THE ALKALOIDS

Chemistry and Physiology

Volume XIX

THE ALKALOIDS
Chemistry and Physiology

Founding Editor
R. H. F. Manske

Edited by
R. G. A. Rodrigo
Department of Chemistry
Wilfrid Laurier University
Waterloo, Ontario, Canada

VOLUME XIX

1981

ACADEMIC PRESS

NEW YORK • LONDON • TORONTO • SYDNEY • SAN FRANCISCO

A Subsidiary of Harcourt Brace Jovanovich, Publishers

ACADEMIC PRESS, INC.
111 Fifth Avenue, New York, New York 10003

United Kingdom Edition published by
ACADEMIC PRESS, INC. (LONDON) LTD.
24/28 Oval Road, London NW1 7DX

Library of Congress Cataloging in Publication Data

Manske, Richard Helmuth Fred, Date.
The alkaloids; chemistry and physiology.

Vols. 8-16 edited by R. H. F. Manske; vols. 17- edited by R. H. F. Manske, R. G. A. Rodrigo.
1. Alkaloids. 2. Alkaloids--Physiological effect.
I. Holmes, Henry Lavergne, joint author. II. Title.
QD421.M3 547.7'2 50-5522
ISBN 0-12-469519-1 (v. 19) AACR1

PRINTED IN THE UNITED STATES OF AMERICA

81 82 83 84 9 8 7 6 5 4 3 2 1

CONTENTS

LIST OF CONTRIBUTORS

Numbers in parentheses indicate the pages on which the authors' contributions begin.

I. Ralph C. Bick (193), Chemistry Department, University of Tasmania, Hobart, Australia

Peter W. Jeffs (1), Department of Chemistry, Paul M. Gross Laboratory, Duke University, Durham, North Carolina 27706

Helmut Ripperger (81), Institute of Plant Biochemistry of the Academy of Sciences of the GDR, DDR-402 Halle/Saale, German Democratic Republic

Klaus Schreiber (81), Institute of Plant Biochemistry of the Academy of Sciences of the GDR, DDR-402 Halle/Saale, German Democratic Republic

Wannee Sinchai* (193), Chemistry Department, University of Tasmania, Hobart, Australia

* Present address: Chemistry Department, Faculty of Science, Prince of Songkla University, Haadyai, Thailand.

PREFACE

Advances in the chemistry of three groups of alkaloids are reviewed in this volume. The *Sceletium* and Phenanthroindolizidine–Phenanthroquinolizidine groups were previously reviewed in Volume IX of this series and a chapter on the *Solanum* alkaloids appeared in Volume X. The present reports deal with the considerable research activity in these areas over the intervening years.

The editor wishes to thank the authors for their cooperation and welcomes advice or criticism from our readers.

R. G. A. Rodrigo

CONTENTS OF PREVIOUS VOLUMES

Contents of Volume I

Contents of Volume II

Contents of Volume III

Contents of Volume IV

Contents of Volume V

Contents of Volume VI

Contents of Volume VII

Contents of Volume VIII

Contents of Volume IX

Contents of Volume X

Contents of Volume XI

Contents of Volume XII

Contents of Volume XIII

Contents of Volume XIV

Contents of Volume XV

Contents of Volume XVI

Contents of Volume XVII

Contents of Volume XVIII

—CHAPTER 1—

SCELETIUM ALKALOIDS

PETER W. JEFFS

Department of Chemistry, Paul M. Gross Laboratory,
Duke University, Durham, North Carolina

I. Historical Background

The plant genus *Sceletium* is made up of about 20 species which are found in the southern regions of Africa, principally in the South-West Cape Province. An interest in these plants stems from early reports (*1*, *2*) of the use of certain species of these plants by the Hottentots and the Bushmen of Namaqualand in the preparation of a drug known as *Channa*. The effects of this drug were described by Kolben (*2*) in the following manner:

> I have seen the effects of Kanna upon the Hottentots. They chew and retain it a considerable time in their mouths. But taking generally too much of it at a time, it

ISBN 0-12-469519-1

> drowns 'em in intoxications. They chew it not long, before their spirits visibly rise, their eyes brighten, their faces take a jovial air, and they sport and wanton under a thousand gaieties of imagination. But in the end it strips 'em of their senses, and throws 'em into the wildest deliria.

Lewin has expressed doubt that such effects are correctly attributed to *Channa* and has suggested that such behavior is more in accord with the use of *Cannabis* (*3*). Later reports have described the effect of chewing the drug as producing a narcotic effect similar to cocaine (*4*). The use of the drug in this way is in keeping with its alternative name, *Koegoed* (*5*, *6*) ("to chew"), which is derived from the language of early Dutch settlers. Unfortunately, there is no recent literature on the physiological effects of extracts of these plants to substantiate the early reports.

The first reported chemical examination of *Koegoed* was by Meiring (*7*) in 1896. His preliminary findings suggested that alkaloids were present. The first definitive evidence of the presence of alkaloids in *Sceletium* species was provided by Zwicky (*8*). In a detailed study of approximately 40 species of plants which were classified as belonging to the genus *Mesembryanthemum*, more than half gave a positive response for the presence of alkaloids. Since that time, the genus *Mesembryanthemum* has been abandoned and a number of the species formerly in this genus have been reassigned to the genus *Sceletium* (Fam. Aizoaceae, *syn*. Ficoidaceae), including *M. expansum*, *M. tortuosum*, and *M. anatomicum*, which are the plants that have been most strongly linked to *Channa* and *Koegoed* (*6*). A summary of the results of Zwicky's survey in which the modern taxonomic classification (*9*) of each species is presented alongside the original designation is given in Table I. The currently accepted names will be used in this review rather than those designated in the original references.

Zwicky's survey for alkaloids was followed by further studies on *S. expansum* and *S. tortuosum* which led to the isolation of a noncrystalline alkaloid, mesembrine, to which was ascribed the formula $C_{16}H_{19}NO_4$ (*10*). Subsequently, Rimington and Roets (*11*) reported the isolation of mesembrine from *Koegoed* and reassigned the molecular formula as $C_{17}H_{23}NO_3$ on the basis of combustion analysis of several crystalline derivatives. More recently, Bodendorf and Krieger (*12*) described the isolation of mesembrine along with two other bases from *S. tortuosum*, and although the ketonic nature of the alkaloid was revealed by its IR spectrum, no significant progress was made in determining the structure of the alkaloid. The structure of mesembrine as *N*-methyl-3a-(3′,4′-dimethoxyphenyl)-6-oxo-*cis*-octahydroindole (**1**) was adduced by Popelak and co-workers (*13*, *14*). The elucidation of the structure of mesembrine has provided the foundation for the studies continued during the past 20 years on the chemistry and biosynthesis of the *Sceletium* alkaloids which are reported in this review.

TABLE I

	Old name	New name	Alkaloid test
Mesembr.	*crystallin.* L	*Crytophytum crystallinum* N. E. Br.	+
	Aitonis Jacq.	*M. aitonis* L. emend. L. Bol.	–
	relaxatum Willd.	*Prenia relaxata* (Willd) N. E. Br.	+
	expans. L.	*Sceletium expansum.* L. Bol.	++
	tortuos. L.	*Sceletium tortuosum* L. Bol.	++
	cordifolium L.	*Aptenia cordifolia* (L.f) Schw	++
	crassulin. spreng.	*Delosperma crassulina* (Haw) L. Bol.	–
	Cooperi Hook. fil.	*Delosperma cooperi* (Hook) L. Bol.	++
	echinqt. Ait	*Delosperma echinatum* (Ait) Schwant	–
	intonsum Haw.	*Trichodiadema intosum* (Haw) Schw	++
	stelligerum Haw.	*Trichodiadema stelligerum* (Haw) Schw	–
	stellatum Mill.	*Trichodiademia stellatum.* (Mill) Schw	++
	bulbosum Haw.	*Trichodiademia bulbosum* (Haw) Schw	–
	floribundum Haw	*Drosanthemum floribundum* (Haw) Schw	++
	hispidum L.	*Drosanthemum hispidum* (L.) Schw	++
	tuberosum Haw.	*Mesotklema tuberosum* (1) N. E. Br.	++
	subincanum Haw.	*Delosperma subincanum* (Haw) Schw	Trace
	Ecklonis Salm-Dyck	*Delosperma ecklonis* (Salm) Schw	Trace
	perfoliat. Mill.	*Ruschia perfoliota* (Mill) Schw	–
	multiflorum Haw.	*Ruschia multiflora* (Haw) Schw	++
	tumidulum Haw.	*Ruschia tumidula* (Haw) Schw	+
	splendens L.	*Nycteranthus splendens* (L.) Schw	++
	umbelliflorum Jacq.	*Nycteranthus umbelliflorus* (Jacq.) Schw	++
	Lehmannii Eckl. & Zeyh.	*Delosperma lehmannii* (Eckl et Zeyh)	+
	inconspicuum Haw.	*Lampranthus inconspicuus* (Haw) Schw	–
	glomeratum L.	*Lampranthus glomeratum* (L.) N. E. Br.	++
	scabrum L.	*Lampranthus scabrum* (L.) N. E. Br.	+
	coccineum Haw.	*Lampranthus coccineus* (Haw) N. E. Br.	–
	curriflorum Haw.	*Lampranthus curviflorus* (Haw) N. E. Br.	–
	aurantiacum Haw.	*Lampranthus aurantiacus* (Haw) Schw	–
	congestum Salm-Dyck	*Ruschia congesta* (S.D.) L. Bol.	+
	lunatum Willd.	*Lampranthus lunatus* (Willd) N. E. Br.	–
	caulescens Mill	*Oscularia caulescens* (Mill) Schwa	Trace
	muricatum Haw.	*Oscularia deltoides* (L.) Schw	–
	lacerum Haw.	*Semanthe lacera* (Haw) N. E. Br.	–
	heteropetalum Haw.	*Ruschia rubicaulis* (Haw) L. Bol.	Trace
	linguiforme L.	*Glottiphylum linguiforme* (L) N. E. Br.	++
	longum Haw.	*Glottiphylum longum* (Haw) N. E. Br.	–

II. The Ring Systems of *Sceletium* Alkaloids

In the space of 10 years, the number of alkaloids isolated from this family has increased from the seven reported by Popelak and Lettenbauer (*15*) in their earlier brief review to more than 25 bases which have fully characterized structures. Consequently, the present chapter is written as a comprehensive review of the subject through 1979. At present, four different ring systems have been identified. These will be referred to as the mesembrine subgroup, the joubertiamine type, the pyridine–dihydropyridone class, and the tortuosamine subgroup. Representative alkaloids belonging to each of these ring systems are shown in Scheme 1. The mesembrine subgroup is by far the largest, comprising 15 bases with known structures. Only two members of the pyridine–dihydropyridine class are known so far; of the remaining alkaloids, eight are in the joubertiamine group and three in the tortuosamine series. The mesembrine alkaloids, which occur in optically active form, are all members of a single enantiomeric series of known absolute configuration. To date, chemical interrelations of the mesembrine group with the other classes of alkaloids in this family are not sufficiently

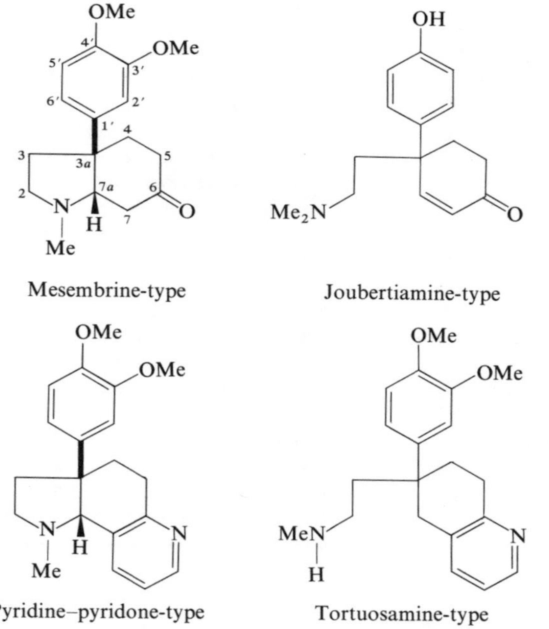

Mesembrine-type

Joubertiamine-type

Pyridine–pyridone-type

Tortuosamine-type

SCHEME 1. *Representative structures of alkaloids belonging to the four basic ring systems exemplifying the mesembrine, joubertiamine, pyridine–pyridone, and tortuosamine subgroups.*

complete to enable one to state with certainty whether the different subgroups are all based on the same absolute stereochemical series. In the few instances where correlations exist between the mesembrine series and the joubertiamine alkaloids, their absolute configurations are directly related. Also, interconversions between the pyridine base, *Sceletium* A_4, and its seco derivatives of the tortuosamine series have established that they belong to the same stereochemical series, but information on the absolute configuration of these two subgroups is not available at present.

III. Isolation and General Methods of Structure Determination

Isolation procedures for *Sceletium* alkaloids have generally relied on column chromatography over alumina and/or silica gel for the separation and purification of the major alkaloids, with repeated preparative-layer chromatography often necessary for separation of the minor bases. In one instance high-pressure liquid chromatography was used for purification of an alkaloid. This latter technique is likely to find increasing application in the future for isolation of the minor alkaloids of this family.

The assignment of an alkaloid to one of the four basic ring systems is now readily accomplished in most instances from mass and ^{1}H-NMR spectral data. The presence of one or two nitrogens in the molecular formula is readily ascertained by high resolution mass spectrometry. Alkaloids containing a single nitrogen belong to either the mesembrine or joubertiamine subgroup. A distinction between these two structural classes, and the identification of the tortuosamine alkaloids, are readily accomplished by the occurrence of characteristic fragmentations which lead to the appearance of diagnostically useful ions. Use of ^{1}H-NMR analysis has proved invaluable for both structural and stereochemical assignments, especially in the mesembrine series where some of the conformational properties are unusual. These structural and conformational assignments have been supported by complementary studies employing single crystal X-ray analysis of several structures in the mesembrine series. The latter technique also has been applied to elucidation of the structures of *Sceletium* alkaloid A_4 and the unusual alkaloid channaine.

Examples of the elucidation of the structures of alkaloids of this family, given in the following section, will illustrate both the chemical approach and the spectral methods which are currently employed. Application of spectral methods alone is often sufficient to derive the structures of new alkaloids; unless new structural classes of alkaloid are found, structure determination by these procedures is at the stage of becoming routine.

IV. Structural Classes

A. Mesembrine Subgroup

This subgroup is the largest representative class and presently consists of fifteen alkaloids. All alkaloids of this subgroup are based on the 3a-aryl-*cis*-octahydroindole skeleton, and the alkaloid mesembrine (**1**) from which this class derives its name was the first member of this family of alkaloids to be structurally characterized.

With one exception, all members of the subgroup contain a 3′,4′-dioxyaryl substituent, either as a veratryl unit or as a 3′-methoxy-4′-hydroxyphenyl substituent. Other variations are found through examples of differences in oxidation state of the hydroaromatic six-membered ring and the presence or absence of a methyl substituent on the nitrogen.

1. (−)-Mesembrine

(−)-Mesembrine (**1**), $C_{17}H_{23}NO_3$, $[\alpha]_D$ −55° (MeOH), is the major alkaloid in *Sceletium namaquense*, occurring in amounts up to 1% of the dried plant. It also occurs in lesser quantities in *S. strictum* (*16*) and has been isolated from *S. tortuosum* as the partial racemate (*17*). The free base is noncrystalline and the alkaloid is most conveniently isolated as its crystalline hydrochloride.

Bodendorf and Krieger (*12*) established the presence of a ketonic carbonyl group from the IR spectrum of mesembrine hydrochloride which contains a carbonyl absorption at 1715 cm^{-1}; this assignment was confirmed by the conversion of mesembrine to an oxime. Analysis of the alkaloid indicated two methoxyls and a tertiary *N*-methyl group. Catalytic hydrogenation of **1** gave a crystalline optically active alcohol, mesembranol (**2**).

The structure of mesembrine was elucidated by Popelak and co-workers (*13*) as follows. Vigorous oxidation of mesembrine gave veratric acid and 3,4-dimethoxyphenylglyoxalic acid, thereby establishing that the two methoxyls were in a veratryl residue. After a number of unsuccessful attempts to obtain useful degradation products by a variety of procedures, a scheme succeeded employing reduction of mesembrine to mesembranol (**2**) followed by dehydration of the latter over P_2O_5 to a mixture of mesembrenes (**3**); on subsequent Hofmann degradation, these yielded a single product (**4**). Further Hofmann degradation of **4** afforded 3,4-dimethoxybiphenyl, ethylene, and triethylamine, which led to the deduction of the basic ring system as shown in Scheme 2. The last-mentioned reaction is an interesting example of a fragmentation reaction which presumably owes much of its driving force to aromatization in the formation of the biphenyl.

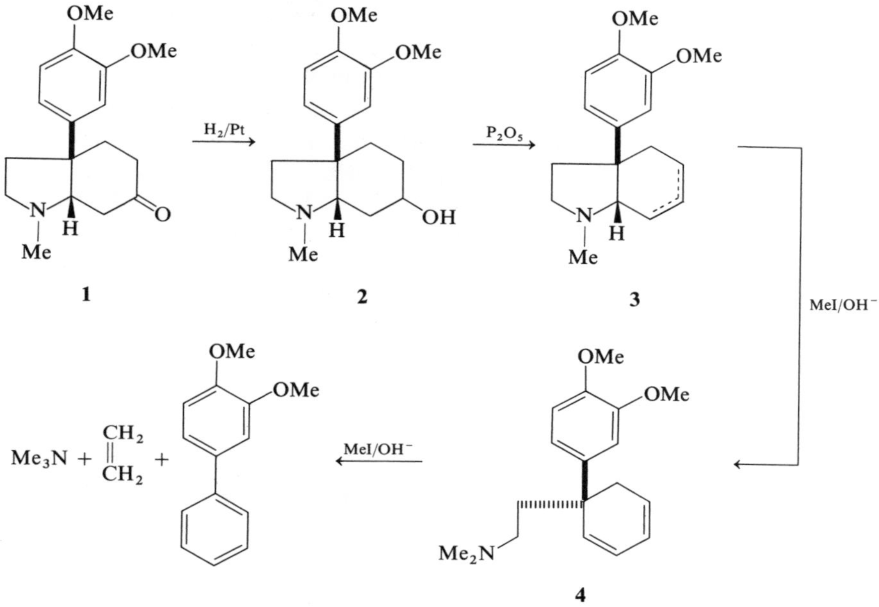

SCHEME 2. *Degradations employed in elucidation of the structure of* (−)-*mesembranol*.

The location of the carbonyl at C-6, which is not established by the foregoing series of reactions, was determined by first reacting mesembrine with phenyllithium to give the alcohol **5**. A sequence of reactions analogous to that described above for mesembranol, involving dehydration of the alcohol **5** followed by two successive Hofmann degradations, was carried out to give 3,4-dimethoxy-*p*-terphenyl (**6**).

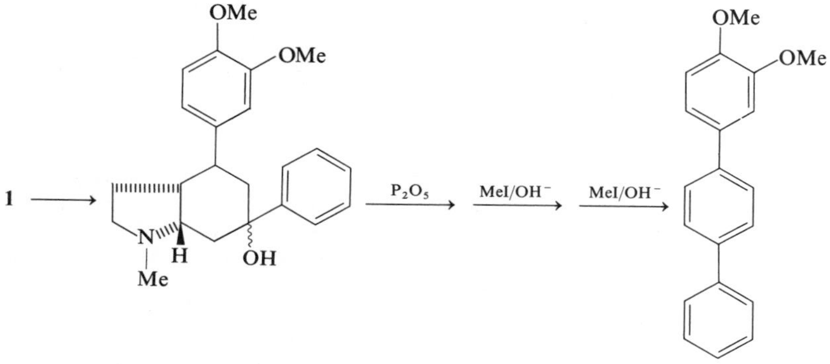

The cis stereochemistry of the ring fusion in mesembrine was elucidated from synthetic studies on a degradation product, (±)-mesembrane (**7**), which was obtained via a companion alkaloid, (±)-mesembrenone, $C_{17}H_{21}NO_3$. The structure of mesembrenone as the α,β-unsaturated ketone **8** was indicated from its IR spectrum (ν_{CO} 1680 cm^{-1}) in conjunction with the observation that catalytic hydrogenation of (±)-mesembrenone gave (±)-mesembrine. A Wolf–Kishner reduction of the latter gave (±)-*N*-methyl-3a-(3′,4′-dimethoxyphenyl)-*cis*-octahydroindole (**7**), which was given the trivial name mesembrane,* and provided the target for a synthesis which established the relative stereochemistry of the basic ring system of this subgroup.

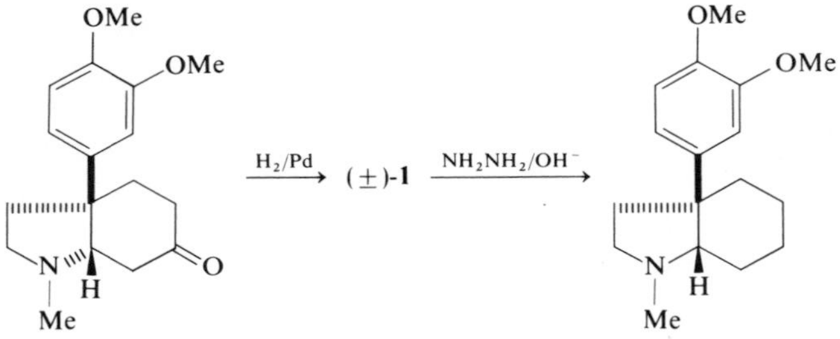

(±)-**8** $\xrightarrow{H_2/Pd}$ (±)-**1** $\xrightarrow{NH_2NH_2/OH^-}$ (±)-**7**

The synthesis (*14*) of (±)-mesembrane, which is presented in Scheme 3, followed closely a route which had been used for a previous synthesis of (±)-crinane, and the assignment of the cis-ring fusion to (±)-mesembrane obtained by this route rests on previous results obtained for catalytic hydrogenations of hexahydroindoles (*18*).

The cis-ring juncture in the octahydroindole skeleton of mesembrine allows for the potential existence of two ground-state conformations (**9** and **10**) in which ring C is in a chair form. A reasonable assumption was made that **9** was the more stable conformation, and on this basis the

* A systematic system of nomenclature based on the adoption of mesembrane as the name for the basic ring system was proposed (*16*) in 1969 and has been used for all alkaloids isolated since. Previously, some confusion existed relating to the naming of compound **8**, which was referred to as both mesembrenine and mesembrinine (*12–14*). Fortunately, the only two other naturally occurring bases described at the time were (−)-mesembranol (**2**), which had been given the name mesembrinol, and (±)-mesembrenone. Although the name mesembrine does not follow this convention, in view of the extensive literature which existed for this alkaloid its trivial name was retained instead of the more systematic name mesembranone. This convention has been followed by subsequent authors except in one instance (*17*) where the name mesembranone has been used instead of mesembrine.

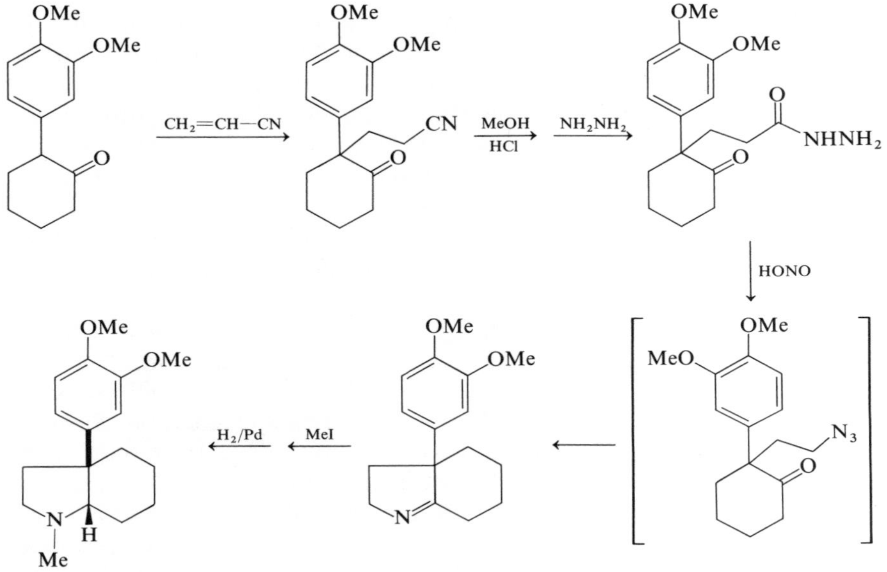

7

SCHEME 3. *Synthesis of* ($\pm$)-*mesembrane.*

negative Cotton effect exhibited by mesembrine was interpreted to indicate that the absolute stereochemistry of the alkaloid is represented by **11** (*19*).

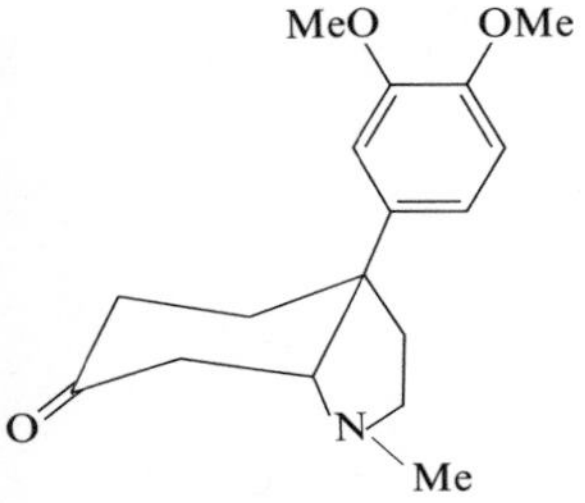

10 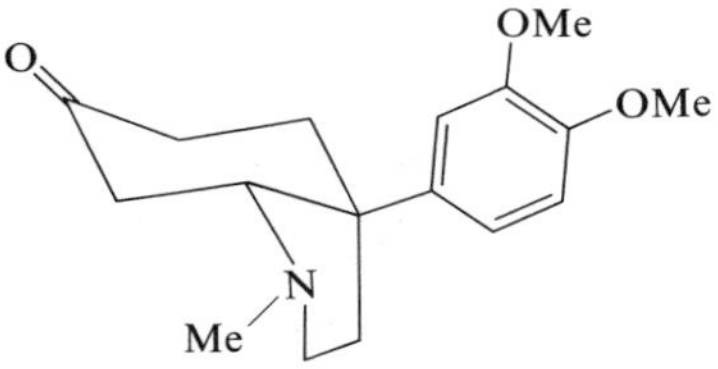 **9** 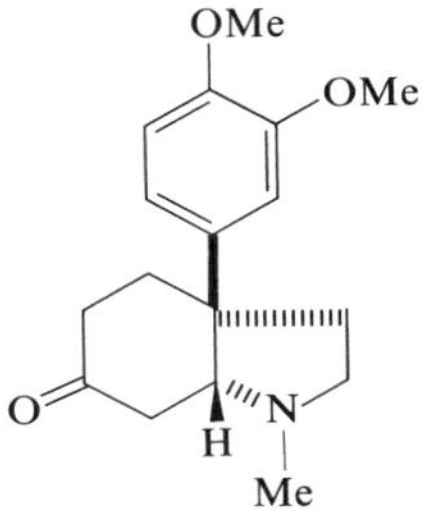 **11**

Later work (*16*), principally emanating from an investigation of the structure of mesembranol, led to the observation that in this alkaloid the more stable ground state conformation possesses ring **B** in a chair form in which the aryl group is in the axial position. Following this observation, extensive investigations of the ^{1}H-NMR spectral properties of mesembrine and related alkaloids led to the assignment of the C-7a hydrogen signal in the spectrum of mesembrine as a triplet at 2.94 δ with an apparent $J = 3.0$ Hz. The triplet nature of this signal and the correspondingly small splitting were of critical importance, for they indicated the equatorial disposition of this proton. This result when considered in conjunction with the complementary studies (*vide infra*) on the conformation of the mesembranols, led to the conclusion that if the cyclohexanone ring of mesembrine exists in a chair form, it should be represented by the conformation in which the aryl group is axial.

Because two nonchair forms of ring B—namely, the twist boat (**12**) and the boat form (**13**)—cannot be distinguished from the chair form from the results of the NMR data, the chiroptical properties of (−)-mesembrine were investigated in an attempt to clarify the conformational situation. The CD spectrum of the alkaloid showed a negative Cotton effect from the $n \rightarrow \pi^*$ band of the carbonyl at 295 nm. A twist-boat conformation was discounted when it was found that the rotational strength of the Cotton effect was weaker than that exhibited by the carbonyl band in the CD spectrum of (−)-dihydrooxocrinine (**14**). The cyclohexanone ring in the latter structure is strongly biased in favor of the chair form by virtue of the nature of the ring fusion. Because a boat form was considered unlikely, it was concluded

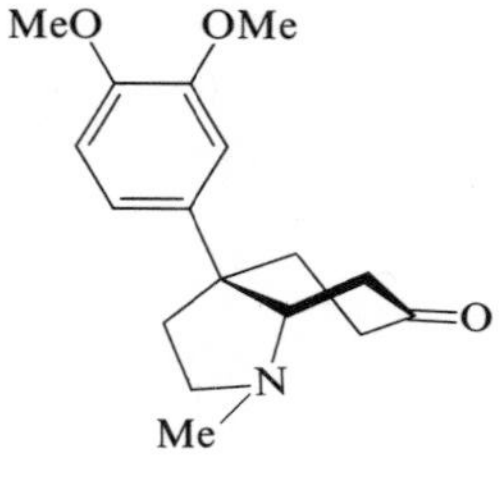

12 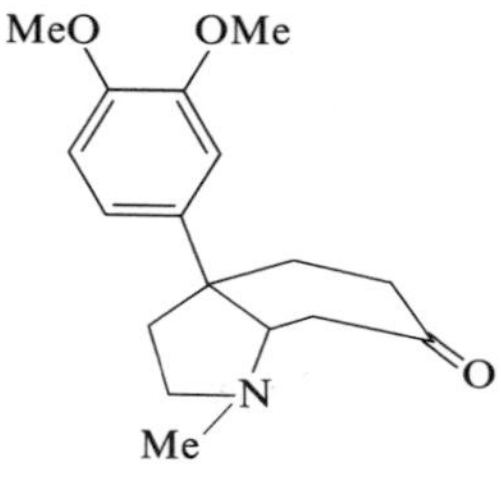 **13** 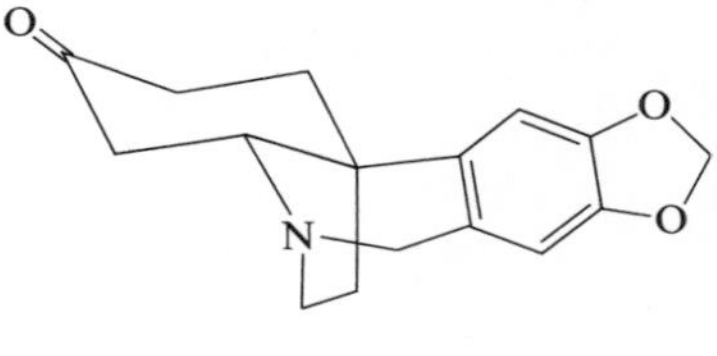 **14**

1 ≡ **15** 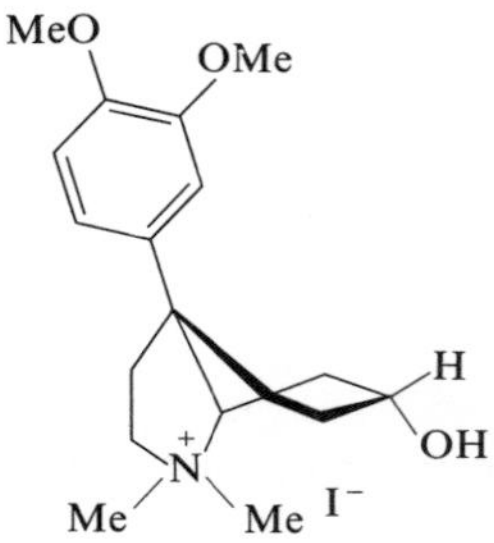 **16**

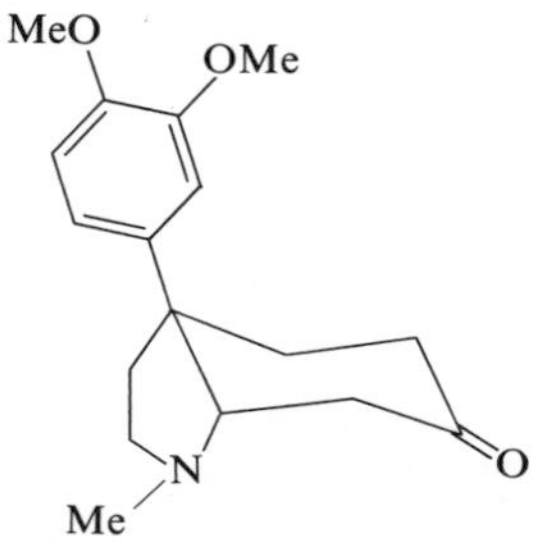

that in solution (−)-mesembrine must exist predominantly in a chair conformation, with the aryl group axial.* This revision of the conformation of the alkaloid invalidated previous conclusions regarding its absolute configuration and required that (−)-mesembrine be correctly represented by the stereostructure **15** on the basis of the Octant rule.

A subsequent X-ray structure determination (*21*) of 6-epimesembranol methiodide (**16**) which had been obtained from (−)-mesembrine confirmed the assignment of the absolute configuration. A more recent (*22*) X-ray structure determination of (−)-mesembrane hydrochloride has provided further substantiation that the absolute configuration proposed on the basis of the CD work which assumed normal Octant behavior is correct. This result is of some interest, as it is in apparent violation of the rules suggested by Hudec (*23*), who has proposed that axial β-aminoketones should exhibit anti-Octant behavior.

With the axial-aryl chair conformation clearly established for all cases of alkaloids in the mesembrine subgroup previously studied, it was surprising to find that the hydrochloride of (−)-mesembrine exists in the solid state with the cyclohexanone ring in a twist-boat conformation, as indicated in **17** (*24*). A study of the CD of mesembrine hydrochloride in H_2O was undertaken in an attempt to seek a possible correlation between the conformation of the hydrochloride in the solid state revealed by the X-ray structure analysis and the solution conformation. Unfortunately, the CD spectrum of mesembrine hydrochloride in H_2O shows the absence of an $n \rightarrow \pi^*$ transition associated with a carbonyl absorption. This evidence, in conjunction with a comparison of the ^{13}C-NMR spectra of the free base in $CDCl_3$ and the hydrochloride in D_2O, clearly shows that the latter exists as the *gem*-diol **18**; the carbonyl resonance which occurs at 210.3 ppm in the spectrum of mesembrine ($CDCl_3$) is replaced by a signal at 94.9 ppm which is consistent with that expected for C-6 in the *gem*-diol. Stabilization of the carbonyl hydrate in mesembrine hydrochloride is possibly due to both the existence of an intramolecular hydrogen bond between the nitrogen and oxygen and the contribution of

* The assumption that mesembrine does not exist in a boat conformation was obviously not justified on the basis of subsequent results obtained from the X-ray structure determination of mesembrine hydrochloride. There is now some doubt that Djerassi's (*20*) earlier finding, indicating that twist-boat conformations of chiral cyclohexanones show abnormally large amplitude Cotton effects, is not without exceptions. Certainly the application of this principle to cyclohexanones with heteroatom substituents presents a quite different situation and in retrospect the conclusion which discounted the boat-conformations in mesembrine must be placed in doubt. Conclusive evidence on the solution conformations of mesembrine and its hydrochloride is still lacking. Despite this uncertainty, the general conformational picture which has emerged for this series, in which the aryl substituent occupies an axial position on a chair form of the substituted cyclohexane ring and the subsequent deductions of absolute stereochemistry derived for the *cis*-octahydroindole series, is unaffected.

dipole effects which are known to favor hydrated forms of carbonyl groups which are *gauche* to ammonium ions (*25*).

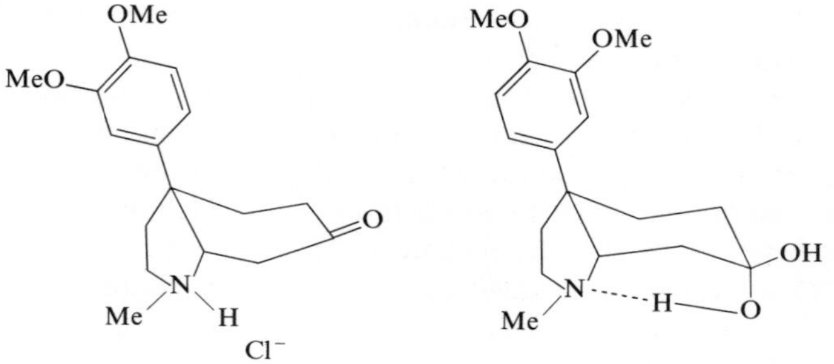

OMe MeO Me N H Cl^- O

17

MeO OMe OH Me N H O

18

2. Mesembrenone

Mesembrenone (**8**), $C_{17}H_{21}NO_3$, has been isolated from both *Sceletium namaquense* and *S. strictum* as the racemate. The free base is a low melting solid, mp 88°, and the alkaloid is most conveniently isolated as its crystalline hydrochloride, mp 192°–193°. The racemic nature of this alkaloid is almost certainly a consequence of the isolation conditions, since (−)-mesembrenone obtained by the careful oxidation of (+)-mesembrenol (**19**) with manganese dioxide under neutral conditions was racemized extremely readily. The ease with which this occurs is illustrated by the observation that a solution of (−)-mesembrenone in ethanol at 25° after 17 hr results in complete racemization of the alkaloid. Presumably, the racemization is the result of an autocatalytic

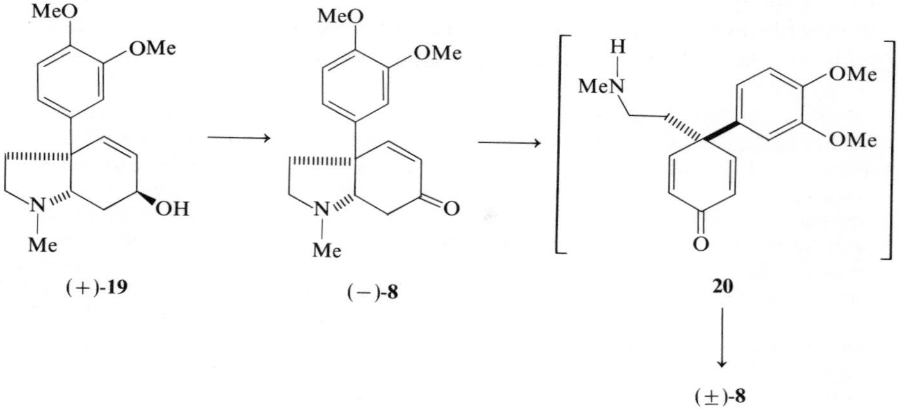

MeO OMe N Me OH

(+)-**19**

MeO OMe N Me O

(−)-**8**

H MeN OMe OMe O

20

(±)-**8**

process involving the alkaloid as a base. Both acid- and base-catalyzed β-elimination reactions can lead to the symmetrical dienone intermediate **20**, which intervenes in the racemization pathway (*26*).

3. (−)-Mesembranol and (+)-Mesembrenol

Besides mesembrine and mesembrenone, the alkaloid (−)-mesembranol (**2**), $C_{17}H_{23}NO_3$, mp 144°–145°, $[\alpha]_D$ −32° ($CHCl_3$), was reported from *S. tortuosum* (*19*). Its structure, without stereochemical details, was evident by its derivation from (−)-mesembrine by catalytic hydrogenation as previously reported (*19*). To provide evidence for the stereochemistry, a detailed study of **2** and the related alcohol, 6-epimesembranol (**21**) was undertaken (*16*). In contrast to catalytic hydrogenation, reduction of mesembrine with either borohydride or LAH provides both **2** and **21**, with the latter generally as the major product. The half-band widths of the C-6 hydrogen in the ^{1}H-NMR spectra of these two epimeric alcohols and their corresponding acetates indicated that mesembranol was an equatorial alcohol and that the 6-epialcohol was its axial counterpart. This implied that the cyclohexanol ring had the same preferred conformation in both alcohols. Of the two possible chair conformations, identification of the C-7a hydrogen signal as an apparent triplet with small splittings of ~3 Hz in the spectra of both alcohols and in the spectra of mesembrine suggested that these three compounds existed with the cyclohexane ring in the chair form in which the 3,4-dimethoxyphenyl substituent was in the axial position. Consideration of the hydroxyl stretching frequences in mesembranol and 6-epimesembranol provided strong support for the foregoing stereochemical assignments. Whereas mesembranol shows a free O—H stretching frequency at 3625 cm^{-1} in dilute solution, the O—H band in 6-epimesembranol exists as a broad band at 3380 cm^{-1} over a wide concentration range in CCl_4 solution. The characteristics of the OH stretching frequency in 6-epimesembranol are indicative of a strong intramolecular OH---N hydrogen bond and accord fully with the *cis*-1,3-diaxial arrangement of the *N*-methyl and the C-6 hydroxyl as represented in conformational structure **22**. Further evidence for the stereochemical details of the structures of **2** and **21** was obtained from kinetic studies. The saponification rates of acetates of cyclohexanols are related to their stereochemical environment, acetates of equatorial alcohols undergoing hydrolysis more rapidly than their axial counterparts. The observation that *O*-acetylmesembranol was hydrolyzed about twice as rapidly as *O*-acetyl-6-epimesembranol was in agreement with the structures and conformations assigned from the spectral studies. Although rates of acetylation of epimeric cyclohexanols are usually found to parallel the order of the hydrolysis rates of the derived acetates, acetylation of 6-epimesembranol was found to be four times faster than the formation of *O*-acetylmesembranol. In fact, this result provided added support for the structural proposals, because the rate enhancement in acetylation for the conformational structure **22**, in which neighboring group participation of the tertiary nitrogen, acting as a base or a nucleophile, was expected in view of its cis-1,3-diaxial relationship to the C-6 hydroxyl (see below).

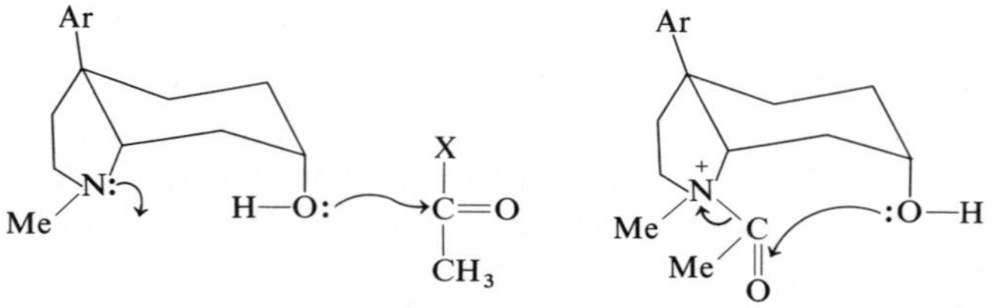

A single crystal X-ray structure determination of (−)-mesembranol subsequently provided full confirmation of the structure of the alkaloid. Furthermore, its conformation as **23** in the solid state was found to correspond to that deduced by NMR analysis for its conformation in solution.

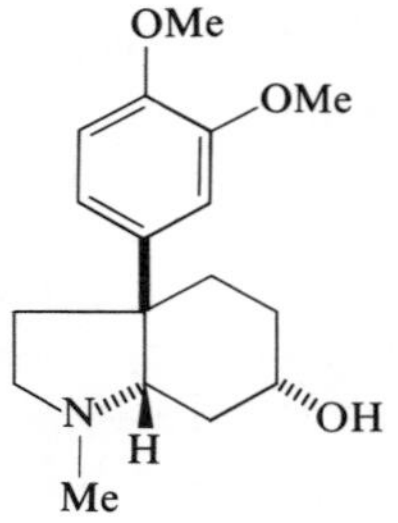

21 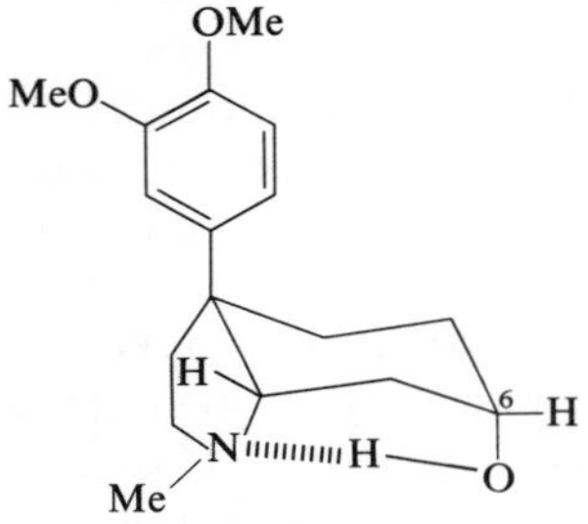 **22**

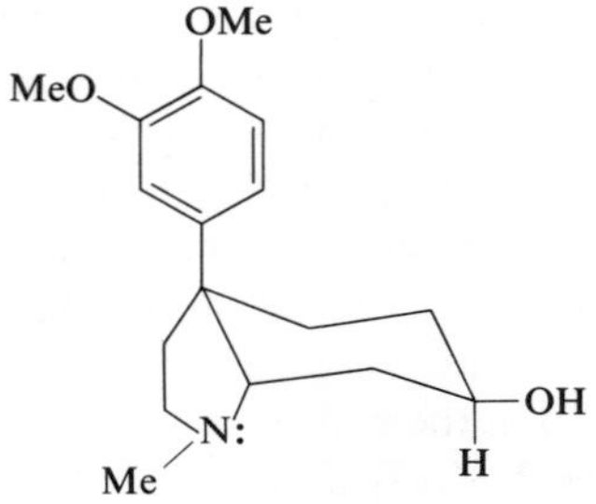

23

The major alkaloids produced by *S. strictum* are (−)-mesembranol and a second optically active, crystalline base, (+)-mesembrenol (**19**), $C_{17}H_{23}NO_3$, mp 142°, $[\alpha]_D + 91°$ ($CHCl_3$). The structure of this latter compound was indicated from its spectral properties (*26*). The IR spectrum of the alkaloid exhibited alcoholic hydroxyl stretching at 3630 and 3450 cm^{-1}, and the NMR spectrum showed typical resonance signals associated with an *N*-methyl group, a veratryl unit, and a disubstituted double bond in a six-membered ring. The mass spectrum of mesembrenol, like the spectra of mesembrine and mesembranol, contains intense ions at *m/e* 219 and *m/e* 70 associated with the fragments, shown in Scheme 4. The location of both the double bond and the position of the hydroxyl group, which are placed in ring B by the NMR and MS results, was established by the conversion of (+)-mesembrenol to (±)-mesembrenone on oxidation with Jones reagent. The racemic nature of the product in this reaction is readily explained by racemization of mesembrenone under the acidic conditions of the oxidation; see, e.g., **8** → **20**.

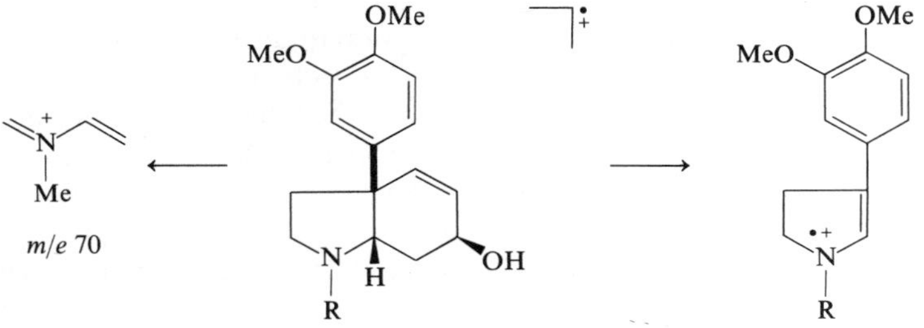

m/e 70

m/e 219 R = Me
m/e 205 R = H

SCHEME 4. *Typical fragments from the two predominant cleavage pathways encountered in the mass spectra of alkaloids in the mesembrine subgroup.*

The stereochemistry of the C-3 hydroxyl and the absolute configuration of (+)-mesembrenol was established by virtue of conversion of the alkaloid to (−)-mesembranol on catalytic hydrogenation.

With the exception of sceletenone and its close relatives, which will be discussed in Section IV,A,7, the remaining extant members of this subgroup are simple relatives of the four alkaloids mesembrine, mesembrenone, mesembranol, and mesembrenol, and include the known base (−)-mesembrane, which has been demonstrated to occur in *S. namaquense* (*22*). The majority of these bases are either *O*- or *N*-demethylated compounds derived from the four above-mentioned alkaloids; discussion of these compounds will be confined to representative examples which illustrate general features of the methods used in the elucidation of their structures.

4. 4′-*O*-Demethylmesembranol and Related Phenolic Alkaloids

Phenolic alkaloids are present in both *S. strictum* and *S. namaquense*. Three of these bases have been shown to belong to the mesembrine subgroup and have been characterized as 4′-*O*-demethylmesembranol (**24**) (*26*), 4′-*O*-demethylmesembrenol (**25**) (*26*), and 4′-*O*-demethylmesembrenone (**26**) (*27*). The essential features of the structures of each of these phenolic alkaloids were apparent by their conversion to known bases by methylation with diazomethane. Compound **25** was converted to **24** by catalytic hydrogenation, indicating that the phenolic hydroxyl occupied the same site in both compounds.

Since the limited availability of the phenolic bases precluded using the usual chemical approaches to determining whether the 3′ or the 4′ position was occupied by the phenolic hydroxyl, a radiolabeling procedure was devised to circumvent the difficulty. Methylation of **24** with [^{3}H]diazomethane provided radiolabeled mesembranol. Drastic oxidation of the alkaloid gave radiolabeled veratric acid in submilligram quantities which was readily isolated by addition of inactive carrier. This technique provided adequate quantities of material to determine the site of the label in the veratric acid as shown in Scheme 5 through selective cleavage of the 3′-methoxyl to give radiolabeled isovanillic acid and by the independent conversion of the labeled veratric acid to unlabeled protocatechuic acid. This sequence, carried out on 5 mg of the original alkaloid, established that the site of the phenolic hydroxyl group was the 4′ position.

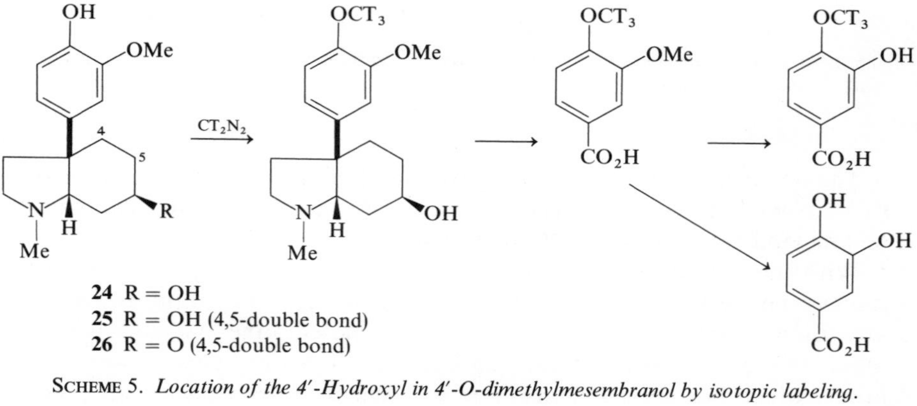

24 R = OH
25 R = OH (4,5-double bond)
26 R = O (4,5-double bond)

SCHEME 5. *Location of the 4′-Hydroxyl in 4′-O-dimethylmesembranol by isotopic labeling.*

Complementary evidence was obtained for the location of the phenolic hydroxyl at the 4′-position in 4′-*O*-demethylmesembrenone. In this case, the sample obtained was biosynthetically labeled at the methoxyl and *N*-methyl groups using [*S-methyl*-^{14}C]methionine as the precursor in experiments with *S. strictum*. Methylation of the labeled alkaloid with diazomethane obtained from this experiment was followed by degradation of the resulting mesembrenone to give unlabeled isovanillic acid, and served to establish the position of the phenolic hydroxyl group as shown in **26**.

5. (+)-*N*-Demethylmesembrenol and (−)-*N*-Demethylmesembranol

The alkaloids (+)-*N*-demethylmesembrenol (**27**), $C_{16}H_{21}NO_3$, mp 63°–65°, $[\alpha]_D$ +86°, and (−)-*N*-demethylmesembranol (**28**), $C_{16}H_{23}NO_3$, mp 178°–185°, $[\alpha]_D$ −13°, have been isolated from *Sceletium strictum* (*28*).

The structural details of both bases were evident from their NMR spectra, which parallel closely the corresponding spectra of mesembrenol and mesembranol with the notable exception of the absence of an *N*-methyl signal. The mass spectral fragmentation pattern provided additional support for the proposed structures in that the 3-aryl-*N*-methylpyrolidinium ion, which occurs at *m*/*e* 219 in the 3′,4′-dimethoxyphenyl-substituted alkaloids of the *N*-methyl series of this subgroup, was replaced by an ion at *m*/*e* 205 (see Scheme 4). Acetylation of *N*-demethylmesembrenol gave an *O*,*N*-diacetate which exhibited both ester (1730 cm^{-1}) and amide (1635 cm^{-1}) carbonyl absorptions, in agreement with the proposed structure.

OMe OMe 4 5 OH N H H H

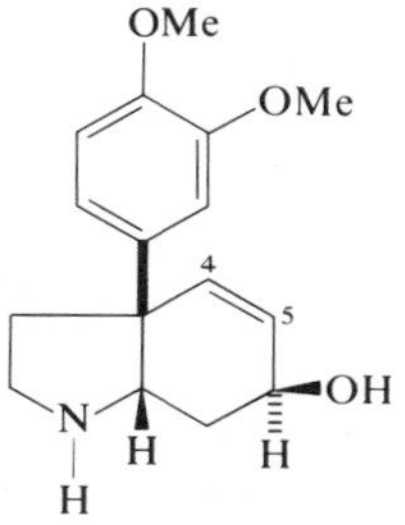

27
28 (No 4,5-double bond)

Catalytic hydrogenation of *N*-demethylmesembrenol afforded a product identical with the companion base **28** and served to establish the structural and stereochemical relationships between the two alkaloids. While there seems little doubt that the structure of these two alkaloids is correctly represented as shown, it is unfortunate that the simple chemical correlation with either mesembrenol or mesembranol through N-methylation was not effected.

6. Channaine

The substance channaine, mp 179°–180°, was first mentioned by Bodendorf and Krieger (*12*) during a study of the alkaloids of *Seletium tortuosum*. The possibility that the molecular formula attributed to channaine, $C_{16}H_{19/21}NO_3$, was incorrect was later stated by Popelak and Lettenbauer (*15*). The latter authors provided additional spectral evidence to indicate that channaine contained two methoxyl groups in a veratryl residue and that *N*-methyl and carbonyl functions were absent. The alkaloid was also stated to be optically inactive, and the suggestion was made that it might be a dimer with the empirical formula $C_{16}H_{19}NO_3$.

More recent work (*29*) has resulted in the isolation of channaine from *S. strictum*, and although the molecular formula $C_{32}H_{38}N_2O_6$ was established and a partial structure could be deduced from ^{1}H-NMR and MS data, the lack of material precluded a definitive chemical investigation of the structure of the alkaloid. Consequently, a single crystal X-ray analysis of channaine, which crystallizes as the hexahydrate from moist ethyl acetate, was performed and showed that it is represented by the racemate of structure **29**.

The structure of channaine suggests its genesis from (±)-normesembrenone by the reactions outlined in Scheme 6.

The origin and occurrence of channaine deserve some comment. With the exception of channaine and the bases mesembrenone, 4′-*O*-demethylmesembrenone, and sceletenone (**30**), all other *Sceletium* alkaloids are known to occur as single enantiomers.* The occurrence of these structurally related bases as racemates is readily explained by racemization of the alkaloids under the conditions of their isolation. However, direct evidence for the formation of these three bases in single enantiomeric forms by the plant lacks firm experimental verification. Still, because each of these alkaloids has been shown to serve as an efficient precursor to (+)-mesembrenol from biosynthetic experiments, it is likely that only a single enantiomeric form of these bases is being utilized for these conversions. Therefore, it is reasonable to assume that only the enantiomeric configuration of these alkaloids which corresponds in absolute configuration to (+)-mesembrenol is being produced, but this point is not unequivocally established.

Given these arguments, which rest on an assumed enantiomeric specificity for the enzymes responsible for the conversion of sceletenone and related bases to (+)-mesembrenol, if channaine were not an artifact it would have been expected to occur in optically active form. Presumably, the natural base is either (+)- or (−)-normesembrenone which is first racemized under the

* There is one report in which mesembrine, mesembranol, and *Sceletium* alkaloid A_4 are reported to occur as the partial racemates in *S. tortuosum* (*17*).

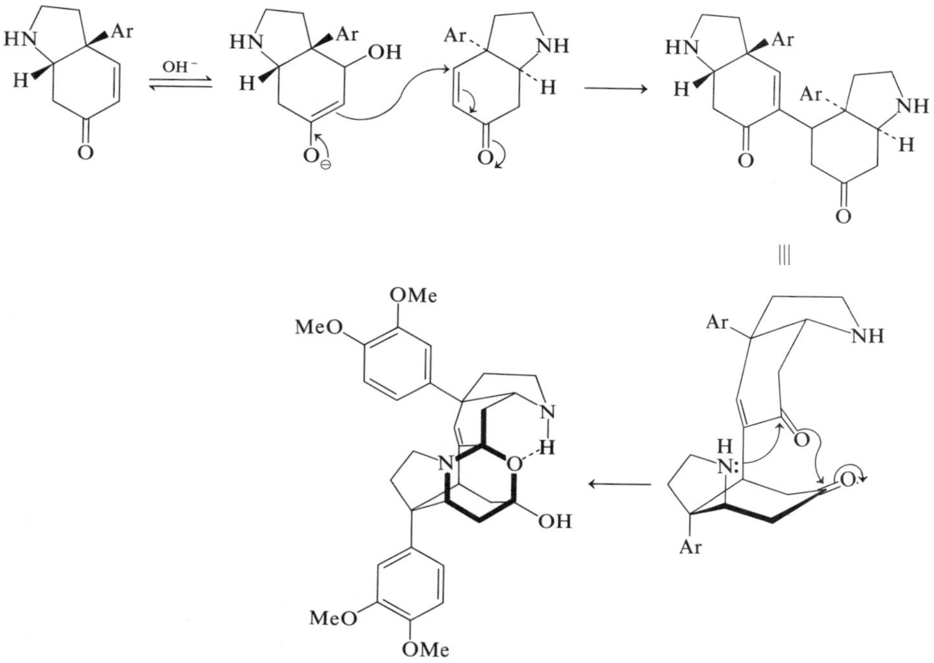

29

SCHEME 6. *Suggested mechanism for the formation of channaine from* (±)-*N-demethylmesembrenone.*

isolation conditions and then converted to channaine according to the pathway shown by the basic medium encountered either during isolation or chromatographic purification.

The stereochemistry of channaine indicates that it is formed by condensation of two molecules of normesembrenone having the same absolute stereochemistry, i.e., (+)(+) and (−)(−). No evidence is available indicating that the formation of the diastereoisomeric racemate, the (+)(−) and (−)(+) combination, occurs under these isolation conditions.

The driving force for the rather remarkable transformation of normesembrenone to channaine has to derive in large measure from the stability conferred by the cage-like spiroaminol–hemiacetal system and the intramolecular N—H---O(6) hydrogen bond, because mesembrenone shows no tendency to undergo self-condensation under similar conditions.

An oxidative pathway leading to the formation of *N*-demethylmesembrenone from mesembrenone is indicated by the isolation of *N*-demethyl-*N*-formylmesembrenone (**31**) (*30*) from *S. strictum.*

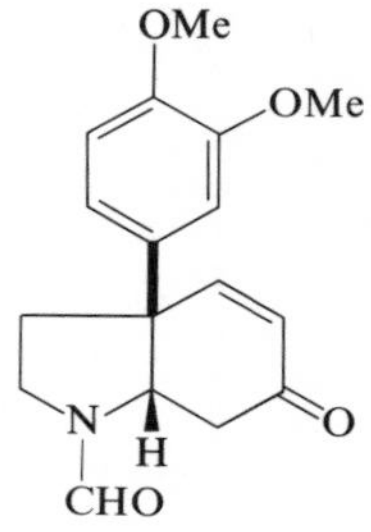

OMe OMe

N H CHO O

31

7. Sceletenone

(±)-Sceletenone (**30**), $C_{15}H_{17}NO_2$, oil, was isolated after extensive chromatographic purification of the phenolic alkaloid fraction of *S. namaquense* (*27*). The structure of this compound was deduced largely from spectral and chemical comparisons with mesembrenone. The key features of the structure

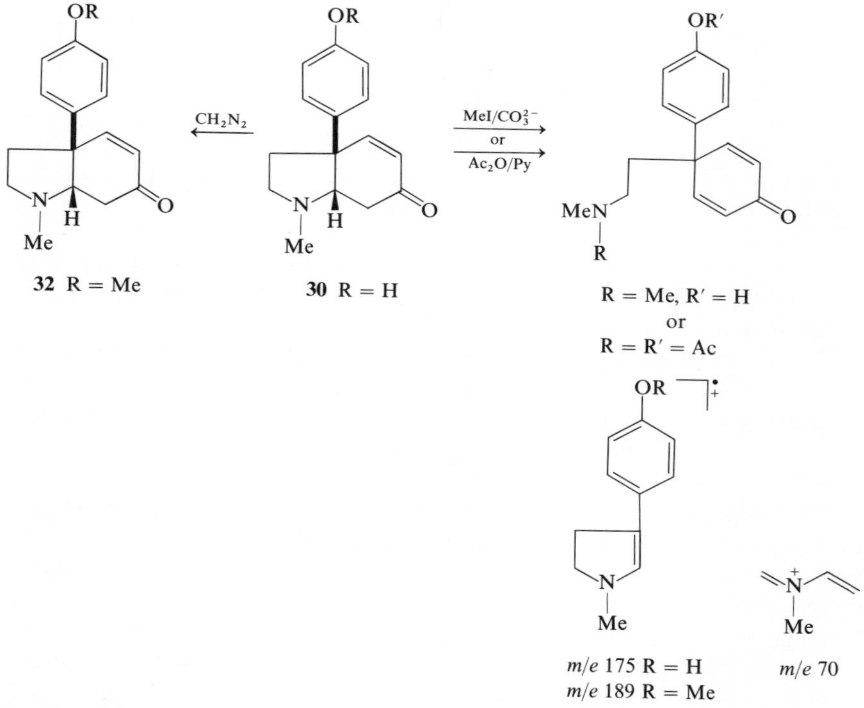

32 R = Me $\xleftarrow{CH_2N_2}$ **30** R = H $\xrightarrow[Ac_2O/Py]{MeI/CO_3^{2-} \text{ or}}$ R = Me, R′ = H or R = R′ = Ac

m/e 175 R = H
m/e 189 R = Me

m/e 70

SCHEME 7. *Reactions utilized in the determination of the structure of* (±)-*sceletenone.*

were revealed by its IR spectrum, which contained phenolic hydroxyl and α,β-unsaturated ketone absorptions at 3592 cm^{-1} and 1678 cm^{-1}, respectively, and the ^{1}H- and ^{13}C-NMR spectra of the alkaloid. The essential features of the ^{1}H-NMR spectrum were embodied in the presence of an *N*-methyl signal, an AB olefinic pattern from the α,β-unsaturated ketone and the AA′BB′ spin pattern in the aromatic region. The mass spectral fragmentation of sceletenone and its *O*-methyl derivative (**32**) both exhibited an *m/e* 70 ion accompanied by the corresponding 3-aryl-*N*-methylpyrrolidinium ions at *m/e* 175 and *m/e* 189, respectively, which provided confirmation that sceletenone was based upon the 3a-aryloctahydroindole ring system. A comparison of the carbon shifts in the ^{13}C-NMR spectra of sceletenone with the corresponding signals in the spectrum of mesembrenone showed remarkably good agreement, except for the aromatic carbon signals, in which the expected differences reflecting the different aromatic substitution patterns were observed. Further evidence for the structure of sceletenone was provided by the chemical transformations to the two 4,4-disubstituted cyclohexadienones, illustrated in Scheme 7.

B. Joubertiamine Subgroup

The alkaloids of this subgroup are based on the N—C(7a) secomesembrane skeleton exemplified by the structure of joubertiamine (**33**), from which this class of alkaloids derives its name. The principal source of the alkaloids of this subgroup has been *Sceletium joubertii* (*31*, *32*) which has yielded joubertiamine, dehydrojoubertiamine (**34**), dihydrojoubertiamine (**35**), and the dioxyaryl base joubertinamine (**36**). Also, some dioxyaryl secomesembrane alkaloids have been found as minor bases in *S. tortuosum* (*22*).

1. Joubertiamine, Dihydrojoubertiamine, and Dehydrojoubertiamine

In an initial study of the alkaloids of *Sceletium joubertii*, Arndt and Kruger (*31*) isolated three new alkaloids. The essential features of the structures of these noncrystalline bases were evident from their spectral properties. The molecular formula of joubertiamine as $C_{16}H_{21}NO_2$ was established by high-resolution mass spectrometry, and fragment ions at *m/e* 58 and *m/e* 78 indicated that the alkaloid contains an *N*-dimethylaminoethyl side chain. A carbonyl absorption in the IR at 1680 cm^{-1} and the UV spectrum were both in accord with the presence of an α,β-unsaturated carbonyl group. A bathochromic shift of the absorption maxima in the UV spectrum of the alkaloid in alkaline solution supported the presence of a phenol which was readily placed in a para-disubstituted ring by the AA′BB′ pattern of the aromatic hydrogen signals in the ^{1}H-NMR spectrum of the alkaloid. The presence of an *N*,*N*-dimethylamino group and the 4,4-substitution pattern of the cyclohexenone ring were deduced from the NMR spectrum by the occurrence of a six-proton singlet at 2.25δ and two one-proton olefinic hydrogen doublets at 6.08δ and 7.04δ ($J = 10$ Hz).

The above data were interpreted in terms of structure **33** for joubertiamine. This structure has been subsequently confirmed by several syntheses of its *O*-methyl derivative **38** (see Section V,A).

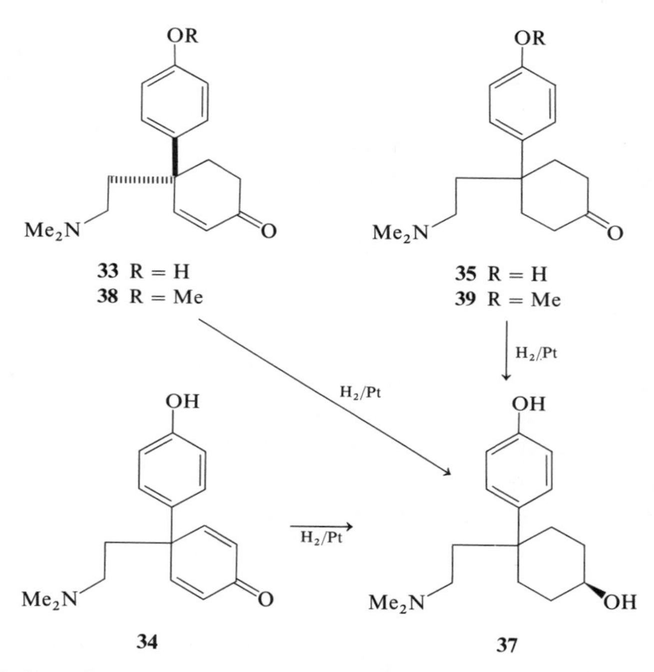

SCHEME 8. *Interrelationship of joubertiamine, dihydrojoubertiamine, and dehydrojoubertiamine.*

The spectral details of dihydrojoubertiamine (**35**) and dehydrojoubertiamine (**34**) showed strong similarities to those of joubertiamine. Correlation of the structures of dihydrojoubertiamine and dehydrojoubertiamine was obtained via catalytic hydrogenation of each of these compounds to the cyclohexanol **37**, as summarized in Scheme 8.

Unfortunately, the chiroptical properties of joubertiamine were not reported, and the absolute configuration rests on an unconfirmed report (*32*) that both *O*-methyljoubertiamine (**38**) and *O*-methyldihydrojoubertiamine (**39**) have been isolated from *S. joubertii* and that the former is represented by the structure shown. Also, evidence on the relative stereochemistry of the reduction product (**37**) is lacking.

2. Joubertinamine

A recent report (*33*) has disclosed the presence of a noncrystalline base, joubertinamine, $C_{17}H_{25}NO_3$, $[\alpha]_D$ $-18°$, in *Sceletium joubertii.* The salient features of the spectral details of this base which led to the postulated structure **36** are fragment ions in the mass spectrum at *m/e* 44 and *m/e* 58 attributed to an *N*-methylaminoethyl side chain; UV absorptions at 231, 280, and 284 nm; and the appearance in the NMR spectrum of a 3H multiplet at 6.7–7.0 and two methoxyl singlets at 3.84 and 3.85 δ, all of which indicated a veratryl residue. A 1H multiplet at 4.0–4.3 δ and 2H singlet at 5.90 δ in the NMR spectrum of **36** were interpreted to indicate the presence of a secondary hydroxyl group and a cis-disubstituted double bond in a cyclohexene ring. The presence of a hydroxyl group was supported by absorptions at 3600 cm^{-1} and 2500–3500 cm^{-1} in the IR spectrum of the alkaloid. The formation of an *O,N*-diacetate (**40**) and oxidation of the alkaloid with MnO_2 to give (−)-mesembrine (**1**) via the intermediate enone **41** provided conclusive evidence for the structure of joubertiamine as **36** (see Scheme 9). The stereochemistry of the hydroxyl group was not assigned by the South African workers, but further unpublished work in the reviewer's laboratory has established the cis-1,4 relationship of the hydroxyl and the dimethoxyphenyl ring as depicted in **36** (*vide infra*).

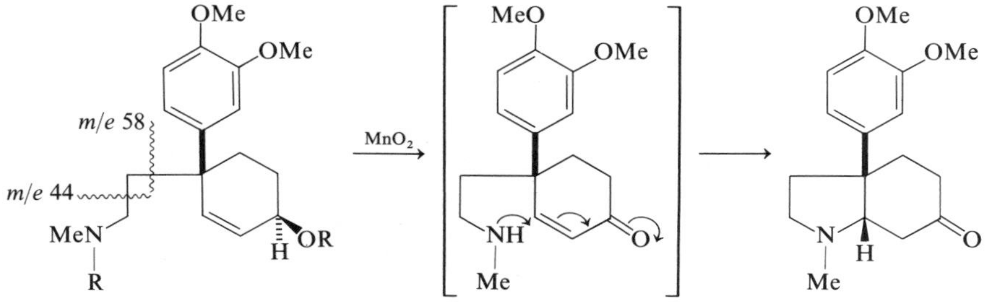

36 R = H
40 R = Ac

41

1

SCHEME 9. *Conversion of joubertinamine to* (−)*-mesembrine.*

3. (−)-3′-Methoxy-4′-*O*-methyljoubertiamine and 3′-Methoxy-4′-*O*-methyljoubertiaminol

Two structurally related bases of the secomesembrane series have been isolated from *Sceletium namaquense* after extensive column and preparative layer chromatography. 3′-Methoxy-4′-*O*-methyljoubertiamine (**42**), $C_{18}H_{25}NO_3$, oil, CD $[\theta]_{340}$ $-6420°$, was readily identified from its spectral properties, which indicated the presence of an α,β-unsaturated ketone fragment from the occurrence of a pair of one-proton doublets at 6.12 δ and 7.09 δ ($J = 10$ Hz) in the NMR spectrum of **42** and a carbonyl band at 1680 cm^{-1} in the IR spectrum of the alkaloid. An unresolved multiplet at 6.80 δ (3H) and a singlet at 3.83 δ (6H) in the NMR spectrum were in keeping with the presence of a 3,4-dimethoxyphenyl group. The mass spectrum contained a molecular ion at *m/e* 303 corresponding to $C_{18}H_{25}NO_3$; more importantly, it exhibited a base peak at *m/e* 58 and an abundant ion at *m/e* 72. The latter ions indicated the presence of an *N*-dimethylaminoethyl side chain, and this was supported by the occurrence of a 6H singlet at 2.25 δ in the ^{1}H-NMR spectrum of the alkaloid.

The spectral data suggested that the alkaloid was represented as **42**. Confirmation of this structure was obtained by establishing the identity of the alkaloid with the product obtained from treating (−)-mesembrine methiodide (**43**) with aqueous base. Because at the time of isolation this represented the first member of the joubertiamine type having a dioxyaryl ring, the possibility that it was an artifact, being produced from a (−)-mesembrine metho salt as a result of the basic isolation conditions, could not be discounted. This eventuality would now seem to be unlikely, for several other dioxyarylsecomesembranes have been isolated subsequently from *S. namaquense*.

The CD spectra of the Hofmann product of (−)-mesembrine methiodide and the new alkaloid were identical, thereby providing the absolute stereochemistry as depicted in structure **42** (*22*).

The negative maximum from the Cotton effect of the $n \rightarrow \pi^*$ band of the enone chromophore at 340 nm suggested that the cyclohexenone ring has left-handed chirality. However, a single crystal analysis of (−)-3′-methoxy-4′-*O*-methyljoubertiamine hydrochloride, while providing independent evidence for the absolute configuration, also showed that the conformation of the cyclohexenone ring exists with the aromatic ring in an axial position and that the enone system has right-handed chirality, i.e., that the 5-6-7-7a torsion angle is positive; *cf.* **44**. The discrepancy between the conclusions regarding the chirality of the enone chromophore from the CD spectrum and the X-ray results may be due to the existence of different conformations in solution and in the solid state. The energy difference between the two conformations having different enone chirality is estimated to be rather small (ca. 3 kcal) and, as such, it is probable that crystal packing forces may be more than sufficient to induce a conformational change.

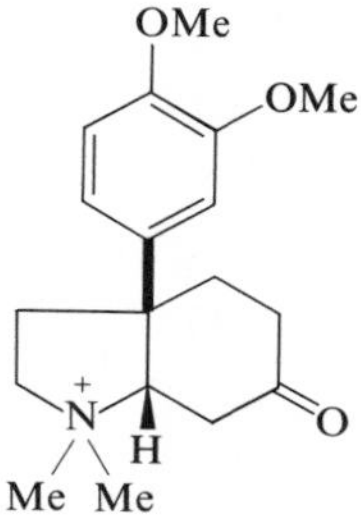

43 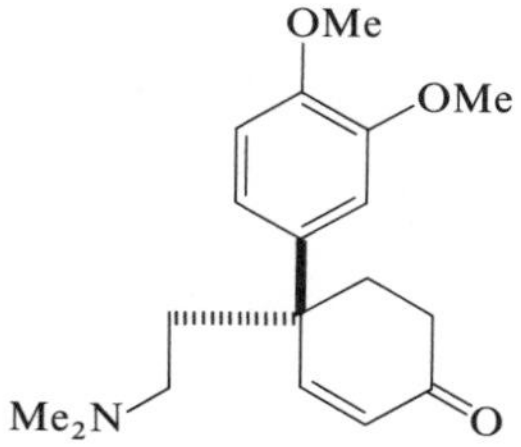 **42** 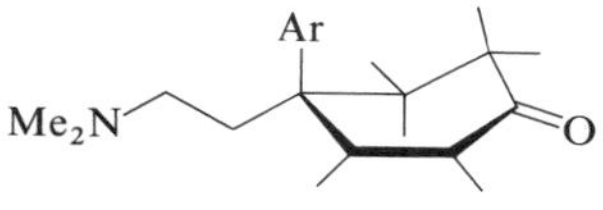 **44**

Preparative-layer chromatography of an ether-soluble alkaloid fraction from *S. namaquense* afforded on optically active crystalline base which has been characterized as (−)-3′-methoxy-4′-*O*-methyljoubertiaminol (**45**) (*34*), mp 100°, CD $[\theta]_{275}$ −2050°. A strong hydroxyl absorption at 3410 cm^{-1} in the IR spectrum of the alkaloid combined with the appearance of a broad one-proton multiplet at 4.30 δ ($W_{1/2} \sim 22$ Hz) indicated an equatorial alcohol. A six-proton singlet at 2.2 δ in conjunction with the occurrence of m/e 58 and m/e 72 ions in the mass spectrum when, considered together with ^{1}H-NMR signals indicative of a veratryl ring, suggested that the alkaloid should be represented by structure **45** (without stereochemical details being implied). Confirmation of the structural assignment was obtained by borohydride reduction of (−)-3′-methoxy-4′-*O*-methyljoubertiamine (**42**) to pair of epimeric alcohols (**45** and **46**). The former proved identical to that of the natural product.

The stereochemistry of (−)-3′-methoxy-4′-*O*-methyljoubertiaminol as depicted in structure **45** was ascertained by the conversion of (−)-mesembranol to the new base by Hofmann degradation.

With the determination of both the relative and the absolute stereochemistries of (−)-3′-methoxy-4′-*O*-methyljoubertiaminol by this conversion, a comparison of the ^{1}H-NMR spectra of the epimeric alcohols **45** and **46** proved informative. The olefinic hydrogen signals in the β-alcohol (**45**) appeared as a two-proton singlet at 6.04 δ, whereas the olefinic hydrogen signals exhibited a more typical AB quartet pattern in the α-alcohol (**46**). Because the spectral results reported for joubertinamine indicate that the two olefinic proton signals appear as a singlet, one can assign the β-hydroxyl configuration to the hydroxyl in joubertinamine, as indicated in structure **36**. A summary of these and other reactions relating to the assignment of the structure of 3′-methoxy-4′-*O*-methyljoubertiamine appears in Scheme 10.

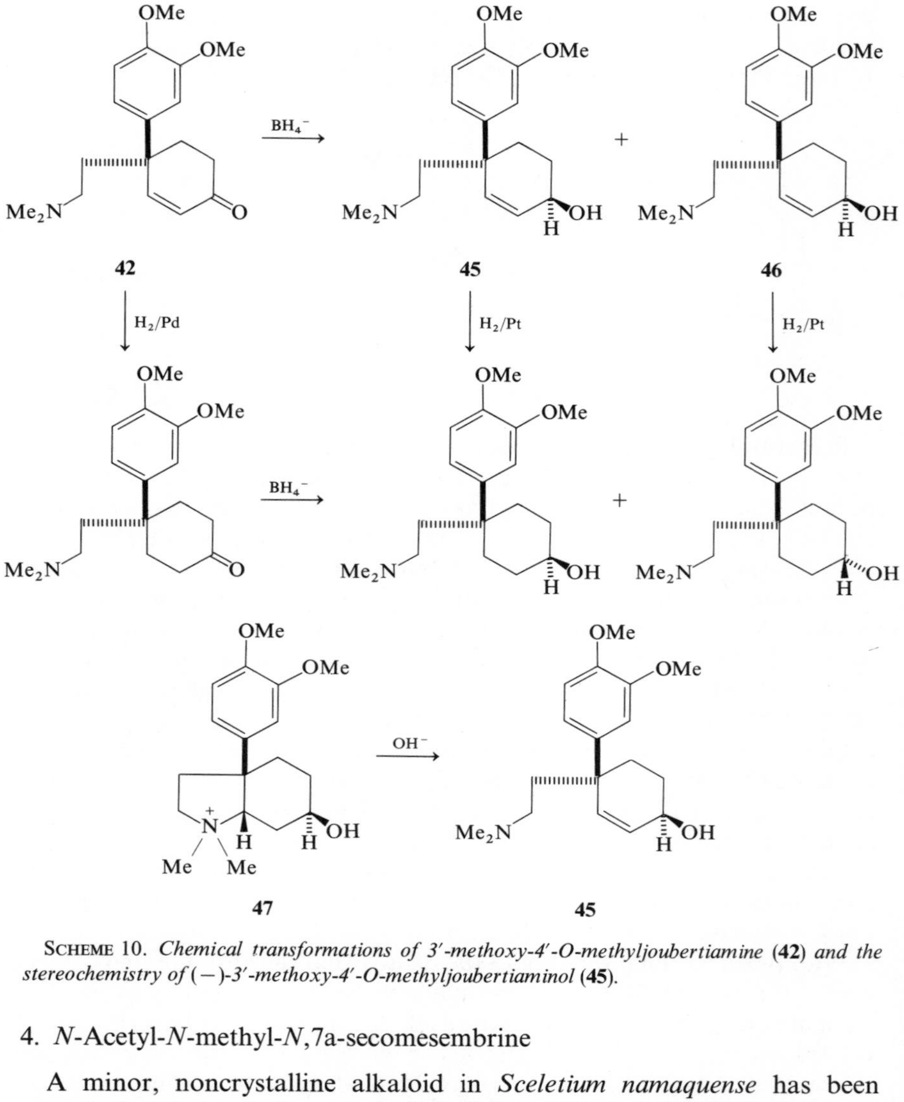

SCHEME 10. *Chemical transformations of 3′-methoxy-4′-O-methyljoubertiamine* (**42**) *and the stereochemistry of* (−)-*3′-methoxy-4′-O-methyljoubertiaminol* (**45**).

4. *N*-Acetyl-*N*-methyl-*N*,7a-secomesembrine

A minor, noncrystalline alkaloid in *Sceletium namaquense* has been recently isolated and characterized as *N*-acetyl-*N*-methyl-*N*,7a-secomesembrine (**48**) (*34*). The IR spectrum indicated the presence of both amide and ketone functions, with carbonyl absorptions at 1658 and 1750 cm^{-1}, respectively. The ^{1}H-NMR spectrum of the alkaloid, in addition to possessing absorptions characteristic of the veratryl group, showed signals attributable to an *N*-methyl-*N*-acetyl group from the occurrence of pairs of singlets at

2.86 and 2.90 δ and at 1.90 and 1.95 δ. These latter signals showed the expected temperature dependence of an amide, each pair of signals collapsing to a single resonance at 80°.

The foregoing spectral data indicated that the alkaloid is represented by structure **48**. Additional support was provided by the mass spectrum of the alkaloid, which displayed a fragmentation behavior yielding a prominent ion at m/e 233 (**49**) corresponding to cleavage of the substituted ethanamine side chain and the complementary ion m/e 100 (**50**).

The structure of the alkaloid was confirmed by its synthesis from mesembrine by the two-step sequence illustrated in Scheme 11.

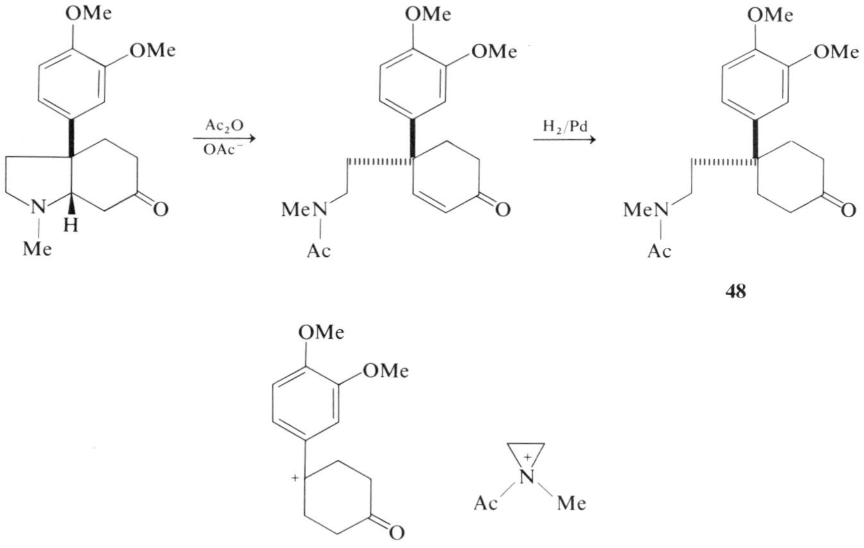

SCHEME 11. *Conversion of* (−)*-mesembrine to N-acetyl-N,7a-secomesembrine* (**48**) *and the structures of the predominant mass spectral ions from* **48**.

5. *N*-Acetyl-4′-*O*-demethyl-*N*,7a-secomesembradienone

The structural features embodied in the *Sceletium* alkaloids coupled with evidence from biosynthetic studies suggest that cyclohexadienone intermediates may be involved in the biosynthesis of these alkaloids. In an attempt to isolate such intermediates, a procedure was devised for the isolation of phenolic alkaloids from *S. namaquense* which avoided the use of acidic conditions which might be expected to induce rearrangement of acid-sensitive cyclohexadienone systems. This procedure involved washing a chloroform extract of the dried plant material with aqueous sodium bicarbonate to remove acidic material, followed by extraction and recovery of a phenolic fraction with aqueous sodium hydroxide and subsequent extraction with solvent after adjusting the solution to pH 8.0.

Repetitive column chromatography gave an alkaloid fraction containing several new components. Further purification by preparative-layer chromatography led to the isolation of a new crystalline base, *N*-acetyl-4-*O*′-demethyl-*N*,7a-secomesembradienone (**51**), $C_{18}H_{21}NO_4$, mp 107.5°. The IR spectrum of the alkaloid contained absorptions in keeping with the presence of a phenolic hydroxyl group (3540 cm^{-1}), a dienone (1665 cm^{-1}), and an amide (1637 cm^{-1}). The NMR spectrum of the alkaloid was found to be temperature dependent. At 26° pairs of signals associated with an *N*-methyl (2.47 and 3.01 δ), *N*-acetyl (2.10 and 2.04 δ), and *O*-methyl (3.82 and 3.86 δ) were observed to collapse to singlets when the temperature was increased to 50°. The appearance of two 2-proton doublets centered at 6.35 and 6.98 δ ($J = 10$ Hz) was in keeping with the presence of a 4,4-disubstituted cyclohexadienone. The mass spectral fragmentation pathways of the alkaloid, which are summarized in Scheme 12, provided additional evidence for the structure of the alkaloid as represented by **51** (*34*).

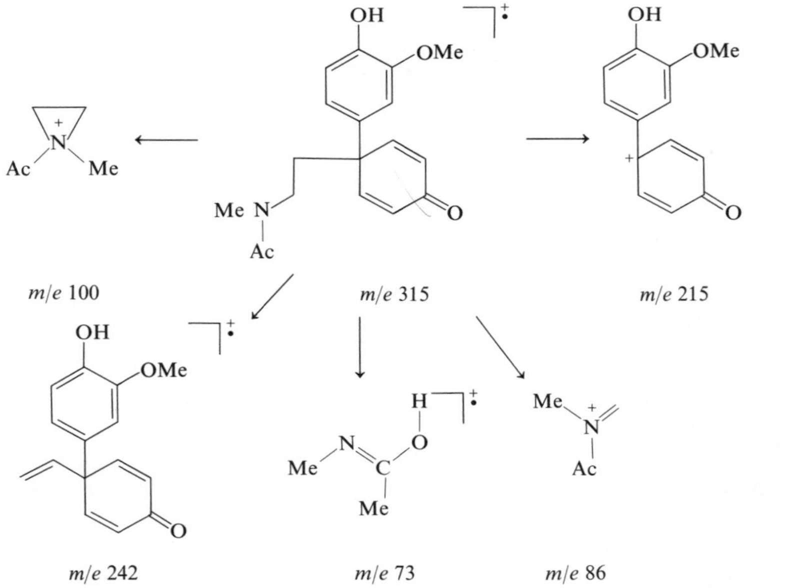

m/e 100 *m/e* 315 *m/e* 215

m/e 242 *m/e* 73 *m/e* 86

SCHEME 12. *Mass spectral fragment ions derived from N-acetyl-4′-O-demethyl-N,7a-seco-mesembradienone.*

Confirmation of this structure was provided by the transformations leading from (−)-4′-*O*-demethylmesembrenone, which are depicted in Scheme 13.

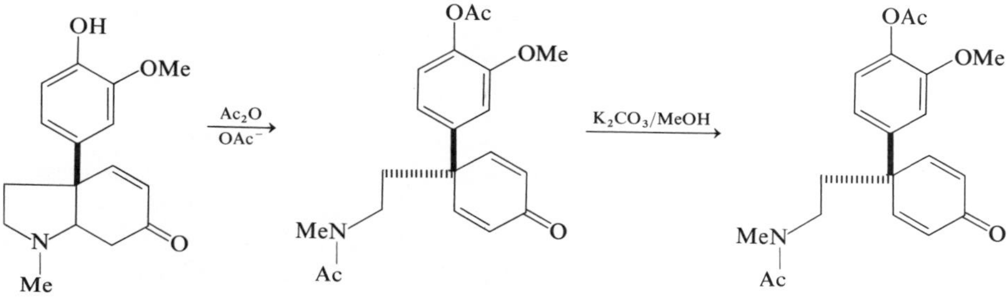

51

SCHEME 13. *Conversion of 4′-O-demethylmesembrenone to N-acetyl-4′-O-demethyl-N,7a-secomesembradienone* (**51**).

C. PYRIDINE–DIHYDROPYRIDONE BASES

At the time of writing there are only two members, *Sceletium* alkaloid A_4 (**52**) (*35*) and the unnamed pyridone base **53** (*36*) constituting this subgroup. A comparison of the structures of these compounds with the structure

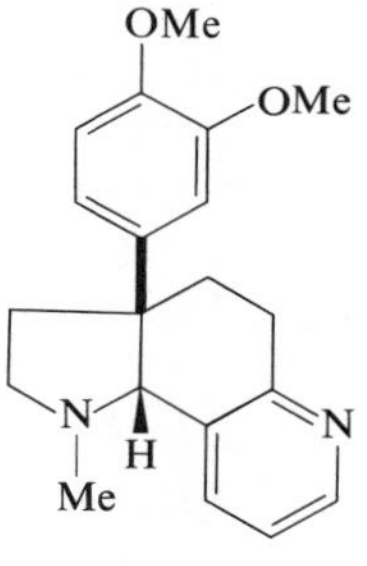

52 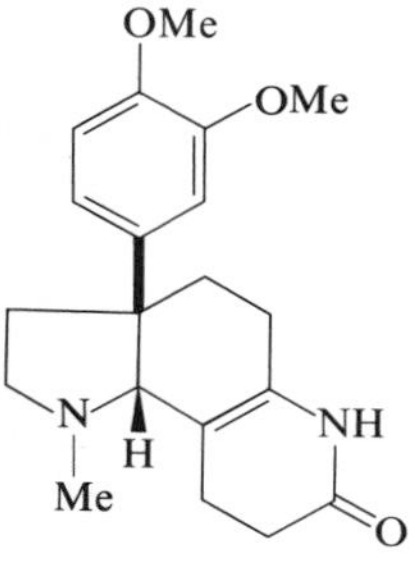 **53**

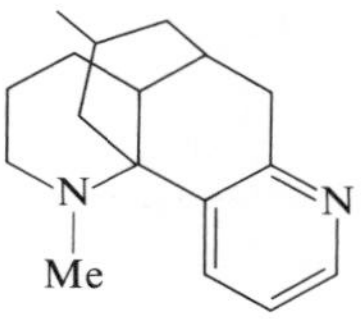

54 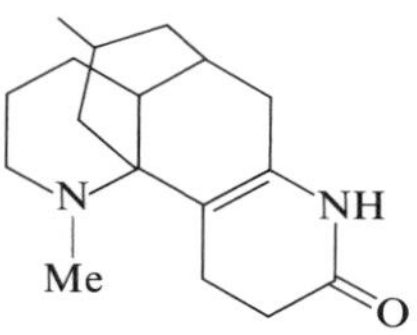 **55**

of mesembrine reveals the presence of an additional six-membered ring containing a single nitrogen atom. It is interesting to note that the structural relationship between these two alkaloids has a direct parallel in the *Lycopodium* series, as may seen from the structures of *N*-methyllycodine (**54**) and α-obscurine (**55**) (*37*), although biogenetically it is difficult to conceive of any common parallels in their origins.

1. *Sceletium* Alkaloid A_4

Sceletium alkaloid A_4, a crystalline base, mp 153°–154°, was first reported by Popelak and Lettenbauer (*15*). While the authors provided no significant structural details other than the assignment of a veratryl unit, and an *N*-methyl group together with the reported absence of groups which are reduced by hydrogen over platinum, the molecular formula of $C_{20}H_{24}N_2O_2$ indicated that this alkaloid was a member of a new structural class.

Further studies (*35*) on the alkaloids of *S. namaquense* led to the isolation of alkaloid A_4 from this plant. The alkaloid was optically active, [α] +131°, and several important structural features became evident from the 250 MHz ^{1}H-NMR spectrum. Two AMX spin systems in the aromatic region were in accord with the presence of both the previously designated veratryl unit and a 2,3-disubstituted pyridine ring. The UV spectrum of alkaloid A_4 was shown to agree well with the summation of isolated chromophores originating from a veratryl ring and a 2,3-dialkylpyridine ring. Although the mass spectrum showed little similarity with the fragmentation behavior of alkaloids of the joubertiamine or mesembrine series, an ion occurs at *m/e* 219 with the same elemental composition, $C_{13}H_{17}NO_2$, as the ion of corresponding mass (see Scheme 4) which is diagnostically useful in characterizing 3,4-dimethoxyphenyl alkaloids of the mesembrine series. This and other mass spectral data provided evidence for alkaloid A_4 containing a 3,4-dimethoxyphenyl unit attached to position 3 of an *N*-methylpyrrolidine ring.

The spectral information reduced the number of possible structures to six, and although biogenetic arguments (see Section VII) led to a clear preference for structure **52** (without stereochemical details), no definitive evidence which would distinguish this structure from the other possibilities was available. Consequently, a single crystal X-ray structure analysis was undertaken, and this established the structure and stereochemistry of *Sceletium* alkaloid A_4 as **52** (*35*, *38*). Because the X-ray structure analysis was effected by the nonheavy-atom direct-methods approach, the absolute stereochemistry was not provided. Structure **52** is represented rather than its mirror image simply because it corresponds to the absolute stereochemistry of the alkaloids of the mesembrine series. The chemical correlation between the mesembrine bases and A_4 which would establish the absolute configuration of the latter has not yet been effected.

An independent report (*39*) has indicated that *Sceletium* alkaloid A_4 occurs as the partial racemate in *S. tortuosum*.

2. Dihydropyridone Base

A noncrystalline, optically active base, $[\theta]_{241}$ $-10{,}500^\circ$, subsequently characterized as the dihydropyridone **53**, has been isolated from the polar fractions resulting after alumina chromatography of the nonphenolic alkaloids of *Sceletium namaquense* (*36*).

The molecular formula $C_{20}H_{26}N_2O_3$ was established by high-resolution mass spectrometry, and the fragmentation pattern was quite dissimilar to the spectra of A_4 or alkaloids of the tortuosamine series. The base peak is a fragment ion with the elemental composition C_2H_5NO; this finding, in conjunction with the observation of other fragments resulting from the loss of CO, M—C_2H_5NO, M—CHNO, and C_2H_3NO, suggested the presence of a dihydropyridone ring. This conclusion was supported by the IR spectrum which contained a strong carbonyl absorption at 1675 cm^{-1}, an N—H stretching vibration at 3420 cm^{-1}, and a carbon double-bond stretch at 1695 cm^{-1}. The ^1H-NMR spectrum, in addition to containing signals attributable to a veratryl moiety, showed an *N*-methyl singlet at 2.50 δ and a singlet at 7.5 δ assigned to an N—H resonance.

The spectral data, in conjunction with comparisons with the structural features of *Sceletium* A_4, suggested that the new base was represented by structure **53**. A racemic compound possesing this structure had been previously synthesized by Stevens and co-workers (*40*) in their studies leading to the synthesis of the A_4 alkaloid. A direct comparison of samples of the two alkaloids revealed their identity and vindicated Professor Stevens' prediction that the dihydropyridone alkaloid would indeed occur as a natural product (*40*).

D. Tortuosamine and Related Bases

The alkaloids of this subclass presently consist of three bases: tortuousamine (**56**) and its *N*-formyl (**57**) and *N*-acetyl (**58**) derivatives. The structures of these compounds bear the same relationship to *Sceletium* A_4 as do the joubertiamine alkaloids to the mesembrine series.

Tortuosamine, a noncrystalline, optically active base, was isolated together with the partial racemate of the A_4 alkaloid from *S. tortuosum* by Wiechers and co-workers (*39*). The molecular formula $C_{20}H_{26}N_2O_2$ was established by high-resolution mass spectrometry, which also provided evidence for an *N*-methylaminoethyl side chain from the observation of the base peak at *m/e* 44 corresponding to the ion **59**. The presence of a 2,3-disubstituted pyridine ring and a veratryl group, which were indicated from the UV spectrum, was confirmed by the ^{1}H-NMR spectrum which at 60 MHz exhibited a low-field AMX system associated with the pyridine ring protons; a 3-proton multiplet in the aromatic region; and three 3-proton singlets at 3.77, 3.81, and 2.32 δ associated with two methoxyls and an *N*-methyl group, respectively.

The secondary amine functionality was confirmed by the formation of an *N*-acetate (**58**) and by methylation of tortuosamine to the tertiary amine **60**. Although the evidence was not sufficient to propose a firm structure for tortuosamine, the spectral data indicated a close structural relationship with *Sceletium* alkaloid A_4. The presence of an *N*-methylethanamine side chain in tortuosamine was in keeping with this alkaloid being an *N*,C-7a seco derivative (mesembrine numbering) of the A_4 base as represented by structure **56**. This structure was confirmed by the conversion of *Sceletium* A_4 (**52**) to tortuosamine by catalytic hydrogenolysis.

Both *N*-formyltortuosamine (**57**) (*27*) and *N*-acetyltortuosamine (**58**) (*34*) have been isolated form *S. namaquense*. Although their initial characterization rested on mass and ^{1}H-NMR spectral evidence, confirmation of the structures of these two compounds was obtained by observation of their interconversion with tortuosamine. The interconversion sequence, *Sceletium* $A_4 \rightarrow$ tortuosamine $\rightleftarrows$ *N*-formyl and *N*-acetyl tortuosamine, has also established that these compounds all belong to the same absolute stereochemical series. Unfortunately, determination of the absolute configuration of these compounds has not been accomplished, although it would be very surprising if their configurations differ from those which are portrayed.

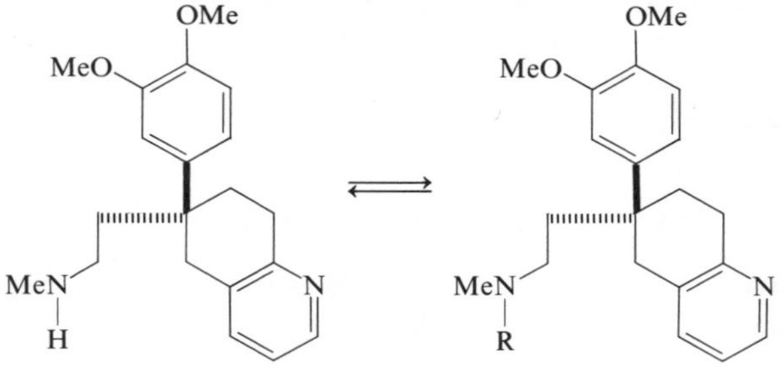

OMe MeO MeN H N

56

OMe MeO MeN R N

57 R = CHO
58 R = COMe
60 R = Me

$MeN^+H{=}CH_2$

59 *m/e* 44

V. Synthesis of *Sceletium* Alkaloids

The area of synthesis of *Sceletium* alkaloids has been one of constant endeavor over the past 15 years, since the report of the first synthesis of (±)-mesembrine by Shamma and Rodriguez (*41*). Following this noteworthy but rather lengthy synthesis, a number of more expeditious routes ensued. Like many other natural products, the *Sceletium* alkaloids have played a role in stimulating the creative endeavors of the organic chemist in devising new reactions to meet the challenge of total synthesis of these alkaloids. While the structures of the alkaloids alone have no doubt provided sufficient challenge for the synthetic organic chemist, the fact that the mesembrine subgroup bears a close structural similarity to the Amaryllidaceae alkaloids of the crinine series, *cf.* crinine (**61**) vs **1**, has undoubtedly provided added impetus for the development of synthetic routes with both structural classes as targets.

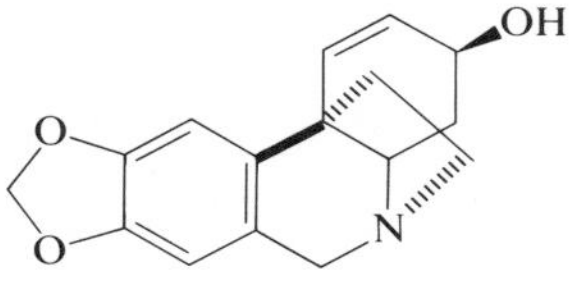

61 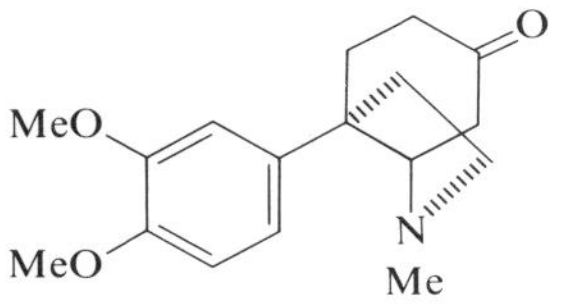 **1**

A. Synthesis of Alkaloids of the Mesembrine and Joubertiamine Classes

1. The Shamma and Rodriquez Synthesis of (±)-Mesembrine

The first total synthesis of (±)-mesembrine was reported in 1965 (*41*). A summary of the reaction scheme employed is presented in Scheme 14. There are several points of interest in the synthesis which should be mentioned. The application of the seldom-used Nef reaction for the introduction of the carbonyl group was not without initial problems. Poor yields of the ketone were obtained until it was recognized that the sodium aci-salt of the nitro compound was sensitive to oxygen and heat. Introduction of a suitable two-carbon fragment necessary for the construction of the pyrrolidine ring was also not without difficulty. Attempts to alkylate the ketone (step 4) with bromoacetate, chloroacetonitrile, and *N*-methylethylenimine were unsuccessful. The solution to this problem was found in using the more reactive alkylating agent, allyl bromide.

Two further points of interest are, first, the use of the mild brominating agent *N*-trimethylanilinium perbromide (step 9) for the selective α-bromination of the ketone in the presence of an activated aromatic ring, and second,

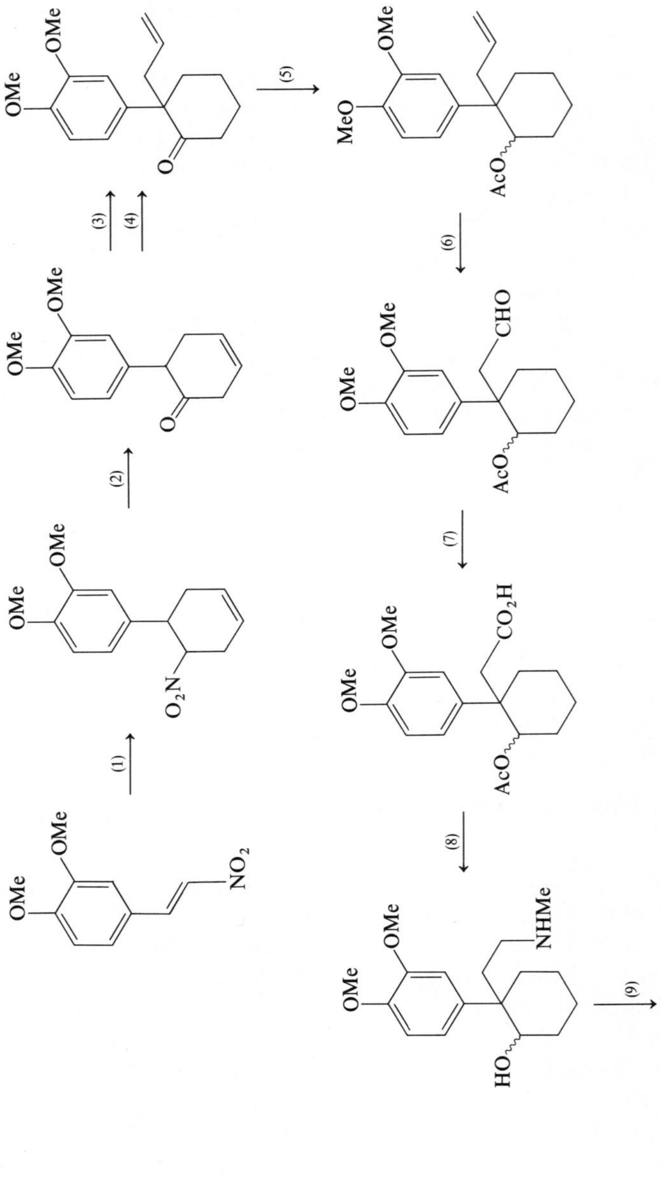
OMe
OMe
NO$_2$
(1)
O$_2$N
(2)
O
(3)
(4)
(5)
MeO
AcO
(6)
CHO
(7)
CO$_2$H
(8)
NHMe
HO
(9)

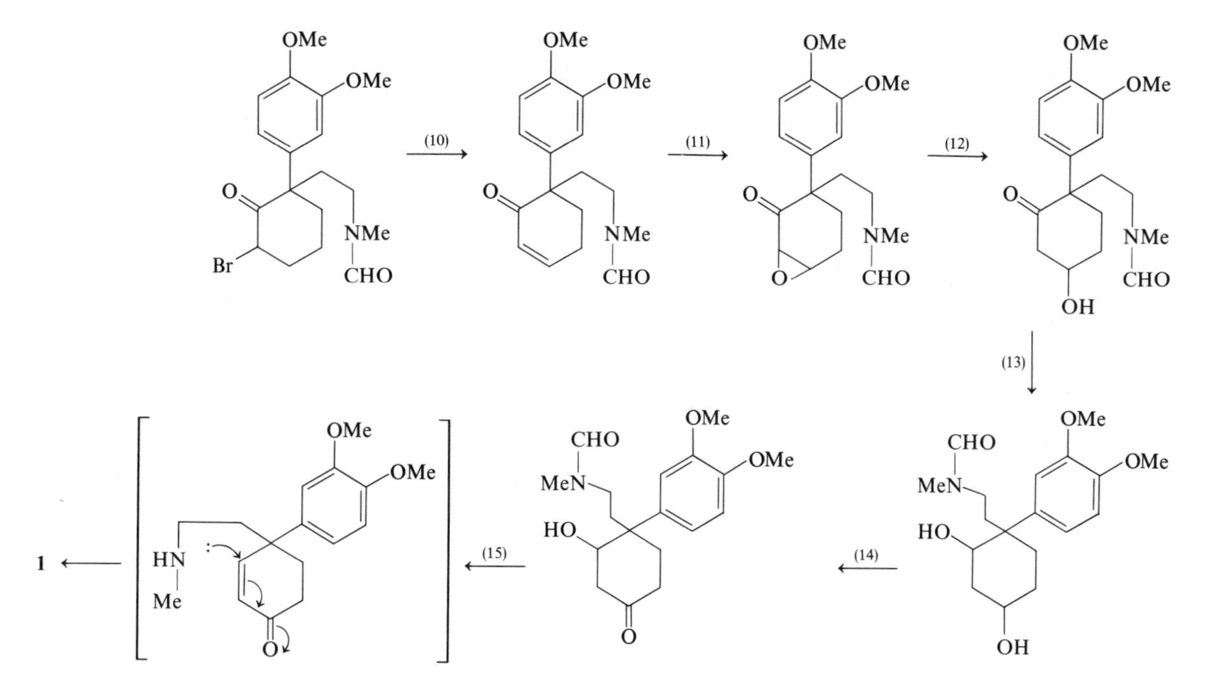

SCHEME 14. *Synthesis of* ($\pm$)*-mesembrine by Shamma and Rodriguez. Legend:* (*1*) *is* $\diagup\!\!\diagdown\!\!\diagup\!\!\!\diagup$; (*2*) *NaOEt, HCl;* (*3*) H_2/Pt; (*4*) $CH_2{=}CH{-}CH_2{-}Br$, *NaH;* (*5*) $LiAlH_4$, Ac_2O; (*6*) OsO_4; (*7*) Ag_2O; (*8*) $COCl_2$, $MeNH_2$, $LiAlH_4$; (*9*) HCO_2Et, $HCrO_4$, $Ph\overset{+}{N}Me_3Br_3^-$; (*10*) $CaCO_3$, *DMF;* (*11*) OCl^-, *Pyridine;* (*12*) $Cr(OAc)_2$; (*13*) $NaBH_4$; (*14*) *Pt,* O_2; (*15*) *HCl, EtOH.*

the transposition of functionality in the ketol via catalytic oxidation (step 14) of the diol in which the least hindered hydroxyl group is selectively oxidized. Treatment of the product ketol from this reaction with 10% HCl effects both *N*-deformylation and β-elimination of the β-hydroxyketone to provide the enone (as the hydrochloride salt) which undergoes a Michael addition on generation of the free base to afford ($\pm$)-mesembrine. The development of the correct cis ring fusion in this latter reaction is predicted as the kinetically controlled product from stereoelectronic considerations, and is also the expected product on thermodynamic grounds.

2. Annelation of Endocyclic Enamines

Utilization of the endocyclic enamine **62** for the synthesis of mesembrine and mesembrenone was reported almost simultaneously from three different

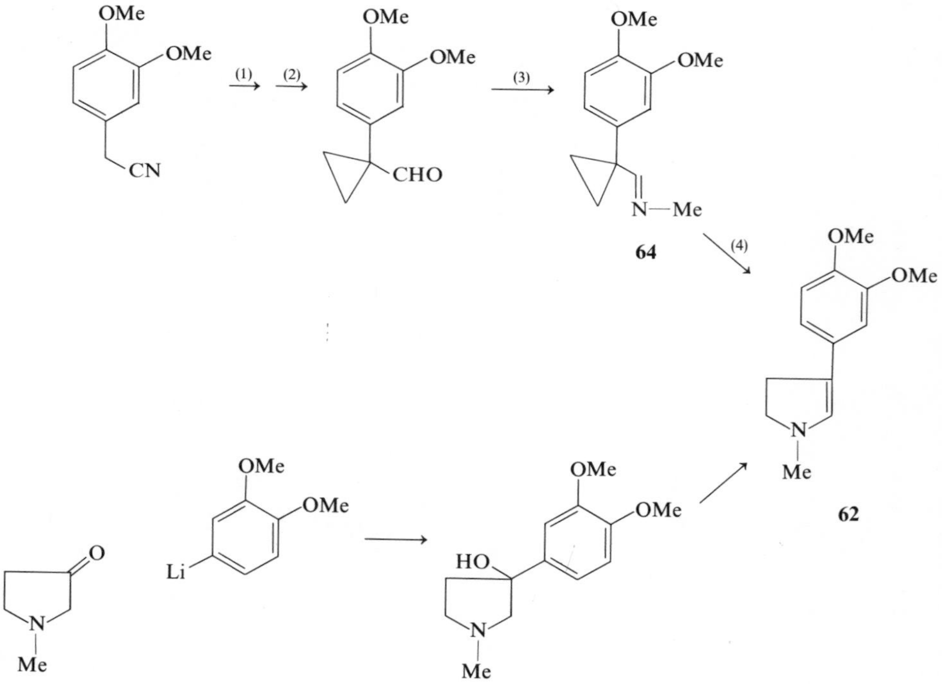

SCHEME 15. *Two routes to the endocyclic enamine mesembrine synthon* (**62**). *Legend: (1) is* $BrCH_2CH_2Br$, $LiNH_2$; *(2) Dibal-H; (3)* $MeNH_2$; *(4)* NH_4Cl, Δ. *From Keely and Tahk (44), Stevens and Wentland (43), and Curphey and Kim (42).*

laboratories (see Scheme 15). Curphey and Kim (*42*) prepared the enamine by the reaction of 4-lithioveratrole with *N*-methyl-3-pyrrolidone (**63**) and subsequent acid-catalyzed dehydration of the product. The other two groups, led by Stevens (*43*) and Tahk (*44*), employed acid-catalyzed, thermally induced rearrangement of a 1-arylcyclopropylaldimine (**64**) for their synthesis of the enamine **62**.

Each of the respective research groups effected the transformation of the endocyclic enamine **62** to (±)-mesembrine by means of a Michael addition with methyl vinyl ketone (Scheme 16). The yields for this reaction reported

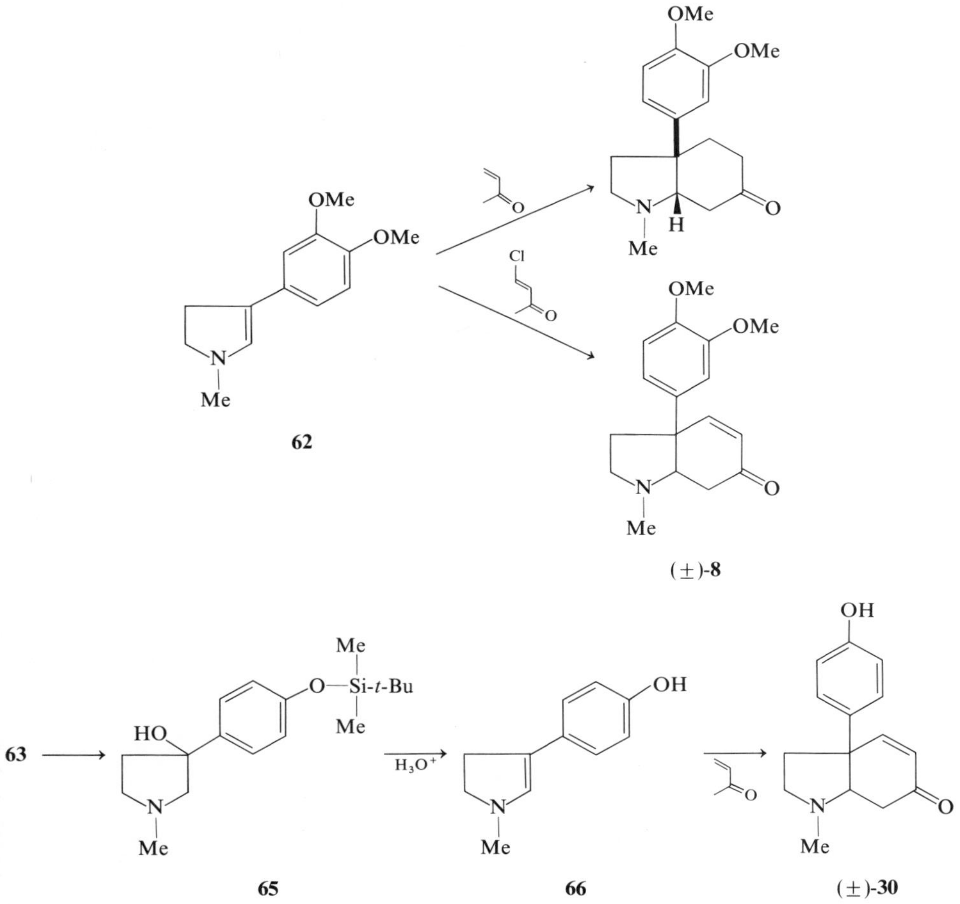

OMe OMe N H Me O OMe OMe Cl O O N Me **62** OMe OMe N Me O (±)-**8**

Me O—Si-*t*-Bu Me HO **63** N Me H_3O^+ OH N Me O OH N Me O **65** **66** (±)-**30**

SCHEME 16. *Utilization of the enamine* **62** *and* **63** *for the synthesis of mesembrine, mesembrenone, and sceletenone.*

by the three groups in these initial attempts were quite variable, ranging from 85% to 5%. Curphey and Kim were able to demonstrate that their high yield was the result of the realization that the Michael addition of methyl vinyl ketone to the endocyclic enamine was strongly subject to acid catalysis. A similar reaction of **62** with β-chlorovinyl methyl ketone was shown to afford the $\alpha\beta$-unsaturated ketone ($\pm$)-mesembrenone (**8**). Later work led to an extension of this approach to the synthesis (*45*) of ($\pm$)-sceletenone (**30**), albeit in poor yield, by an analogous reaction sequence **63** → **65** → **66**; see Scheme 16.

The synthesis of 1-arylcyclopropylimines has been improved (*34*), and their conversion to Δ^1- or Δ^2-pyrrolines for further elaboration with methyl vinyl ketone and methyl vinyl ketone analogs has been brilliantly employed by Stevens and his co-workers for efficient syntheses of not only *Sceletium* alkaloids and the structurally related alkaloids of the *Amaryllidaceae* family, but alkaloids of diverse structure representing other alkaloid families (*46*).

The application of the annelation of endocyclic enamines to the *Sceletium* alkaloids is exemplified in the synthesis of ($\pm$)-joubertiamine (**33**), ($\pm$)-*O*-

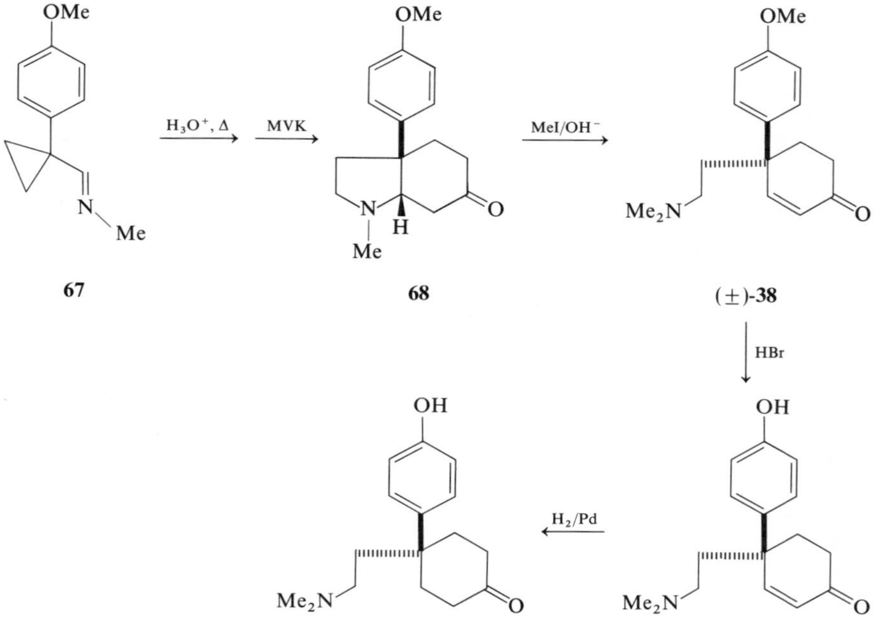

SCHEME 17. *Synthesis of joubertiamine* (**33**) *and related bases* **38** *and* **35** *by the endocyclic enamine route.*

methyljoubertiamine (**38**), and (±)-dihydrojoubertiamine (**35**) (*47*). Transformation of the *p*-methoxyphenyl cyclopropylaldimine **67** to the octahydroindole intermediate **68** was accomplished by reactions which paralleled those previously described in Schemes 15 and 16. The conversion of **68** to the secomesembrine skeleton of the alkaloids of the joubertiamine series represented by structures **38**, **33**, and **35** was accomplished readily by standard reactions, as portrayed in Scheme 17.

3. Acid-Catalyzed Cyclization of 3-Acetyl-3-(3′, 4′-dimethoxyphenyl)adiponitrile

The development of an alternate route (*48*) to the mesembrine ring system which has led to the synthesis of (±)-mesembrine is based on the interesting observation that, on treatment with H_2SO_4, each of the compounds **69–71** affords the lactam–enamide **72**. The mechanism of these conversions has been established and involves a series of complex equilibria (*49*) the details of which will not be discussed here. A point of interest concerns the mode of cyclization which, under the reaction conditions (65% H_2SO_4 at 140° for 5 min), gave exclusively the lactamenamide **72** with none of the isomer containing a six-membered lactam ring being observed. Unfortunately, the conversion of **72** to mesembrine via **73** could not be effected satisfactorily. Despite a study of the reduction of **73** with a variety of reagents under different conditions, the most expenditious method employing catalytic reduction over Raney nickel gave a mixture of products containing only 30–46% of the stereoisomeric 2-oxomesembranols (**74**) together with the hyhrogenolysis products **75** and **76**. Jones oxidation of **74** afforded a 1:1 mixture of 2-oxomesembrine (**77**) and the isomer **78** containing a trans ring fusion. After separation of this mixture by chromatography on alumina, ketalization of **77** to give **79** followed by reduction with LAH and subsequent deketalization afforded (±)-mesembrine (**1**). The diastereoisomer of mesembrine, 7a-epimesembrine (**80**), was obtained from **78** by the same sequence. This synthesis, which is summarized in Scheme 18, has little to commend it from the viewpoint of either yield or stereoselectivity.

4. Synthesis of Mesembrine and Joubertiamine Alkaloids via Routes Involving Geminal Aklylation at a Carbonyl Center

Recent work by Martin and co-workers (*50*) has provided examples of two elegant synthetic routes to (±)-mesembrine and (±)-*O*-methyljoubertiamine.

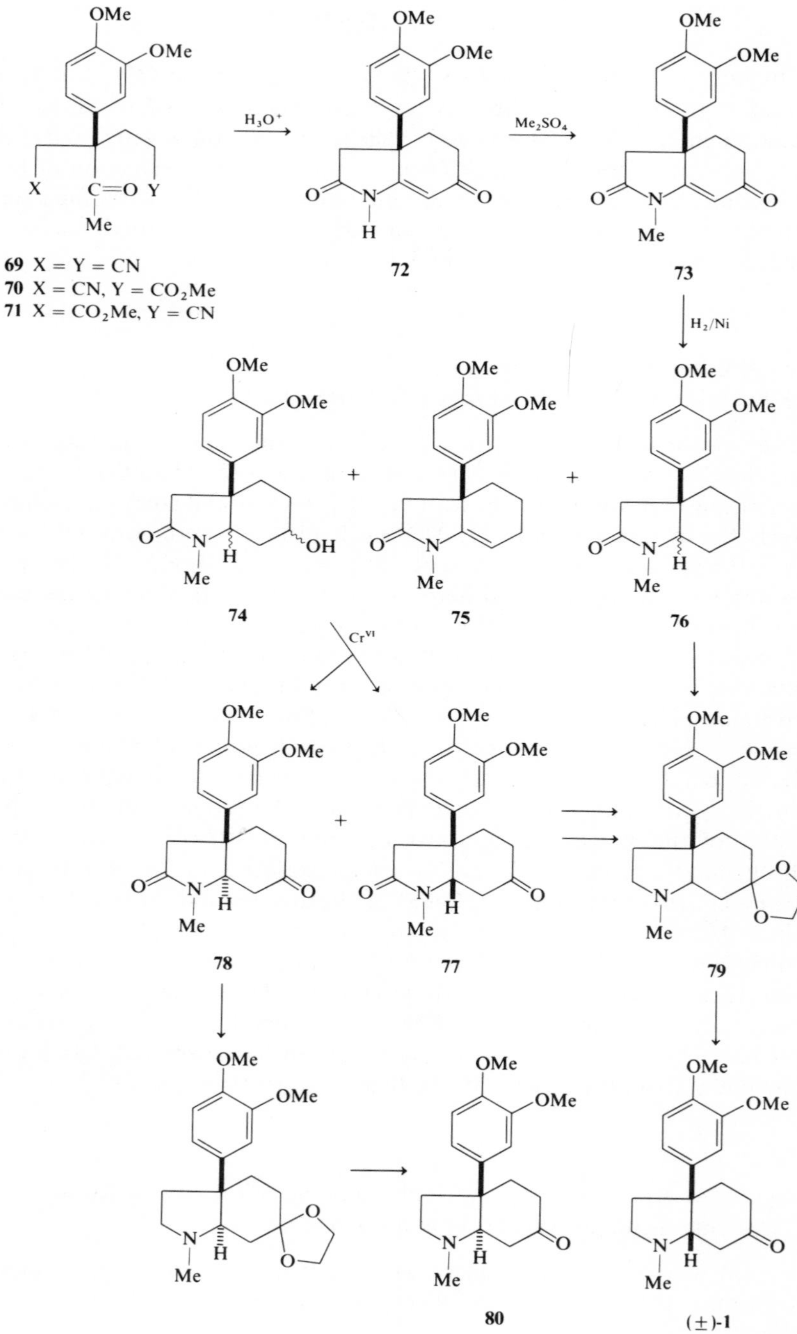

69 X = Y = CN
70 X = CN, Y = CO_2Me
71 X = CO_2Me, Y = CN

SCHEME 18. *Synthesis of* (±)*-mesembrine and* (±)*-7a-epimesembrine* (**80**).

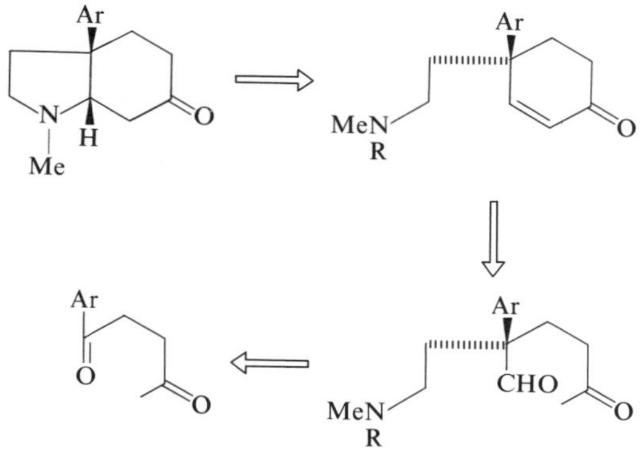

SCHEME 19. *Retrosynthetic analysis of a synthetic scheme for mesembrine.*

The first route employed successfully for the synthesis of (±)-*O*-methyljoubertiamine (**38**) and (±)-*O*-methyldihydrosceletenone (**68**) is based on the retrosynthetic scheme presented in Scheme 19. The creation of the quaternary carbon center in this scheme, which is a key feature of the approach, is accomplished by conversion of the carbonyl center to the enamine **81** using the requisite Wittig reagent (**82**), followed by subsequent alkylation. These reactions and the subsequent synthetic route (±)-*O*-methyljoubertiamine are presented in Scheme 20. The keto aldehyde **83**, which is obtained from alkylation of **81** with allylbromide, undergoes a base-induced ring closure to give a 4,4-disubstituted cyclohexenone (**84**). The latter substance could be transformed most efficiently (75% yield) to (±)-*O*-methyljoubertiamine (**38**) by selective ozonolysis of the vinyl group and reductive amination of the intermediate aldehyde with *N*,*N*-dimethylamine and $NaCNBH_3$. By a slight modification involving protection of the carbonyl in the enone **84** as its ketal, the synthesis of (±)-*O*-methyldihydrosceletenone (**68**) was also achieved.

An attempt to synthesize (±)-mesembrine by a sequence of reactions which paralleled those employed for the synthesis of **68** revealed unexpected difficulty in alkylation of the enamine of 3,4-dimethoxyl phenyl ketone (**85**). The solution to this problem was found by a modified route in which a metalloenamine (**86**) is utilized as depicted in Scheme 21.

Metalloenamines are more nucleophilic than enamines and thus provide greater flexibility in choice of alkylating agents. This flexibility proved advantageous in that a properly functionalized nitrogen-containing side chain could be introduced directly; ultimately, it led to a facile synthesis of (±)-mesembrine in 40% overall yield from veratraldehyde.

The general applicability of the method of using metalloenamines for a regioselective conversion of carbonyl centers to quaternary carbon centers

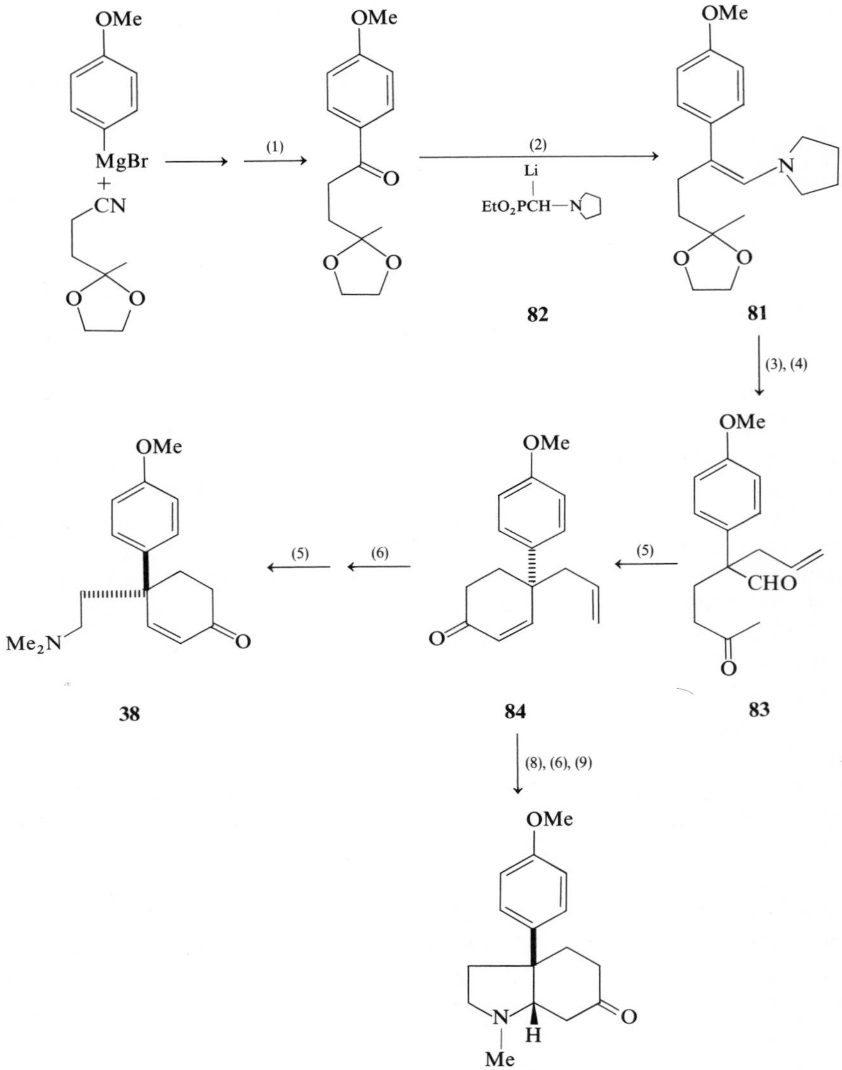

82 81 83 84 38 68

SCHEME 20. *Synthesis of* (±)-*O-methyljoubertiamine* (**38**). *Legend:* (*1*) *is* NH_4Cl/H_2O: (*2*) *THF*, $-78° \rightarrow 25°$; (*3*) $CH_2{=}CH_2Br$, *dioxane*, Δ; (*4*) H_3O^+; (*5*) $KOH/H_2O/MeOH$; (*6*) O_3, CH_3Cl_2, $-78°$; (*7*) Me_2NH, $NaCNBH_3$, *t-BuOH*; (*8*) $Me_2C(OCH_2CMe_2CHO)$, *p-TsOH*, Δ; (*9*) $MeNH_3Cl$, $NaCNBH_3$, *MeOH*.

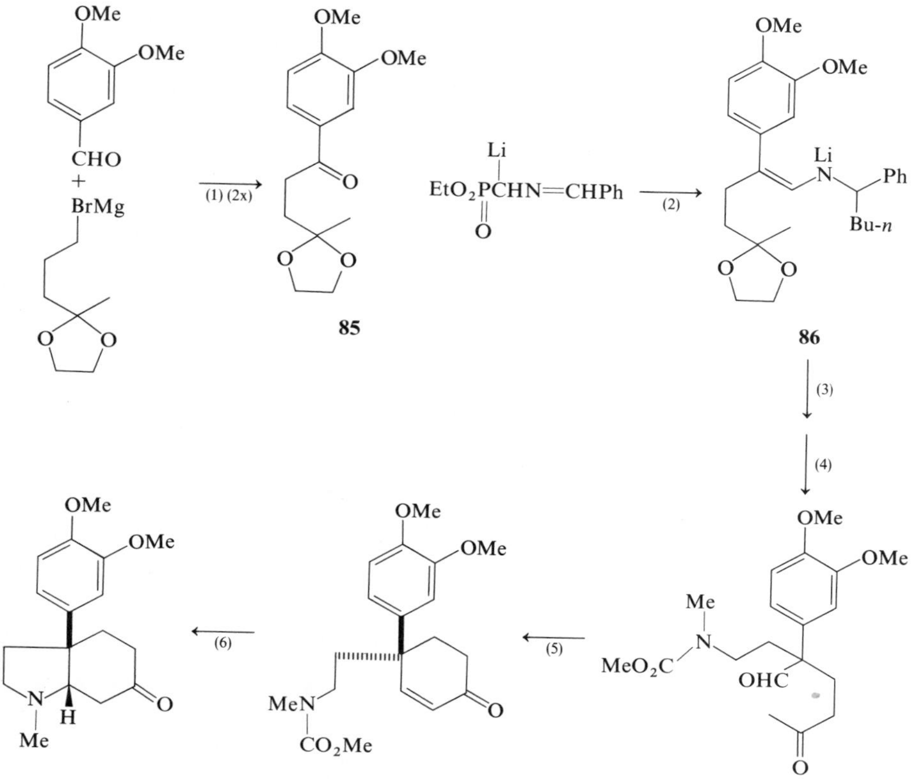

SCHEME 21. *Metalloenamine route to* (±)-*mesembrine. Legend:* (*1*) *is* $Cr(VI)$; (*2*) *n-BuLi*, $-78°$; (*3*) $Br(CH_2)_2NMeCO_2Me$, $-78° \rightarrow 25°$; (*4*) H_3O^+; (*5*) $KOH/H_2O/MeOH$; (*6*) $KOH/H_2O/EtOH$, Δ.

containing different ligands was further demonstrated by an alternate synthesis of (±)-mesembrine by the reaction scheme presented in Scheme 22.

5. Ketene–Alkene Cycloadditions and Aza-Ring Expansion of Cyclobutanones in the Synthesis of Mesembrine and Joubertiamine-Type Alkaloids

The synthesis of alkaloids in the mesembrine and joubertiamine groups by aza-ring expansion of suitably functionalized *cis*-3-arylbicyclo[4.2.0]octanones (**87**) has evolved from earlier studies (*51*) which were designed to provide model systems represented by the general structures **88** and **89**.

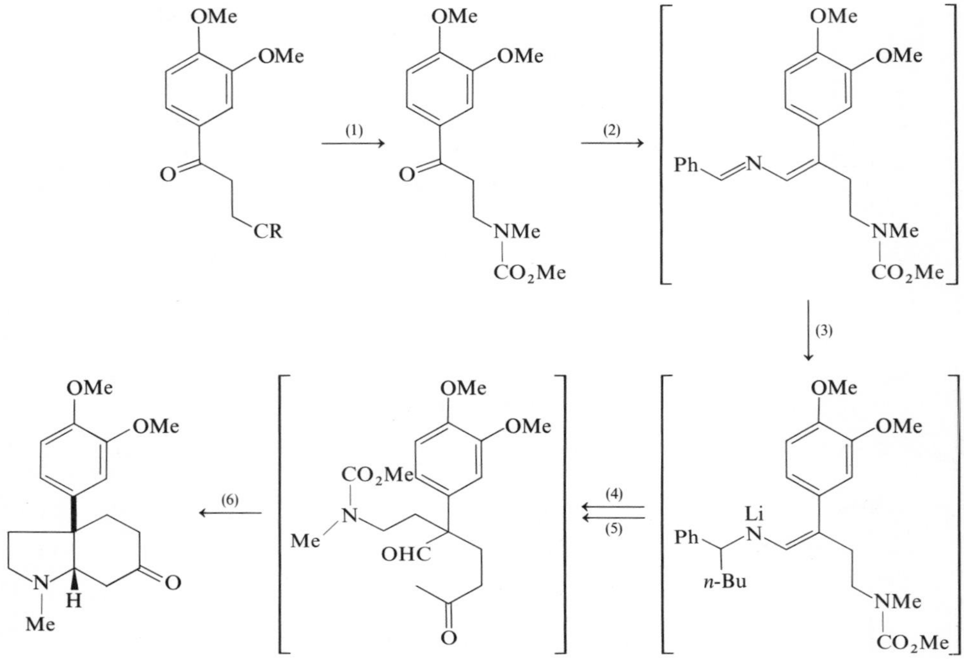

SCHEME 22. *Alternate synthesis of* ($\pm$)*-mesembrine via alkylation of a metalloenamine.*

Legend: (*1*) *is* $MeNHCO_2Me$, $PhCH_3$, $TsOH$, Δ; (*2*) $PhCH{=}NCHLiP(O)(OEt)_2$, *THF*, $-78° \rightarrow \Delta$; (*3*) *m-BuLi*, $-78°$; (*4*) $Br(CH_2)_2C(OCH_2CH_2O)CH_3$, *THF*, *HMPA*, $-78° \rightarrow 25°$; (*5*) H_3O^+; (*6*) $KOH/H_2O/MeOH$.

The strategy of the overall scheme is presented in Scheme 23 and relies on affecting the conversion of a 1-substituted cyclohexene to a cis-1-substituted bicyclo[4.2.0]octanone with a ketene or its equivalent and the subsequent aza-ring expansion of the cyclobutanone to the *cis*-octahydroindole skeleton.

Generation of the desired bicyclo[4.2.0]octanone intermediate from the 1-arylcylohexene requires that the ketene cycloaddition proceed in both a regio- and stereospecifically controlled manner. Initial studies indicated that in keeping with earlier findings, the trisubstituted double bond of l-alkyl- and l-arylcyclohexenes was essentially inert to dichloroketene when it was generated *in situ* from the dehydrochlorination of dichloroacetyl chloride with triethylamine. However, in contrast to this observation, the important finding was noted that the ketenoid reagent obtained from activated zinc and trichloroacetyl bromide when prepared *in situ* in the presence of 1-arylcyclohexenes reacted to provide a single 1,1-dichloro-cyclobutanone possessing the required structural and stereochemical details

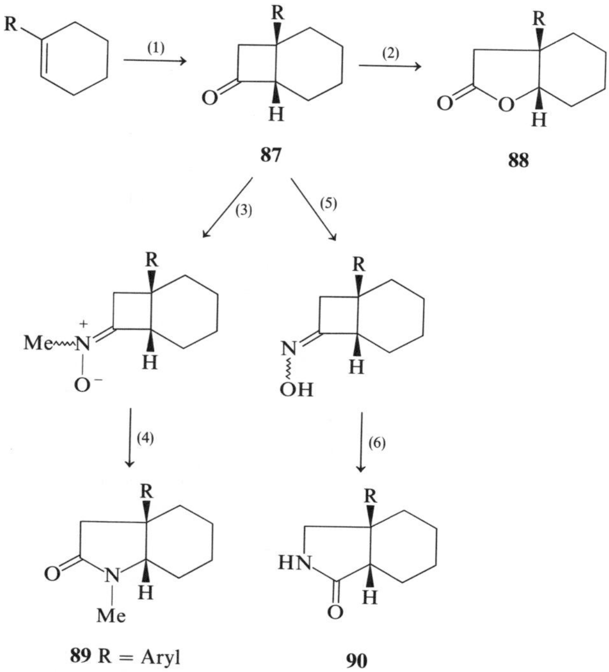

87 **88**

89 R = Aryl **90**

SCHEME 23. *Regiospecific conversion of 1-cyclohexenes to bicyclo[4.2.0]octanones and their aza ring expansion by Beckman and modified Beckman rearrangements. Legend: (1) is Cl_3CCOBr-Zn: (2) H_2O_2, OH^-; (3) MeNHOH, pyridine; (4) tosyl chloride/pyridine; (5) NH_2OH; (6) H_3PO_4,P_2O_5.*

for elaboration to the bicyclic systems **88** and **89**. Although the furanone (**88**), was readily obtained by a Baeyer–Villiger reaction of the corresponding bicyclo[4.2.0]octanones, the aza-ring expansion of *cis*-bicyclo[4.2.0]octanones related to **87** using various modifications of the Beckman rearrangement led only to octahydroisoindolones (cf. **90**). The solution to this problem was found by use of the modified Beckman rearrangement procedure devised by Barton *et al.* involving the solvolytic rearrangement of a nitrone derivative (see Scheme 23), which gave the desired lactam **89** exclusively.

With the successful completion of the model studies, the general approach has been extended to an efficient synthesis of *O*-methyljoubertiamine as illustrated in Scheme 24. The functionalized 1-arylcyclohexene **91** was prepared in high yield from 4-methoxyphenyllithium and the THP ether of 4-hydroxycyclohexanone in four steps by the standard reactions shown. Reaction of **91** as its *O*-acetate with the trichloroacetylbromide–zinc reagent

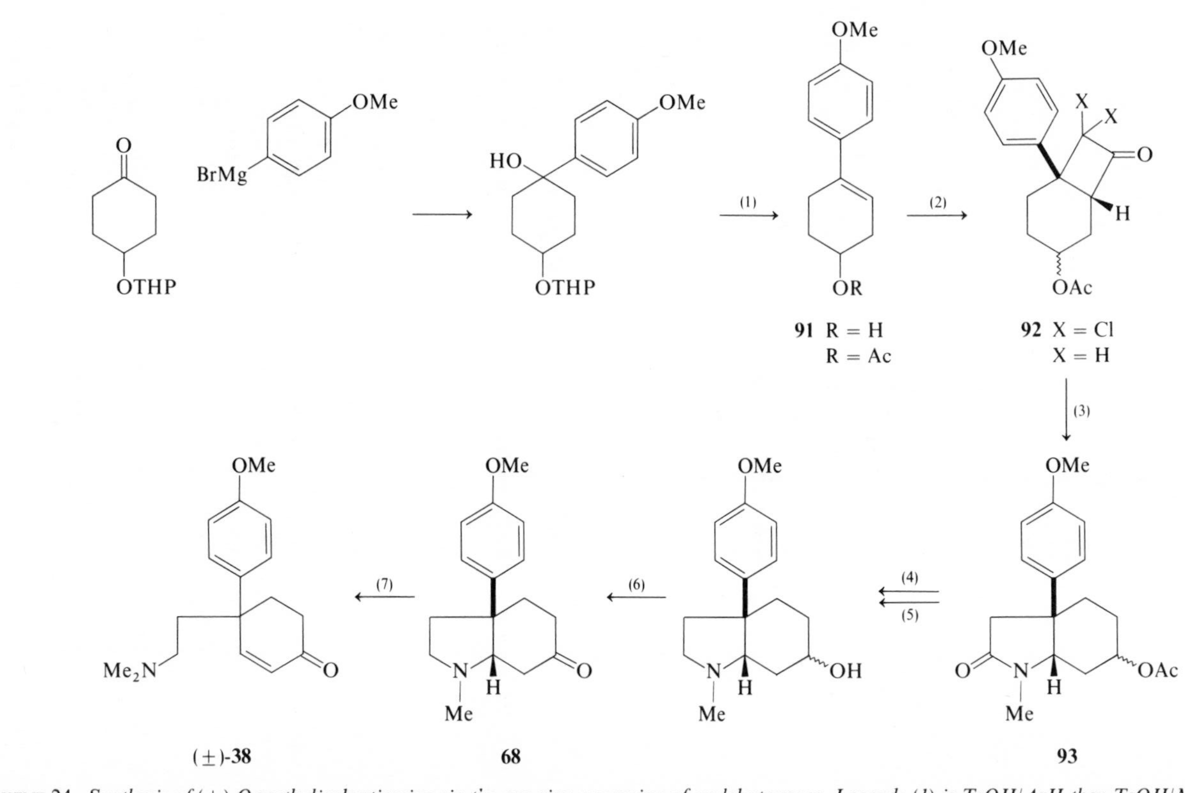

SCHEME 24. *Synthesis of* (±)-*O-methyljoubertiamine via the aza ring expansion of cyclobutanones. Legend:* (*1*) *is TsOH/ArH then TsOH/MeOH,* Ac_2O/Py; (2) Cl_3CCOBr/Zn *then* $Zn/MeOH/NH_4Cl$; (3) *MeNHOH/Py then tosyl chloride/Py,* H_2O; (4) BH_3/THF; (5) $LiAlH_4/THF$; (6) *Cr(VI)/*Me_2CO; (7) *MeI then* $NaHCO_3$.

afforded the dichlorocyclobutanone **92**. Although this compound exists as a pair of diastereoisomers, it was not necessary to effect a separation because the penultimate step in the sequence, involving an oxidation of an alcohol to a ketone, results in the formation of single compound.

Reductive removal of the two chlorines in **92** with zinc in methanol containing ammonium chloride affords the cyclobutanone which when subjected to the Barton modification of the Beckman rearrangement, yields the single regioisomer **93**. Sequential treatment of **93** with diborane followed by LAH without isolation of the intermediates effected the reduction of the lactam and ester groups to give a mixture of epimeric alcohols. Oxidation of this mixture afforded (±)-*O*-methyldihydrosceletenone (**68**) which, on treatment with methyliodide and base, gave (±)-*O*-methyljoubertiamine in 79% overall yield from the alkene **91** (*52*, *53*).

This synthetic procedure has been applied to the synthesis of (±)-mesembrine by an analogous series of reactions to those portrayed in Scheme 24 by replacing 3,4-dimethoxyphenyllithium for 4-methoxyphenylmagnesium bromide in the first step (*52*).

6. Cyclization of α-Acylimmonium Ions

A recent synthesis of (±)-mesembrine reported by Wijnberg and Speckamp (*54*) utilizes an interesting solvolytically induced intramolecular cyclization of an acetylene with an α-acylimmonium ion for the construction of the octahydroindole ring system.

The synthesis, illustrated in Scheme 25, is initiated with the 3-arylsuccinimide **94**. Alkylation of this compound with 4-iodobutyne, as shown, gives the 3,3-disubstituted derivative **95**. Previous studies by the Dutch group had shown that borohydride in acid media effected the regioselective reduction of the more hindered 2-carbonyl in these systems to give the 2-carbinolaminolactam as the major product. When this reaction was employed with amide **95**, it afforded the desired regioisomer **96** in 67% yield. On stirring **96** at room temperature in HCO_2H, it underwent dehydration to the α-acylimmonium ion **97** and concomitant solvolytic cyclization to afford the keto-lactam **77** in near quantitative yield. The conversion of the keto-lactam to (±)-mesembrine was effected by the sequence of reactions described previously in Scheme 18.

Extension of this synthetic procedure to the synthesis of the crinine alkaloid dihydromatridine was also demonstrated. An interesting sidelight in the latter synthesis was the observation that the corresponding *N*-benzyl-α-acylimmonium ion (**98**) containing an olefinic double bond underwent solvolytic cyclization in HCO_2H to give a 4:1 mixture of the formate esters

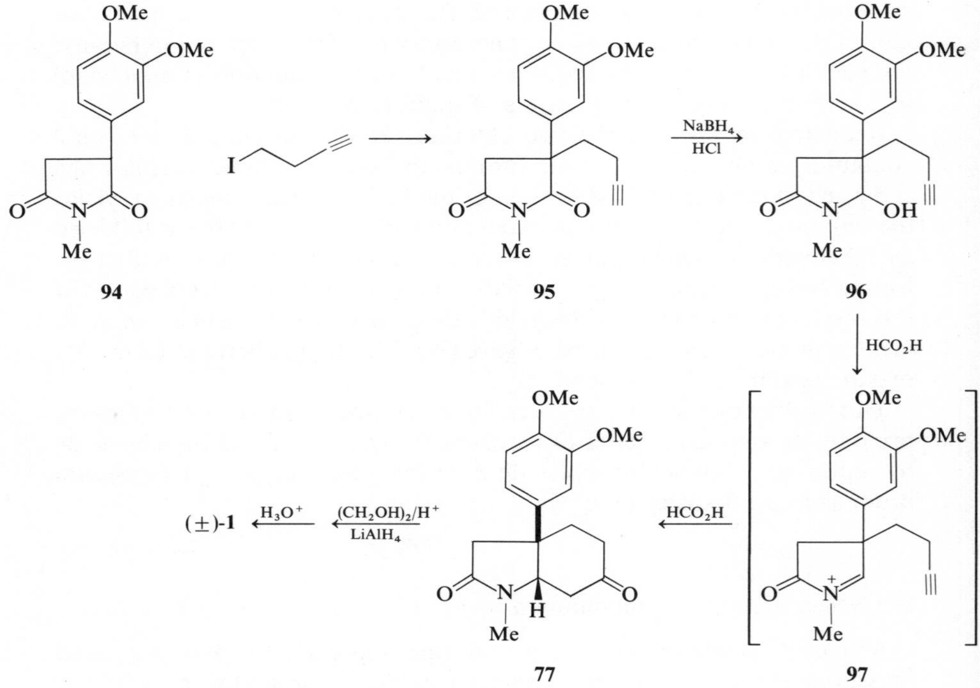

SCHEME 25. *Intramolecular solvolytic cyclization of an alkyne with an α-acylimonium ion as a route to mesembrine.*

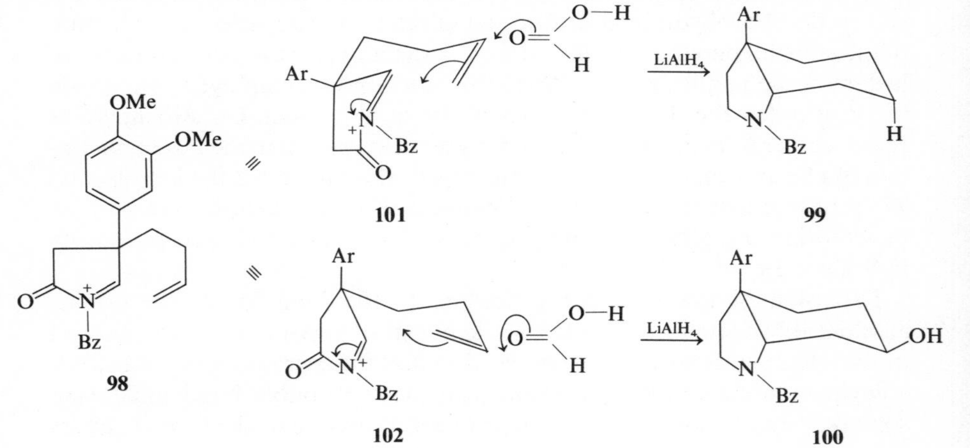

of the C-6 epimeric alcohols **99** and **100**. This finding implies that transition state **101**, possessing an equatorial aryl conformation, is more stable than its axial aryl counterpart, **102**. This situation is in contrast to the preferred ground-state conformations of these and other compounds in the series which, as mentioned in Section IV,A,1, exist in the axial aryl conformation. A possible explanation of this result is that the orbital overlap of the incoming nucleophile and the two π systems in conformation **101** is significantly better than it is in **102**.

7. Enantioselective Synthesis of (+)-Mesembrine

An interesting synthesis of (+)-mesembrine has been described by Otani and Yamada (*55*). The approach follows from earlier studies (*56*) by these authors on the enantioselective synthesis of chiral 4,4-disubstituted cyclohexenones (see Scheme 26) in which construction of the cyclohexenone ring is effected with Michael addition of methyl vinyl ketone to an enamine (**103**) formed from L-proline pyrrolidide and the aldehyde **104**. In the reactions presented in Scheme 26, the *R*-(+)-isomer of 4-methyl-4-phenylcyclohexenone was obtained in 50% enantiomeric excess.

Ph Me H CHO (±)-**104** H N H O N → Me Ph CH—N H O N **103** O HOAc Me Ph O

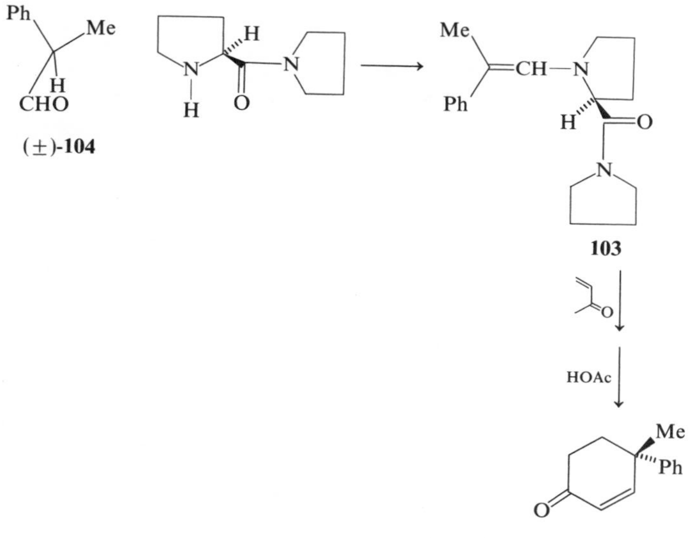

R-(+)-isomers (50% enantiomeric excess)

SCHEME 26. *Enantioselective synthesis of chiral 4-methyl-4-phenylcyclohexenone.*

Extension of this approach to the synthesis of (+)-mesembrine is presented in Scheme 27. In its fundamental approach it is similar in concept to that developed later by Martin's group, in that an intermediate aldehyde **105**,

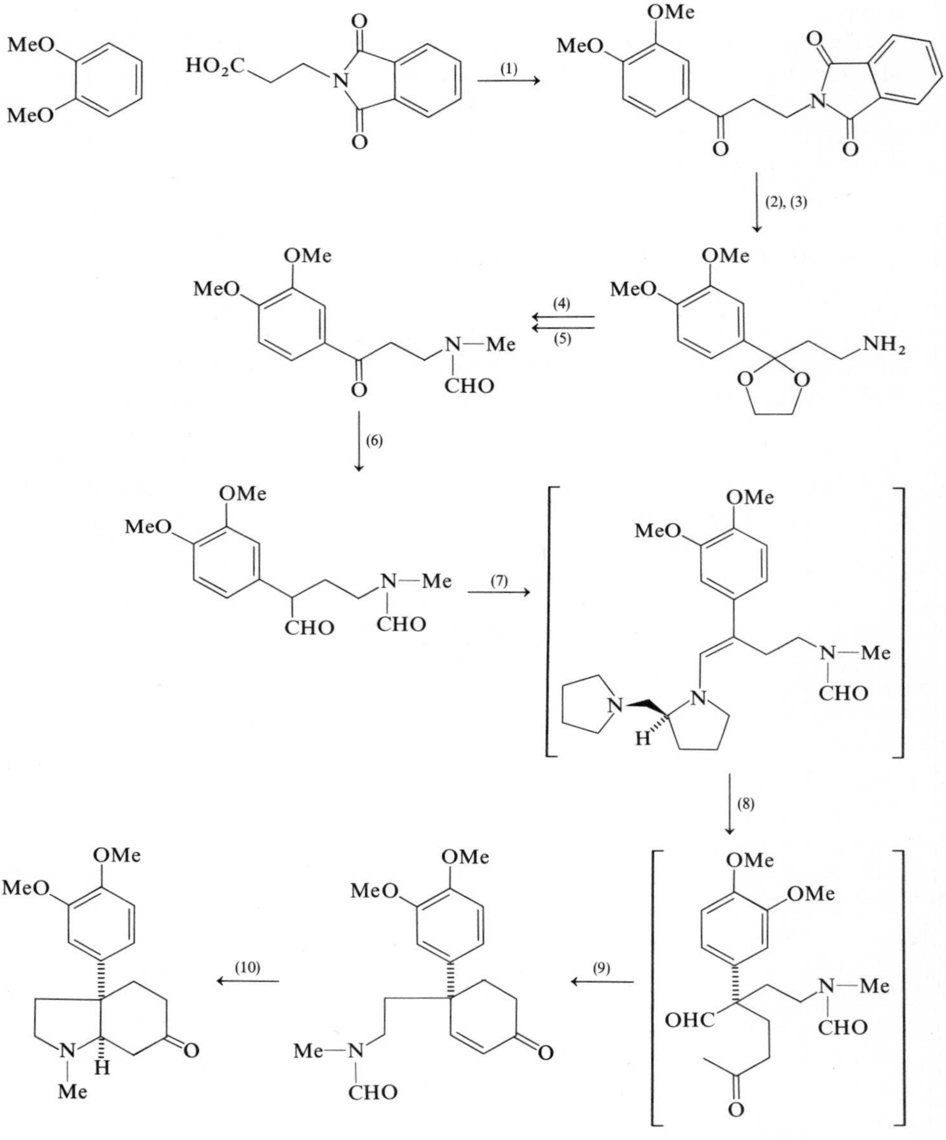

106 **105**

SCHEME 27. *Enantioselective synthesis of (+)-mesembrine. Legend: (1) is polyphosphoric acid; (2)* $(CH_2OH)_2$, *TsOH; (3)* N_2H_4; *(4)* HCO_2Ac *then* $LiAlH_4$; *(5)* HCO_2Ac *then* Me_2CO, *TsOH; (6)* $ClCH_2CO_2Me$, *t-BuO*$^-$ *then NaOH followed by HOAc; (7)* L*(+)-proline pyrrolidide; (8)* $CH_2{=}CH{-}COCH_3$; *(9) HOAc/*H_2O*/pyrrolidine; (10) 10% HCl/EtOH.*

with four disparate groups, is synthesized by an enamine reaction. This procedure differs from that of Martin *et al.* in that a chiral aldehyde is obtained with one enantiomer in excess. Cyclization of the aldehyde gives the protected amine **106** which, on deformylation, leads to mesembrine in 28% overall yield. The specific rotation of the free base of +16° indicated that the unnatural (+)-isomer had been formed in 30% enantiomeric excess. The hydrochloride of this sample fortuitously cocrystallized as prisms and needles. After mechanical separation of the two crystalline modifications, the prisms were discovered to be the racemate while the needles afforded the (+)-isomer in 93% optical purity.

8. (±)-*O*-Methyljoubertiamine: Use of the Claisen–Eschenmoser Rearrangement

Strauss and Wiechers (*57*) have described a synthesis of (±)-*O*-methyljoubertiamine (**38**) which is illustrated in Scheme 28. The approach used by these authors relies on use of the Claisen–Eschenmoser 3,3-sigmatropic rearrangement to create the functionalized 4,4-disubstituted cyclohexenone derivative **107**. This scheme follows closely one utilized successfully by Muxfeldt and co-workers (*58*), reported in 1966 for the synthesis of (±)-crinine. The other reactions employed in the transformation of **107** to **38** via the cyclohexenone **108** are straightforward and need no further comment.

In a later paper (*59*), the South African authors employed the same approach in reporting a synthesis of (±)-joubertinamine **36** (Scheme 28).

The 3,4-dimethoxyphenyl derivative **109** was prepared by a series of reactions which paralleled those of the analogous 4-methoxyphenyl compound **107**. Reductive cleavage of the *N*,*N*-dimethyl amide group in **110** to the aldehyde occurred simultaneously with a stereospecific reduction of the enone to give a single allylic alcohol (**111**) of unspecified stereochemistry. (The stereochemistry depicted in **111** is based on its subsequent transformation to joubertinamine, which has been demonstrated (*34*) to have the relative stereochemistry represented in structure **36**.) Reductive amination of the aldehyde group in **111** with methylamine, and subsequent reduction of the product with borohydride to give (±)-joubertinamine (**36**), completed the synthesis. This synthesis also constitutes a synthesis of (±)-mesembrine, for oxidation of **36**, as previously noted, provides the parent octahydoindole base **1** (see Scheme 9).

B. The Synthesis of *Sceletium* A_4 and Related Bases

Stevens and co-workers (*40*) have extended their fundamental approach from the earlier described synthesis of *Sceletium* alkaloids by employing the

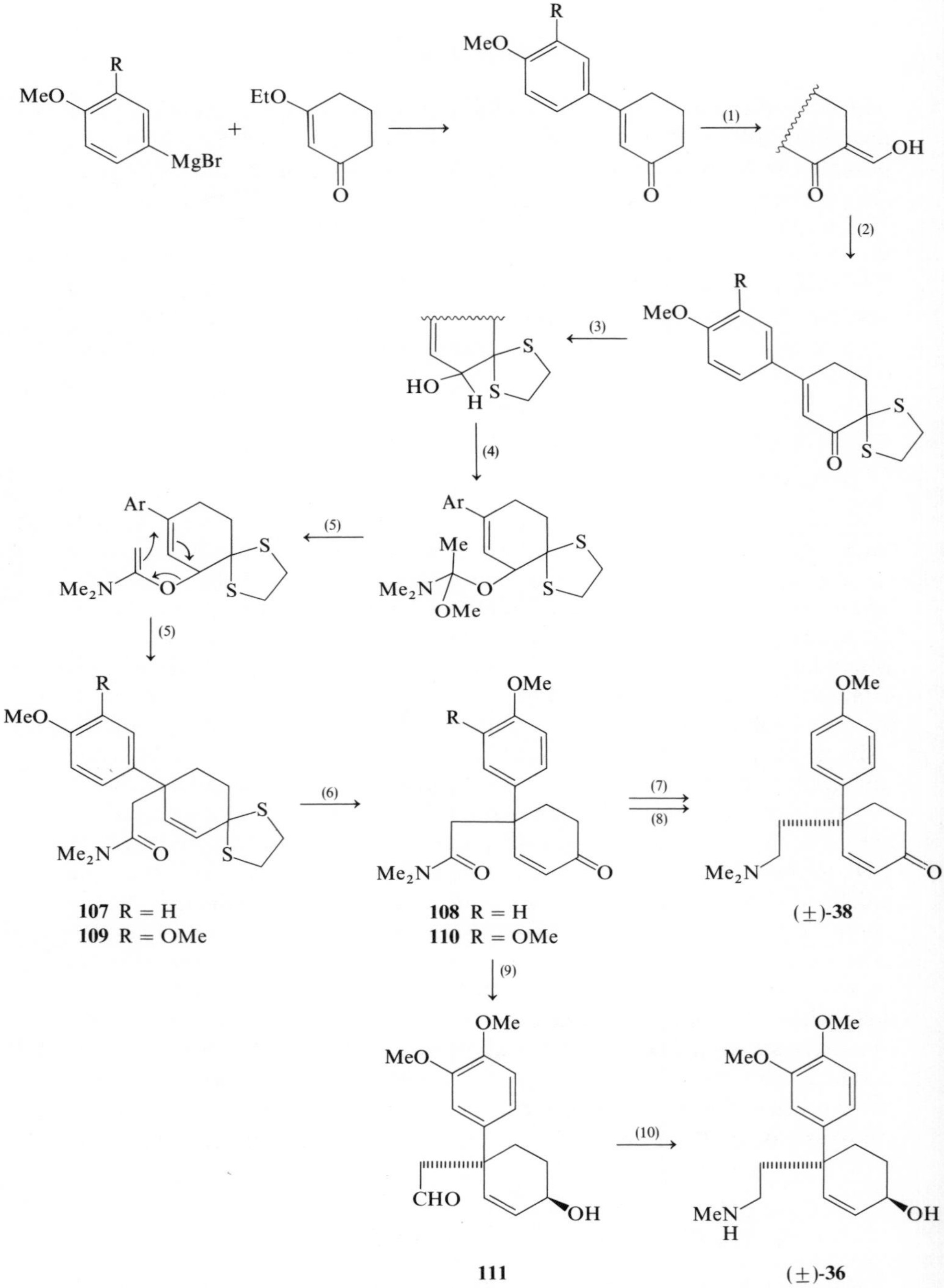

107 R = H
109 R = OMe

108 R = H
110 R = OMe

(±)-**38**

111

(±)-**36**

SCHEME 28. *Employment of the Claisen-Eschenmoser rearrangement in the synthesis of* (±)-*O-Methyljoubertiamine. Legend: (1) is* $HCO_2Et/NaH/benzene$; (2) $CH_2(CH_2S\text{-}Ts)_2\text{-}KOAc$; (3) $LiAlH_4/THF$; (4) $Me_2N\text{-}C(OEt)_2Me$; (5) Δ; (6) MeI/Me_2CO *or* $MeOH/H_2O$; (7) $LiAlH_4$; (8) MnO_2; (9) *Dibal-H/ArMe*; (10) $Me_2NH/ArH/MgSO_4$ *then* $NaBH_4$.

endocyclic enamine synthon **62** in a synthesis of the dihydropyridone **53** and *Sceletium* A_4 (**52**) as summarized in Scheme 29.

An attempt to synthesize the A_4 alkaloid by reacting **62** and 2-vinylpyridine was not successful. However, reaction of **62** with the enone **112** provided intermediate **113**, which was converted in separate reaction paths to

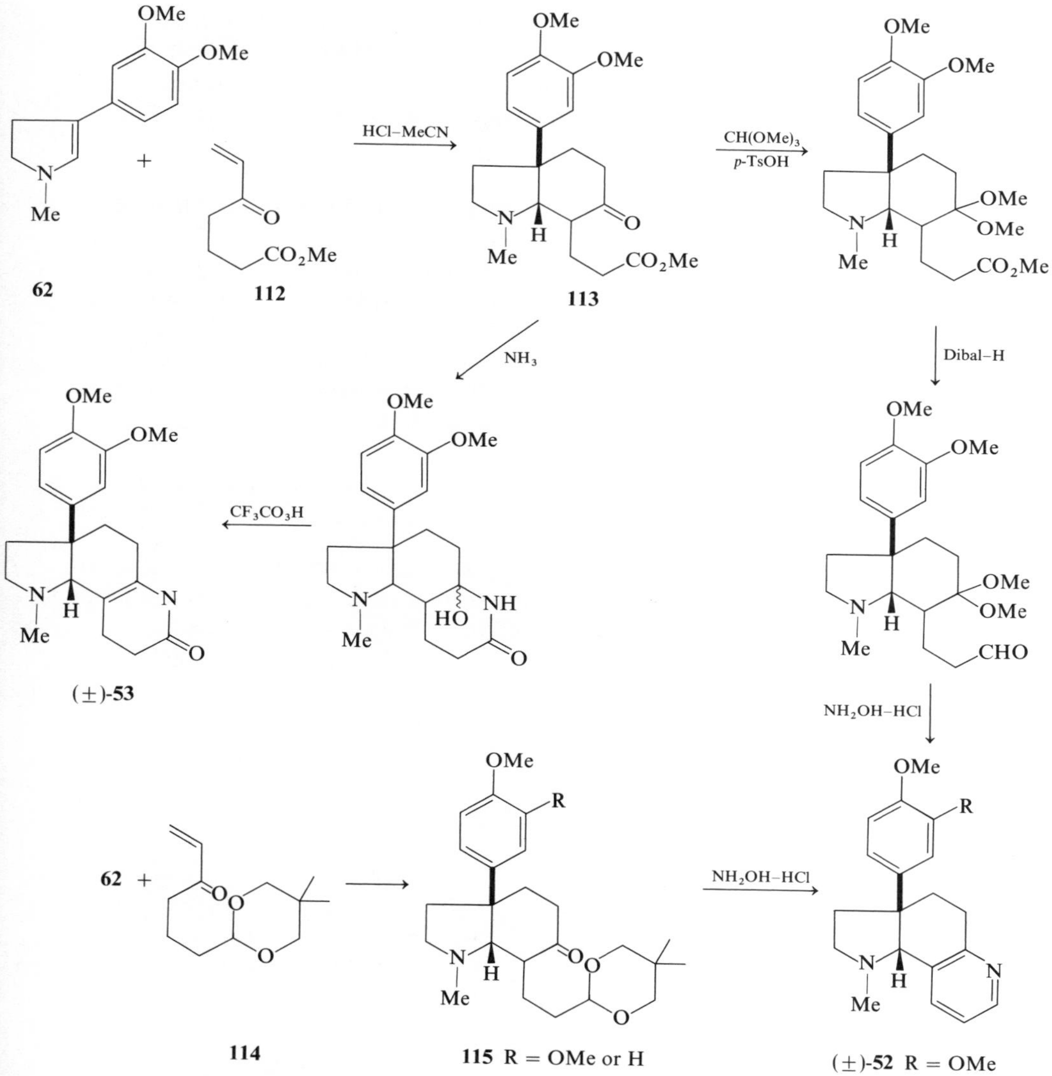

SCHEME 29. *Synthesis of Sceletium alkaloid A_4 and a related dihydropyridone via the endocyclic enamine route.*

the dihydropyridone **53** and *Sceletium* alkaloid A_4 (**54**). The dihydropyridone base, as mentioned previously, was subsequently isolated as a natural product.

A modification of the above route to the A_4 alkaloid and its 3′-demethoxy analog has recently been reported by the Wiechers group (*60*). In a series of reactions which directly parallel those reported by Stevens *et al.*, Wiechers and co-workers reacted the synthon **62** and its 3′-demethoxy analog in separate reaction schemes with the enone acetal **114** to give **115** (R = H or OMe); they converted the latter compounds directly to the pyridine alkaloid and its 3′-demethoxy base with hydroxylamine hydrochloride (see Scheme 29).

VI. Miscellaneous Chemical Transformations of Mesembrine Alkaloids

There has been relatively little chemistry undertaken in effecting functional group modification of the alkaloids of this family. Of the reactions discussed in this section, attempts to effect either N- or O-demethylation of mesembrine by previously established procedures is of interest because it has produced unexpected results. These results, together with some additional chemical transformations, will be discussed in this section.

A. Reaction of Mesembrine and Mesembrenone with Diethyl Azodicarboxylate

The reaction of *N*-methyl tertiary alkylamines with diethyl azodicarboxylate is known (*61*) to result in the formation of an intermediate hydrazodiester (**116**), which on acid hydrolysis gives the corresponding secondary amine together with hydrazine 1,2-dicarboxylate diethyl ester and formaldehyde

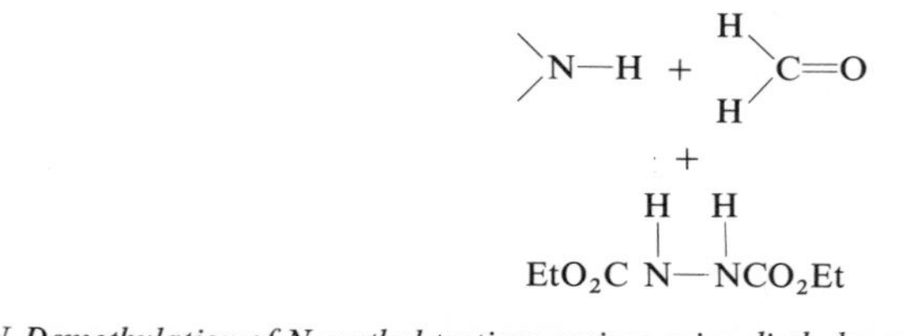

$$>N{-}CH_3 + EtO_2C\,N{=}NCO_2Et \longrightarrow >N{-}CH_2{-}N(CO_2Et){-}N(H)CO_2Et \quad \mathbf{116}$$

$$\mathbf{116} \xrightarrow{H_3O^+} >N{-}H + H_2C{=}O + EtO_2C\,N(H){-}N(H)CO_2Et$$

SCHEME 30. *N-Demethylation of N-methyl tertiary amines using diethylazodicarboxylate.*

(see Scheme 30). The mechanism of this interesting reaction, which probably involves a nitrogen ylide, has been discussed (*62*), and its application to *N*-demethylation in the morphine alkaloid series has been presented (*63*).

The reaction of mesembrine with diethyl azodicarboxylate in refluxing acetone gave a compound which from its spectral properties was not the expected *N*-alkylhydrazo ester. The structure of the product as the β-enaminoketone **117** was deduced from its spectral properties, in which the occurrence of an intense band at 1590 cm^{-1} in the IR spectrum and a low-field *N*-methyl signal at 2.98 δ were particularly informative. Further support for the structural assignment was obtained by its conversion with methyl iodide to the *O*-alkyl derivative **118** and by its reduction to mesembranol and 6-epimesembranol with sodium in isopropanol. The reaction of (±)-mesembrenone with diethyl azodicarboxylate proceeded similarly to give the conjugated β-enaminoketone **119** (*64*).

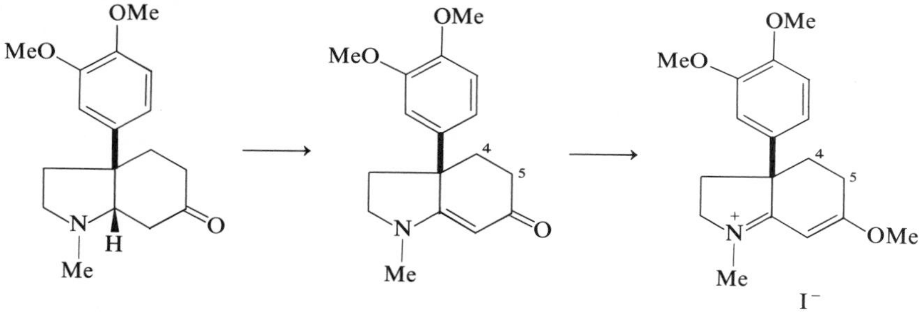

OMe
MeO
N
H
Me
O
⟶
OMe
MeO
4
5
N
Me
O
⟶
OMe
MeO
4
5
$\overset{+}{N}$
Me
OMe
I^-

117
119 (4,5-double bond)

118
120 (4,5-double bond)

Although there are numerous mechanistic schemes which could account for the oxidation of mesembrine and mesembrenone with diethyl axodicarboxylate, an ene pathway via the enol as shown in Scheme 31 is keeping

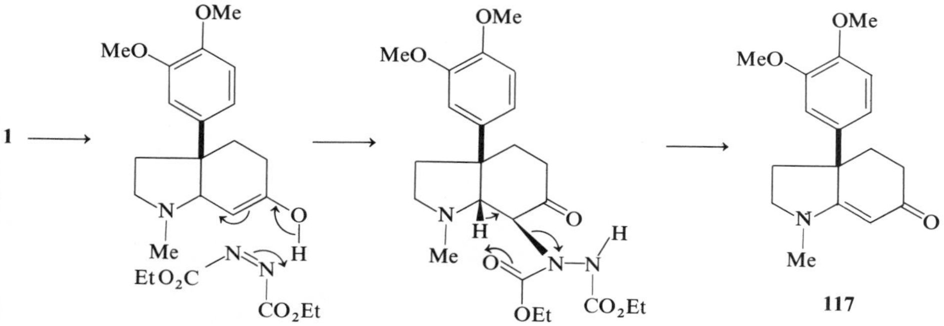

1 ⟶
OMe
MeO
N
Me
O
H
EtO_2C
N
N
H
CO_2Et
⟶
OMe
MeO
N
H
Me
O
O
N—N
H
OEt
CO_2Et
⟶
OMe
MeO
N
Me
O
117

SCHEME 31. *Oxidation of mesembrine to* Δ^7*-mesembrenone with diethylazodicarboxylate.*

with the observation that the oxidation of mesembrine-7,7-d_2 with diethyl azodicarboxylate in CH_2Cl_2 solution affords **117** devoid of deuterium.

A further point of interest concerns the reaction of the iminium salts **118** and **120** with borohydride (*65*). The reduction of **118** or **120** with sodium borohydride gave an identical product mixture in which the major components were 6-*O*-methylmesembranol (**121**) and 6β-*O*-methyl-7a-epimesembranol (**122**) in 78% and 16% yields, respectively (see Scheme 32). Two minor components representing less than 6% of the total products were identified as mesembrane (**7**) and its C-7a epimer **123**. While the structures of the compounds with a trans ring fusion rested on spectral evidence, the identity of 6-*O*-methylmesembranol was established by its synthesis from mesembranol by selective O-methylation of the sodium salt with methyl-*p*-toluenesulfonate, while that of *cis*-mesembrane was identified by comparison with an authentic sample.

The major feature of interest in these borohydride reductions is that the conjugated iminium ions **118** and **120** undergo reduction exclusively by a sequential 1:6 → 1:4 → 1:2 hydride addition pathway. Since a previous report (*66*) on reduction of iminium salts led to the observation of exclusive 1,2-addition and no conjugate addition, further study on these systems is warranted.

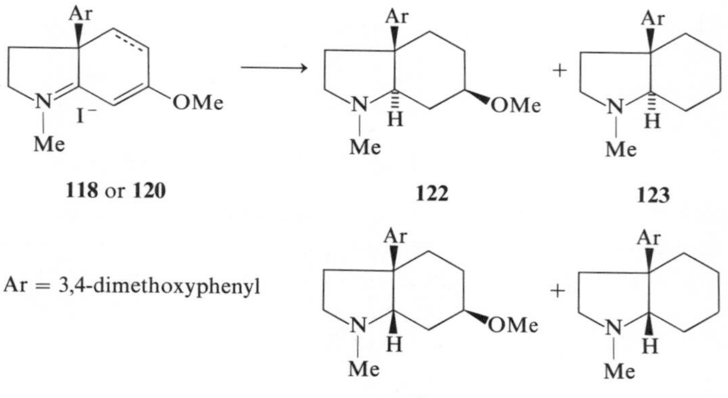

Ar OMe N I⁻ Me H

118 or **120** **122** **123**

Ar = 3,4-dimethoxyphenyl

121 **7**

SCHEME 32. *Sequential conjugate hydride addition to the iminium salts* **118** *and* **120**.

B. N-DEMETHYLATION AND REACTION OF MESEMBRINE WITH BORON TRIBROMIDE

The conversion of mesembrine to *N*-demethylmesembrine has been accomplished by the reaction of the ethylene ketal **124** with *p*-nitrophenylchloroformate. Use of the ketal was necessary because mesembrine underwent β-elimination of the nitrogen substituent when reacted directly with chloroformate esters. The urethane **125**, obtained from the ketal, could be con-

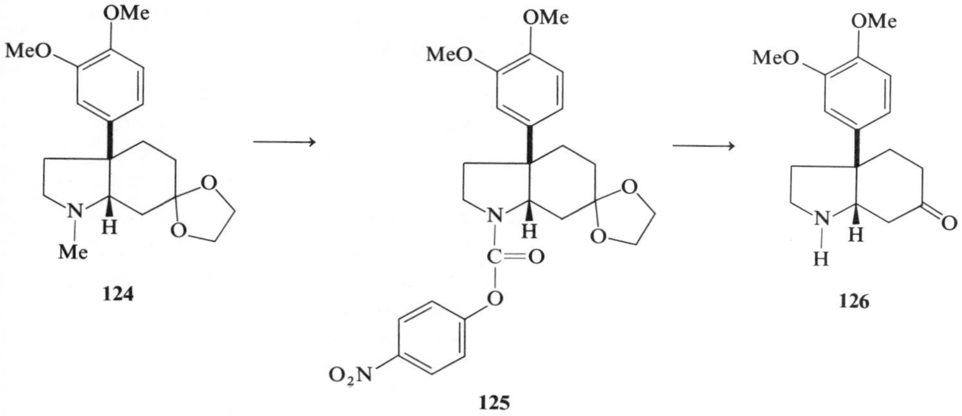

124 → **125** → **126**

verted to *N*-demethylmesembrine (**126**) by sequential hydrolysis procedures employing aqueous acid and then aqueous base. In view of the isolation of the *N*-demethyl bases **27** and **28** from *Sceletium strictum* (see Section IV,A,5), it is expected that further isolation studies will result in the identification of *N*-demethylmesembrine as a natural product.

The reaction of mesembrine with boron tribromide at 0° was carried out in an attempt to effect O-demethylation of the aromatic ether groups. The IR spectrum of the product from this reaction showed no carbonyl absorption, and although the absence of methoxyl groups was evident from the ^{1}H-NMR spectrum, the occurrence of two one-proton singlets at 6.61 and 6.88 ppm indicated that the desired O-demethylation had been accompanied by further reaction involving the aromatic ring. The product from this reaction is accounted for by structure **127**, and results from an intramolecular Friedel–Crafts reaction as shown in Scheme 33. Methylation of the product with diazomethane gave a compound with spectral properties in full agreement with those expected for the dimethyl ether **128**.

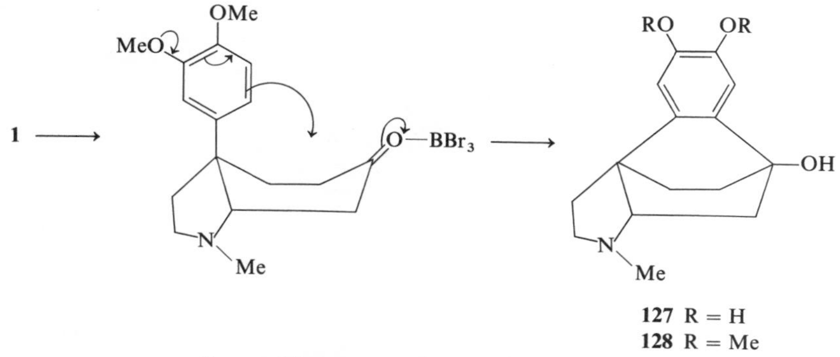

127 R = H
128 R = Me

SCHEME 33. *Reaction of mesembrine with* BBr_3.

C. Reactions of 4,4-Disubstituted Cyclohexadienones and Related Compounds Derived from Octahydroindole Bases

The chemistry of the octahydroindole alkaloids mesembrine, mesembrenone, sceletenone, and related bases as β-aminoketones is characterized by facile β-elimination reactions which occur on N-alkylation or N-acetylation of these compounds.

Earlier studies had shown that the cyclohexadienone **129** derived from the Hofmann degradation of mesembrenone on acid treatment underwent the expected cyclohexadienone–phenol rearrangement to give **130**. Further interest in this system was stimulated by the need to develop controlled degradation schemes to ascertain labeling patterns at the C-2 and C-3 carbons

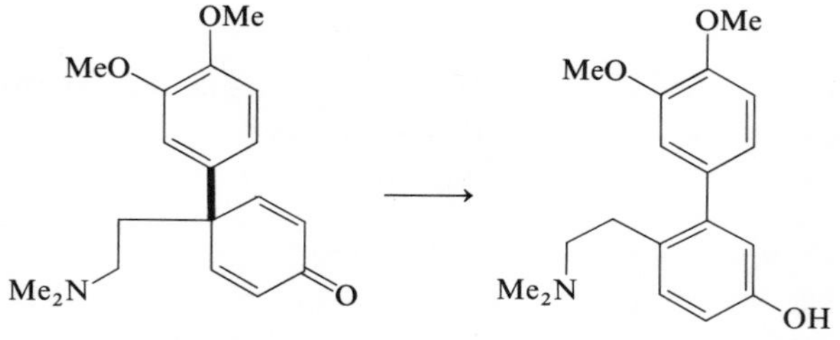

129 **130**

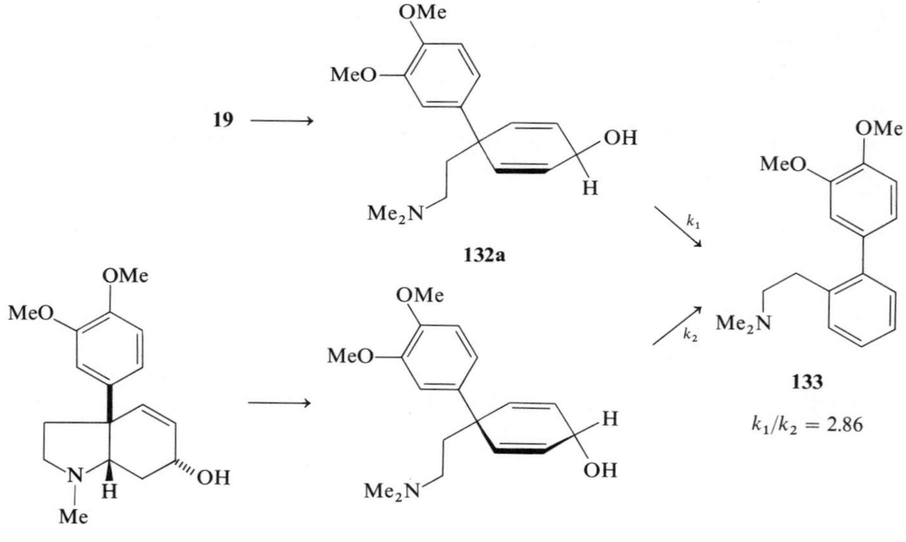

19 **132a** **133**

$k_1/k_2 = 2.86$

131 **132b**

Scheme 34. *Rates of dienol–benzene rearrangement of the epimeric cyclohexadienols* **132a** *and* **132b**.

in connection with biosynthetic studies. Hofmann degradation of mesembrenol (**19**) and 6-epimesembrenol (**131**), the latter obtained from the reduction of mesembrenone with $LiAlH_4$ or $NaBH_4$, gave the corresponding epimeric dienols **132a** and **132b**, respectively. The latter two compounds provided a convenient system for examination of the possible role of stereoelectronic effects in the dienol–benzene rearrangement in this system. Acid-catalyzed rearrangement of dienol **132a** to biphenyl **133** was found to occur at about three times the rate of its epimer **132b** (see Scheme 34). If the reaction is concerted, the transition state for the rearrangement of the dienols is analogous to that of an S_N2' process, which despite some recent controversy would seem to favor a syn pathway. This is in keeping with the relative rates observed. However, the small rate difference would seem to rule out the participation of an arenium ion; consequently, a nonconcerted mechanism with a small steric acceleration by the *cis*-aryl group in **132a**, appears to provide the best explanation of the results.

An efficient scheme for the conversion of saturated ketones in the mesembrine series to their 4,5-unsaturated analogs has been developed; see Scheme 35. The reaction of mesembrine with trifluoroacetic anhydride gives the *N*,7a-

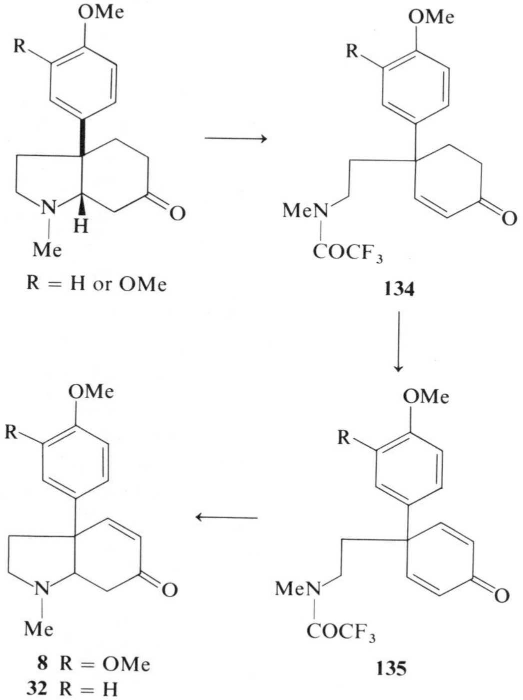

OMe R N H Me O R = H or OMe

OMe R MeN $COCF_3$ O **134**

OMe R N Me O **8** R = OMe **32** R = H

OMe R MeN $COCF_3$ O **135**

SCHEME 35. *Conversion of mesembrine to mesembrenone.*

secoamide **134** (R = OMe), which is readily dehydrogenated with 2,6-dicyano-3,5-dichloroquinone to the symmetrical dienone **135** (R = OMe). Treatment of compound **135** (R = OMe) with aqueous base effects removal of the trifluoroacetyl group to release the secondary amine, which spontaneously cyclizes to mesembrenone. The same sequence has been employed in the 4-methoxyphenyl series to give *O*-methylsceletenone **32** (*53*).

VII. Biosynthetic Studies

A. Introduction

In considering possible biogenetic schemes for the alkaloids of this family, analysis of the structural features of the mesembrine subgroup indicated that the ring system may be constructed from a hydroaromatic C_6—C_2—N unit and an aromatic C_6 unit. Although the occurrence of a C_6—C_2—N unit is encountered in the structures of many alkaloids in different families, the appearance of an isolated aromatic C_6 unit was an unusual feature. An explanation of the origin of the latter unit as deriving from the commonly encountered C_6—C_1 unit was suggested by a possible biosynthetic relationship of the mesembrine alkaloids with the structurally similar crinine alkaloids of the Amaryllidaceae family, *cf.* **61** (see discussion introducing Section V above). Extensive studies on the biosynthesis of the Amaryllidaceae alkaloids which had preceeded the initial studies on the biosynthesis of the *Sceletium* alkaloids provided a detailed picture of the origins and pathways by which this former family of alkaloids were derived.

On the basis of this background work, a scheme for the biogenesis of mesembrine was proposed as shown in Scheme 36 (*67*). The proposed scheme rested on the premise that the mesembrine alkaloids were biosynthesized by an extension of the pathway previously established for the crinine alkaloids in which phenylalanine and tyrosine are used in separate pathways to provide the C_6—C_1 unit and a C_6—C_2—N unit, respectively, to give norbelladine (**136**) as an intermediate. After regiospecific O-methylation to **137**, the latter is converted to the crinine ring system by a phenol oxidative coupling process, as shown in Scheme 36. The proposed conversion of the crinine skeleton to the mesembrine skeleton, which requires loss of a C_1 unit from the C_6—C_1 unit in the former structure, involving benzylic hydroxylation, oxidation, and decarboxylation, appeared reasonable because there were several Amaryllidaceae alkaloids known at the time which contained a benzylic hydroxyl group. This hypothesis for the biogenesis of the mesembrine alkaloids formed the basis of the early experimental tracer studies, described on p. 61.

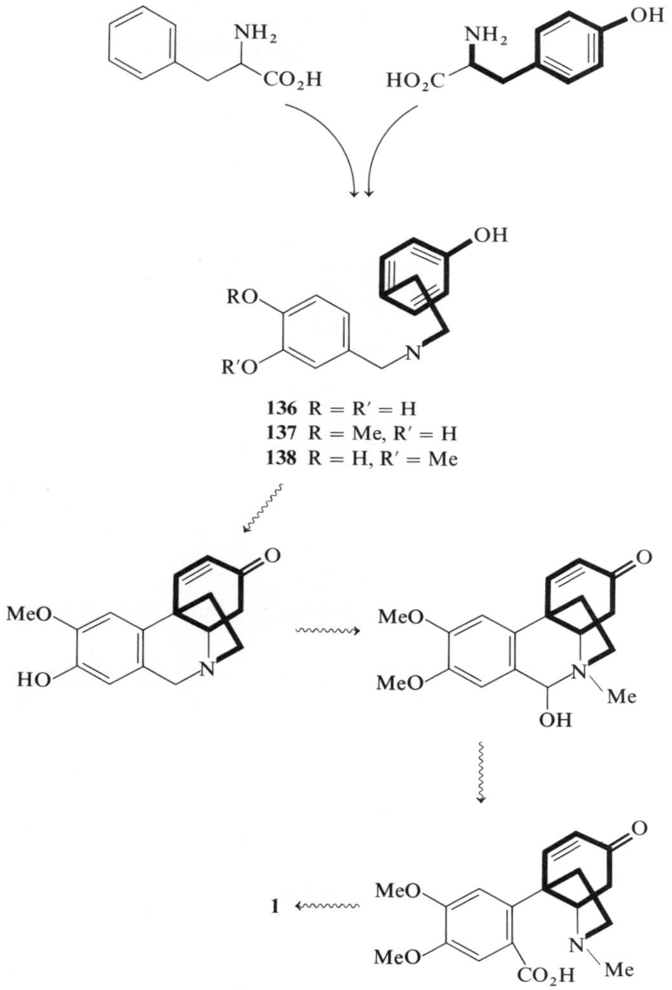

136 R = R′ = H
137 R = Me, R′ = H
138 R = H, R′ = Me

SCHEME 36. *Early biogenetic scheme for mesembrine: extension of the crinine pathway.*

B. THE AMINO ACID PRECURSORS

The basic premise that phenylalanine and tyrosine provided the C_6 and C_6—C_2—N units, respectively, of the mesembrine skeleton was readily established by appropriate radiolabeled tracer experiments. Both DL-[2-^{14}C] tyrosine and DL-[3-^{14}C]tyrosine, when administered to *Sceletium strictum*, gave rise to radiolabeled mesembrine and mesembrenol. Evidence for specific incorporation of this amino acid was obtained by subjecting the labeled mesembrenol from the [2-^{14}C]tyrosine experiment to the degradation shown in Scheme 37.

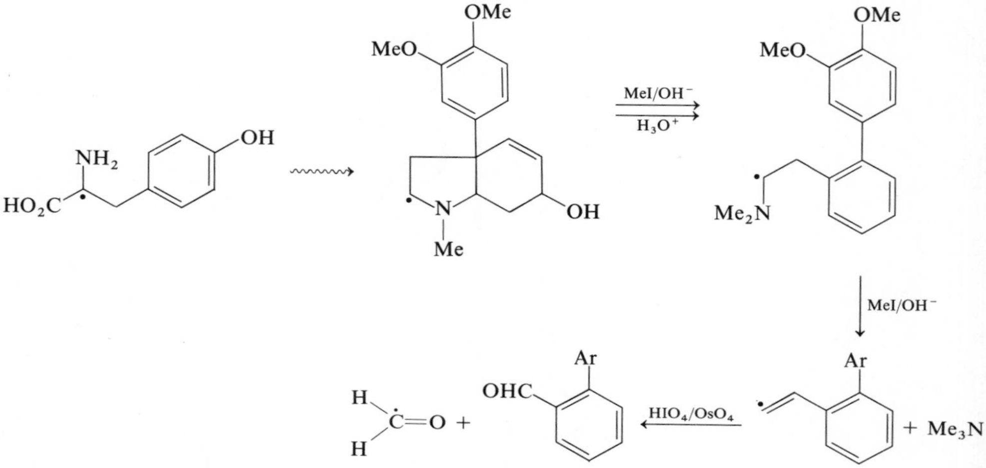

SCHEME 37. *The degradation of mesembrine derived from* L-[2-^{14}C]*tyrosine.*

The observation that DL-[2-^{14}C]phenylalaine and DL-[3-^{14}C]phenylalanine showed no incorporation of radiolabel into mesembrine established that phenylalanine is not converted to tyrosine and that these two aminoacids

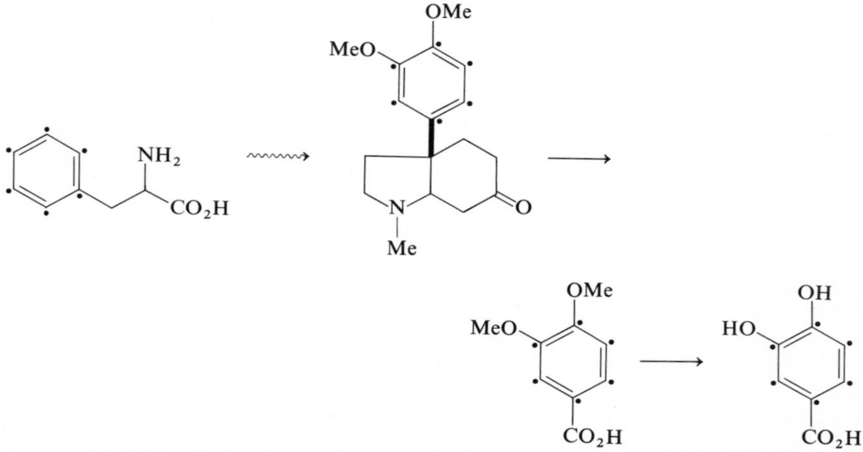

SCHEME 38. *Degradation of mesembrine derived from* DL-[*Ar*-^{14}C]*phenylalanine.*

are metabolized on separate pathways, a finding in consonance with the metabolic role of these two amino acids in the biosynthesis of the Amaryllidaceae alkaloids. The role of phenylalanine as a precursor of the aromatic ring of mesembrine was established initially by the specific incorporation of DL-[Ar-^{14}C]phenylalanine. The location of the radiolabel in the mesembrine derived from this experiment was shown to be restricted to the aromatic ring carbons of the alkaloid by its oxidation to veratric acid and demethylation of the latter to protocatechuic acid (see Scheme 38).

The remaining carbons not accounted for constitute the two aromatic methoxyls and the *N*-methyl group. These were shown to originate from the *S*-methyl group of methionine, presumably by the agency of the ubiquitous C-1 donor, *S*-adenosylmethionine.

These preliminary experiments identified the roles of the three amino acids phenylalanine, tyrosine, and methionine in providing the various structural units encountered in the mesembrine alkaloids, as summarized in Scheme 39.

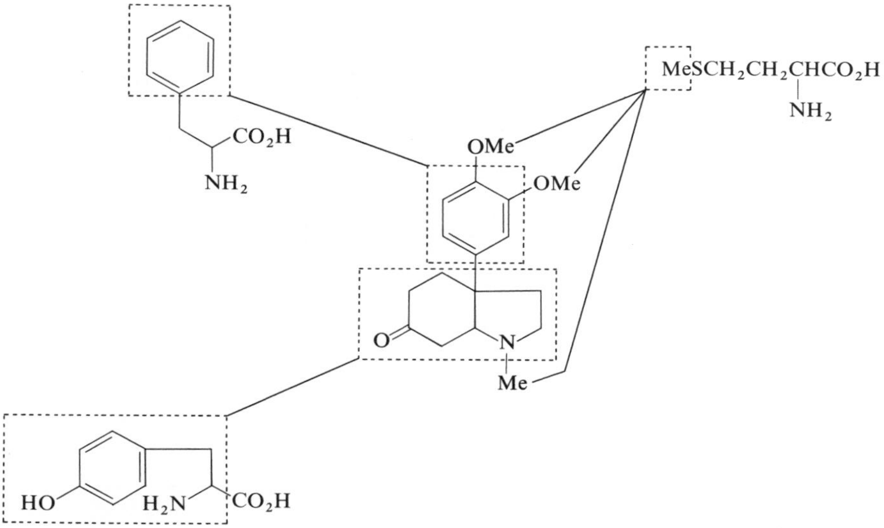

SCHEME 39. *Utilization of phenylalanine, tyrosine, and methionine in the biosynthesis of mesembrine.*

C. EVIDENCE FOR A BISSPIRODIENONE INTERMEDIATE

Following these early studies, which appeared to provide support for the biogenetic hypothesis outlined in Schemes 36 and 39, tracer experiments were carried out with norbelladine (**136**) and its 4′-*O*-methyl and 3′-*O*-methyl derivatives **137** and **138**.

An extended series of experiments with these compounds and related structures provided the first indication that the biosynthesis of the mesembrine and crinine alkaloids proceeded by quite different pathways (*68*). Norbelladine and its two *O*-methyl derivatives when labeled in ring A with tritium or at C-1 with ^{14}C all showed incorporation of radioactivity into mesembrenol when fed to *Sceletium strictum*. The results were unexpected, because 3′-*O*-methylnorbelladine cannot undergo direct phenol–oxidative coupling to generate a crinane intermediate without rearrangement. Further, both sets of experiments demonstrated that the incorporation of radiolabel was significantly higher from the 3′-*O*-methylnorbelladine than from its 4′-*O*-methyl isomer, although the overall level of incorporation in both was low in comparison to that of L-[*S*-methyl-^{14}C]methionine which was fed as a control.

Because the biogenesis of the mesembrine alkaloids was reasonably assumed to have involved a phenol–oxidative coupling process, these results suggested an alternate mode of coupling. In view of the uncertainty of the intermediacy of a norbelladine type intermediate presented by the low incorporation results, the mode of coupling was pursued by an experiment with phenylalanine. Loss of the C-3 side chain from phenylalanine during its biosynthetic conversion into the alkaloid ring system prevents any conclusion being drawn as to which of the original nuclear carbons of phenylalanine correspond to the position of coupling represented by the C-1′ of mesembrine. To circumvent this situation, a sample of DL-phenylalanine containing tritium at the symmetry-equivalent 2′,6′-positions and a ^{14}C label at the C-1′ nuclear carbon was administered to *S. strictum*. Isolation of the labeled alkaloids from this experiment provided radiolabeled mesembrine and mesembrenone with a $^{3}H:^{14}C$ ratio essentially unchanged from that of the labeled phenylalanine from which these alkaloids derived. This result was unexpected, for if the biosynthesis of these alkaloids proceeded by an extension of the crinine pathway as previously described in Scheme 36, a 50% loss of tritium would have occurred as a consequence of the phenol–oxidative coupling step involved in the formation of the crinane system. Further, degradation of the double-labeled mesembrine from this experiment according to the reactions in Scheme 40 established the location of tritium labels at the 2′,6′-positions and pinpointed the site of coupling as the aromatic carbon which previously bore the phenylalanine side chain.

The implications of this singularly important result allowed several conclusions to be made regarding the biosynthesis of these alkaloids. First, if one assumes that phenol–oxidative coupling is indeed involved in linking the two aromatic rings of a post-phenylalanine and a post-tyrosine intermediate,

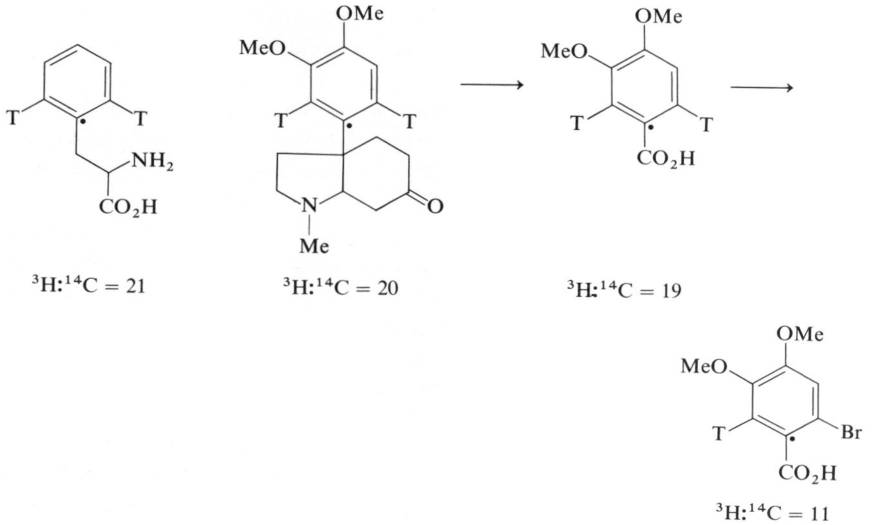

T T NH_2 CO_2H → OMe MeO T T N Me O → OMe MeO T T CO_2H →

$^3H:^{14}C = 21$ $^3H:^{14}C = 20$ $^3H:^{14}C = 19$

OMe MeO T Br CO_2H

$^3H:^{14}C = 11$

SCHEME 40. *Degradation scheme for mesembrine derived from* DL-[2′(6′)-^3H;1′-^{14}C]*phenylalanine.*

then the essential structural features of a product derived by this process are expressed by the bisspirodienone **139** in Scheme 41. Second, the eventual aromatization of the ring A dienone in this coupling product must proceed by a fragmentation process which is mechanistically analogous to that indicated in Scheme 41 to account for the retention of tritium at its original location(s).

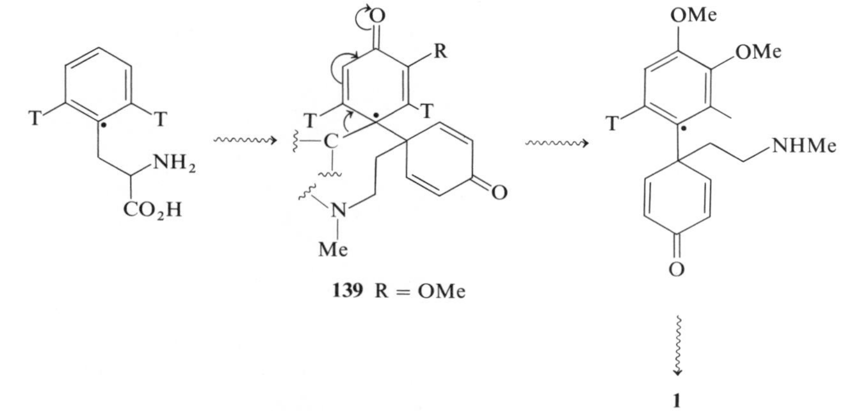

T T NH_2 CO_2H O R T T C N Me O OMe OMe T NHMe O

139 R = OMe

1

SCHEME 41. *Biogenetic scheme involving a bisspirodienone intermediate* (**139**) *and its resulting fragmentation–aromatization.*

It should be noted that a dienone–phenol rearrangement of the bisspirodienone (**139**) to a crinane system could be excluded, because this would require a 50% loss of tritium.

The results of this experiment seemed to provide an explanation of the more efficient incorporation of label from 3′-*O*-methylnorbelladine in comparison to that observed for 4′-*O*-methylnorbelladine, which had been previously assumed the more likely precursor.

Confirmation of the role of 3′-*O*-methylnorbelladine was sought in more decisive experiments with the double-labeled compounds **140** and **141**. Degradation experiments on mesembrenol derived from **140** showed that the ^{14}C label was scattered among the two *O*-methyl groups and the *N*-methyl group. An indication that incorporation of radioactivity from **140** was not occurring by simple 3′-O-demethylation and subsequent incorporation of the derived norbelladine was provided by the marked change in ^3H:^{14}C ratios observed in alkaloids derived from experiments using **141** and the double-labeled norbelladine (**142**). These results demonstrated that the appearance of radioactivity in the alkaloids from the labeled norbelladines occurred as a result of reincorporation of fragments derived from prior degradation of the test precursors.

Subsequent experiments with *N*-methylnorbelladine (**143**), which is an effective precursor of the Amaryllidaceae alkaloid galanthamine (*69*), and with 3′-*O*,*N*-demethylnorbelladine (**144**), failed to show incorporation into the alkaloids.

Although the monooxyaryl alkaloid sceletenone was not known at the time, the occurrence of the monooxyaryl bases of the joubertiamine series

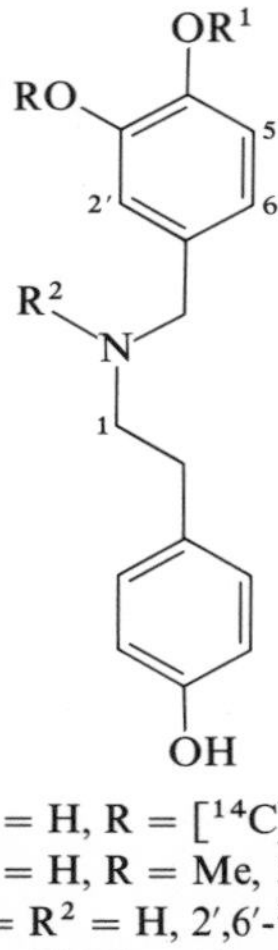

140 $R^1 = R^2 = H$, $R = [^{14}C]Me$, 1-^{14}C
141 $R^1 = R^2 = H$, $R = Me$, 5′-^3H, 1-^{14}C
142 $R = R^1 = R^2 = H$, 2′,6′-^3H, 1-^{14}C
143 $R = R^1 = H$, $R^2 = Me$, 2′,6′-^3H, 1-^{14}C
144 $R = R^2 = H$, $R^1 = Me$, 5′-^3H, 1-^{14}C

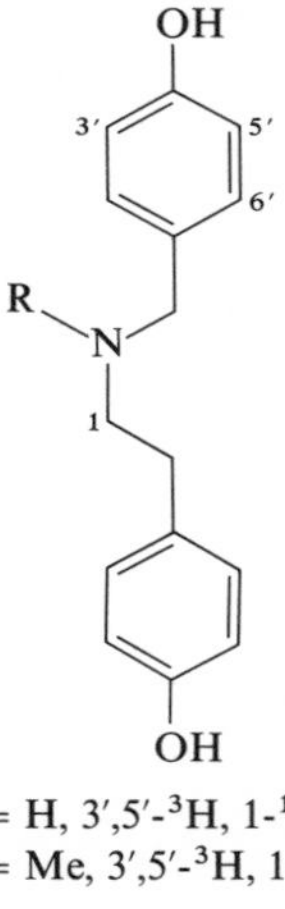

145 $R = H$, 3′,5′-^3H, 1-^{14}C
146 $R = Me$, 3′,5′-^3H, 1-^{14}C

had prompted consideration of 3′-deoxynorbelladine (**145**) and *N*-methyl-3′-deoxynorbelladine (**146**) as possible precursors of the mesembrine alkaloids. When these compounds were administered as the doubly labeled compounds **145** and **146**, again no incorporation into the alkaloids was observed.*

These foregoing studies clearly showed that norbelladine-type compounds containing the Ar–C_1–N–C_2–C_6 system do not appear to be involved in the biosynthesis of the mesembrine alkaloids.

D. Post-Tyrosine and Post-Phenylalanine Intermediates

1. The Tyrosine Pathway

With the failure to demonstrate that norbelladine or its relatives plays a role in the biosynthesis of the mesembrine alkaloids, a reevaluation led to a modified approach in which attempts to identify the sequence of occurrence of the post-tyrosine and post-phenylalanine intermediates were made. There is now a substantial body of information available to suggest that phenylalanine and tyrosine have separate metabolic roles in plants belonging to the order Dictolyoden. Not only do they lack the enzyme phenylalanine hydroxylase (phenylalanine 4-monooxygenase) which is necessary for the conversion of phenylalanine to tyrosine, but the metabolic pathways of these two amino acids are generally quite different in secondary metabolism (*70*). Phenylalanine is involved in initial conversion to cinnamic acid and subsequent transformation to structural units of the so-called phenyl–propanoid pathway, which include Ar—C_3, Ar—C_2, and Ar—C_1 structural entities. On the other hand, the role of tyrosine in the biosynthesis of secondary metabolities is most frequently seen as the precursor of Ar—C_2—N and C_6—C_2—N units, and somewhat less frequently, as Ar—C_2 and C_6—C_2 units.

With regard to the utilization of tyrosine in the biosynthesis of mesembrine, various possibilities existed which pertained to the timing of the decarboxylation and N-methylation of the tyrosine-derived C_6—C_2—N unit during its incorporation into the octahydroindole portion of the mesembrine alkaloid skeleton.

Several experiments provided interlocking evidence for the sequence tyrosine → tyramine → *N*-methyltyramine occurring during the biosynthesis of mesembrine-type alkaloids. These included the observation that both

* All four *N*-methyl compounds **143**, **144**, **145**, and **146** appear to be metabolized by *Sceletium strictum*. The products of this metabolism are aberrant metabolites of unknown structure but are known to contain ^3H:^{14}C ratios which correspond closely to that of the administered precursors.

$[2\text{-}^{14}C]$tyramine and *N*-$[^{14}C]$methyl-$[2\text{-}^{3}H]$tyramine led to radiolabeled mesembrenol and that the incorporation of the latter occurred to give mesembrenol in which the $^{3}H:^{14}C$ ratio was unchanged (*71*).

Confirmatory evidence for the intermediacy of *N*-methyltyramine was secured by the observation that when an equimolar mixture of $[2\text{-}^{14}C]$-tyramine and *N*-methyl-$[2\text{-}^{3}H]$tyramine with a $^{3}H:^{14}C$ ratio of 7.6 to 1 was administered to *Sceletium Strictum*, the radiolabeled mesembrenol obtained had a $^{3}H:^{14}C$ ratio of 39 to 1. This result was interpreted on the premise that if two compounds of similar structure show similar uptake, then intermediates closer to the final product are expected to be incorporated more efficiently than those which appear earlier (*72*).

While the evidence for the sequence tyrosine → tyramine → *N*-methyltyramine as intermediates in the biosynthesis of the mesembrine alkaloids appeared to be secure, more compelling evidence was sought for the utilization of the aromatic ring and the attached C_2—N side chain as an intact C_6—C_2—N unit. The experiments, discussed below, which confirmed this proposal also brought to light several other important and unexpected findings (*72*).

A doubly labeled sample, **147**, consisting of DL-$[2\text{-}^{14}C]$tyrosine and L-$[3,5\text{-}^{3}H]$tyrosine with a $^{3}H:^{14}C$ ratio of 15.7 to 1, was administered to *S. strictum*. The resulting radiolabeled mesembrenol had a $^{3}H:^{14}C$ ratio of 8.6 to 1, corresponding to a 50% loss of tritium. If there had been no loss of tritium, as anticipated, a $^{3}H:^{14}C$ ratio of 31.4 to 1 would have been predicted because the enzymatic decarboxylation of tyrosine to tyramine was expected to be stereospecific, utilizing only the L isomer. Two further experiments clarified the situation. Incorporation of L-$[U\text{-}^{14}C,3,5\text{-}^{3}H]$-tyrosine also occurred with a 50% loss of tritium (after correction for loss of the carboxyl carbon) and served to indicate that the aromatic ring and the attached C_2—N unit is incorporated intact into the octahydroindole portion of the mesembrine skeleton. Location of the sites of tritium labels at the predicted 5,7-positions, as discussed later, further substantiated the utilization of the C_6—C_2—N unit tyrosine in this manner.

The 50% loss of ^{3}H observed during the incorporation of both ^{3}H- and ^{14}C-labeled L-tyrosine and the mixed L-$[^{\beta}H]$tyrosine and DL-$[^{14}C]$tyrosine may be explained by the existence of either a fast interconversion of D and L isomers or alternatively, the rapid conversion of D to L through a tyrosine racemase or epimerase. A third possibility—that the decarboxylase enzyme is nonstereospecific—would appear less likely. An indication that one of the two former possibilities was more plausible was provided by the observation that DL-$[3\text{-}^{14}C,2\text{-}^{3}H]$tyrosine is incorporated into mesembrenol by *S. strictum* with a 56% loss of the tritium label from the chiral center (*40*).

A full interpretation of these results must await further labeling studies with chiral-labeled D-[3-^{14}C,2-^{3}H]- and L-[3-^{14}C,2-^{3}H]tyrosines.

It should be noted that another example of incorporation of both D- and L-tyrosine with apparent identical efficiency has been observed in the work of Kirby in the biosynthesis of the Amaryllidaceace alkaloids norpluviine and lycorine (*73*).

Experiments with the achiral test precursors [2-^{14}C,3′,5′-^{3}H]tyramine (**148**) and *N*-methyl-[2-^{14}C,3′,5′-^{3}H]tyramine (**149**) also showed incorporation with loss of 50% of the tritium. These results demonstrate that the loss of tritium must occur in an intermediate which appears subsequent to *N*-methyltyramine (Scheme 42).

The loss of 50% of the tritium label from each of the test precursors **147**–**149** during their bioconversion to mesembrenol was investigated by a series of degradations. They showed that mesembrenol obtained from these experiments contained an equidistribution of tritium at the 5 and 7 positions (see Scheme 42). Furthermore, the retention of both tritiums in the biphenyl compound, obtained by the acid-catalyzed rearrangement of the Hofmann product, indicated that the tritium at H-7 must occupy the 7α position exclusively. The latter assignment is in keeping with the biosynthetic formation of the pyrrolidine ring by a Michael addition of the amine group of of the *N*-methyl-*N*-ethylamine side chain to C-7a from the 7a *si*; 7 *re* face and subsequent stereospecific protonation of the resulting enolate at C-7 from the opposite 6 *re*, 7 *si* face. (See Scheme 43.)

The reason for the loss of 50% of the tritium from *N*-methyl[3′,5′-^{3}H]-tyrosine and its immediate precursors is not apparent. The loss of half the original tritium label, when considered in conjunction with the observed equipartition of the surviving ^{3}H label between the 5 and 7α positions leaves only one possible interpretation of the sequence of events involved: two aromatic substitutions must occur. The first has to occur at one of the symmetry equivalent 3′ or 5′ positions in the aromatic ring of *N*-methyl-tyramine or in a post-*N*-methyltyramine intermediate in which some group X other than hydrogen is introduced. The group X which is introduced must then undergo substitution by hydrogen (see Scheme 43) in a subsequent step. It is important to note that this second substitution must occur while the ring is still aromatic, for after the formation of a dienone by oxidative coupling of the tyrosine-derived moiety, the 3′ and 5′ positions become enantiotopic and are therefore distinguishable by enzyme-mediated reactions, and substitution at one of these positions at the dienone oxidation level could not give rise to the observed labeling. The necessity for the occurence of this double aromatic substitution prior to phenol–oxidative coupling cannot be discerned at the present time.

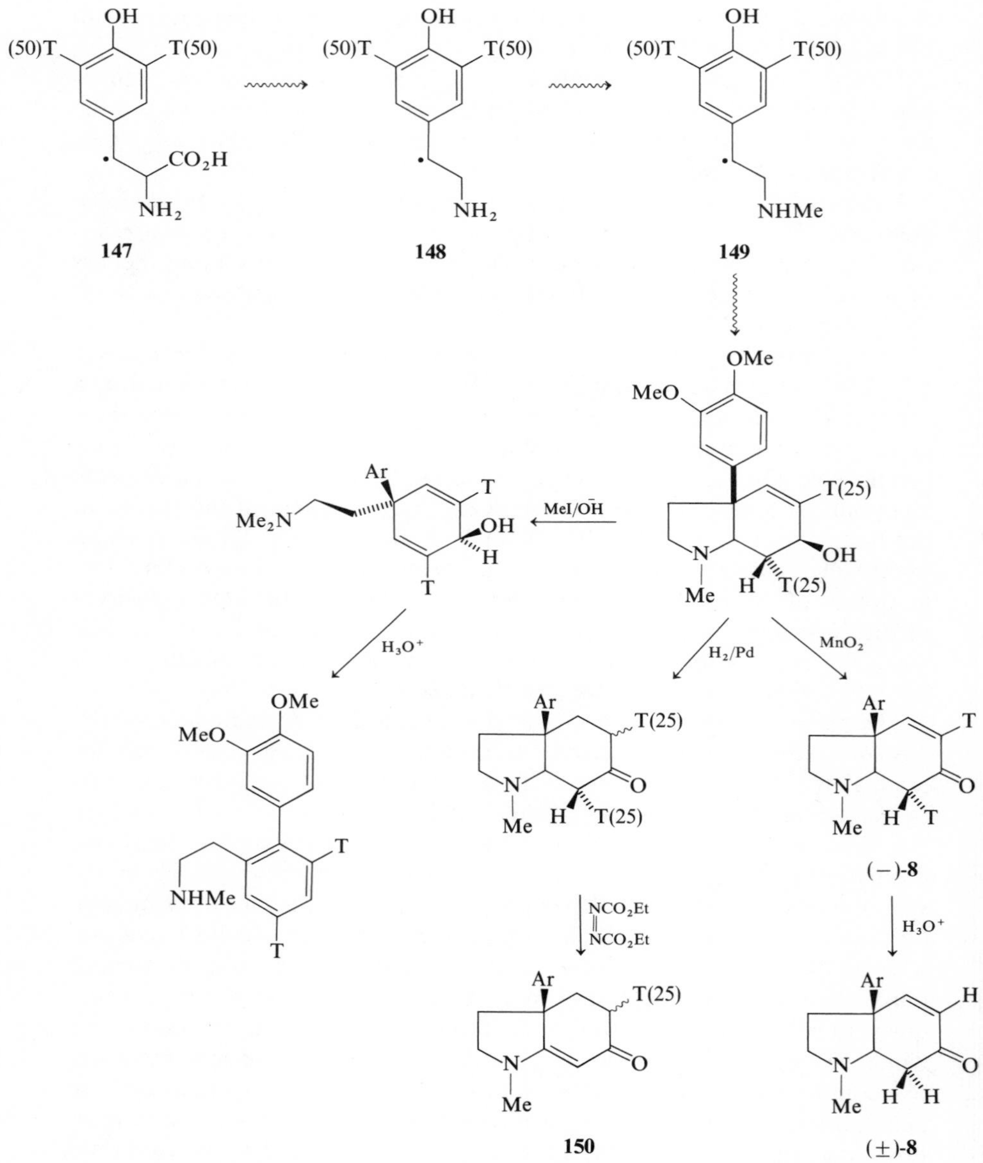

147

148

149

(−)-8

150

(±)-8

SCHEME 42. *Incorporation of double-labeled precusors* **147–149** *into mesembrenol and the location of the labels in the alkaloid by degradation.*

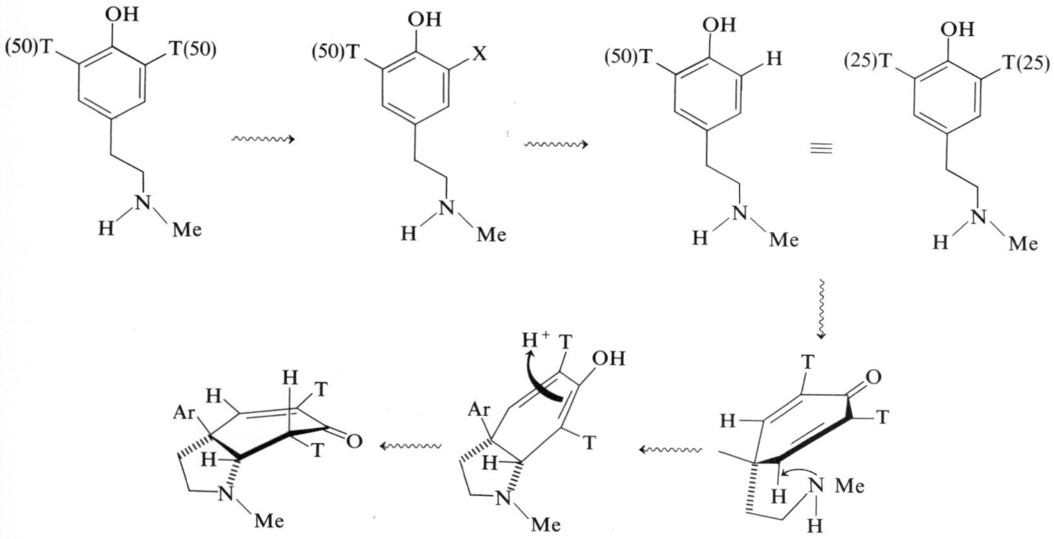

SCHEME 43. *Stereochemistry of C-7a-N bond formation and C-7 protonation.*

2. The Phenylalanine Pathway

The utilization of phenylalanine for the biosynthesis of the mesembrine alkaloids was shown to involve its conversion to cinnamic acid (**151**) and 4-hydroxycinnamic acid (**152**), intermediates which are characteristic of the phenylpropanoid pathway (*74*).

The incorporation of **151** in its 2′-^{3}H,1′-^{14}C-labeled form into mesembrenol occurred with an essentially unchanged ^{3}H:^{14}C ratio and provided additional evidence for the intervention of a bisspirodienone as first indicated from the earlier work with phenylalanine. 4-Hydroxycinnamic acid with a 3′,5′-^{3}H label was incorporated much more efficiently than either 3′,4′-[5′-^{3}H]dihydroxycinnamic acid or 4′-[5′-^{3}H]hydroxy-3′-methoxycinnamic acid. The low levels of incorporation of the latter two dioxyarylcinnamic acids and the subsequent discovery of sceletenone, when considered in conjuction with the occurrence of monooxyaryl alkaloids of the joubertiamine series, suggested that consideration should be given to a pathway in which the 3′-aromatic oxygen of the mesembrine bases is introduced at a late stage in the biosynthesis.

Examination of a number of derivatives of 4-hydroxycinnamic acid as precursors showed that 4′-hydroxydihydrocinnamic acid (**153**) was a very good precursor, giving in one instance 2.3% incorporation of radioactivity from **153** containing a 3′,5′-^{3}H label. Of the other 4′-hydroxycinnamic acid

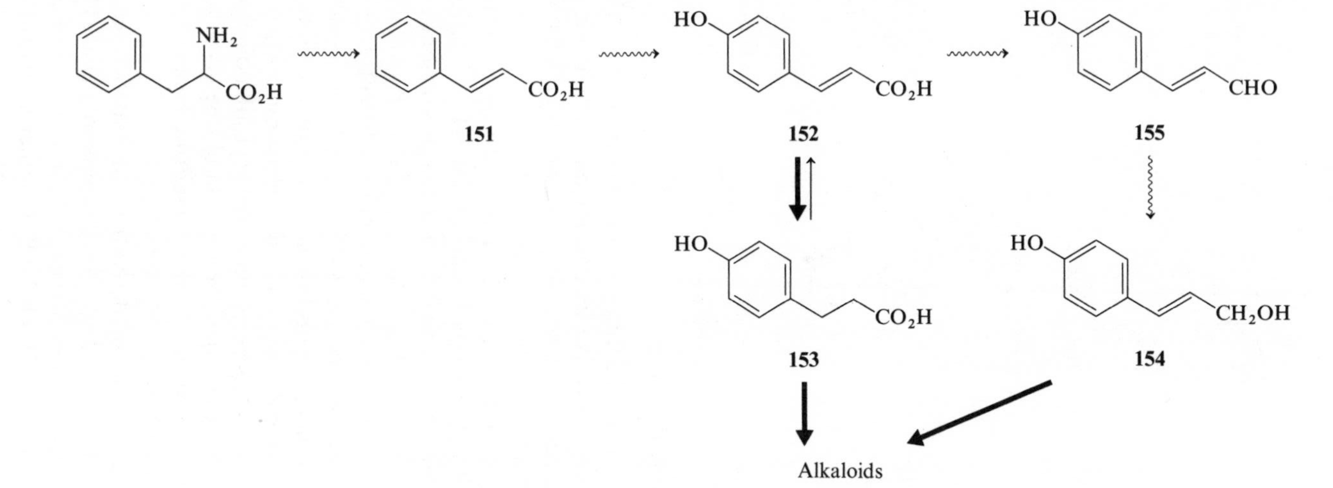

SCHEME 44. *Metabolic grid involving cinnamic acid intermediates.*

derivatives examined, only 4′-hydroxycinnamyl alcohol (**154**) containing a 3′,5′-^{3}H label gave any significant (0.15%) incorporation; the levels of incorporation of ^{3}H-labeled derivatives of 4′-hydroxycinnamyl aldehyde (**155**) and 4′-hydroxydihydrocinnamyl alcohol were low.

The specific incorporation of **154** was demonstrated by combining samples of the ^{3}H-labeled mesembrenol which had been derived from feeding 3′,5′-^{3}H-labeled **154** and [Ar-^{14}C]phenylalanine to *Sceletium strictum* in separate experiments. The ^{3}H:^{14}C ratio of the mesembrenol in this combined sample was essentially unchanged after its oxidation to veratric acid. This clearly demonstrated that the ^{3}H label was restricted to the aromatic ring of the alkaloid, and constituted support for the conversion of 4′-hydroxycinnamyl alcohol to mesembrine.

A series of related experiments (*74*), in which various labeled 4′-hydroxycinnamic acid derivatives (**151**–**155**) were administered in admixture with different members of the series in unlabeled form as trapping agents, led to the following two conclusions:

1. The existence of a facile redox equilibrium between 4′-hydroxycinnamic acid and its dihydro compounds, with one or both being converted directly to an intermediate(s) on the biosynthetic pathway to the mesembrine alkaloids.

2. The involvement of a metabolic grid in which interconversion of various 4′-hydroxycinnamyl derivatives participate. (See Scheme 44).

E. Investigation of Ar—C_3—N—C_2—C_6 and Ar—C_2—N—C_2—C_6 Precursors

The observation that the aromatic ring of phenylalanine is incorporated into the mesembrine system via phenylpropanoid intermediates, when considered in conjunction with the demonstrated role involving the intact incorporation of *N*-methyltyramine, led to the investigation of Ar—C_3—N—C_2—C_6 compounds as precursors. This possibility seemed attractive because precedent exists for the utilization of such compounds in the phenethyltetrahydroisoquinoline alkaloids and related bases, such as colchicine, where experimental data has shown that these compounds are biosynthesized from a phenylalanine-derived Ar—C_3 fragment and tyrosine-derived C_6—C_2—N unit (*75*). Such a scheme seemed to possess a further advantage in that utilization of an Ar—C_3—N—C_2—C_6 unit allows the construction of a general biosynthetic scheme (see Scheme 45) which encompasses all extant structural classes of *Sceletium* alkaloids.

A rigorous examination of this scheme has not been carried out. However, preliminary tracer experiments with the labeled cinnamyl amides **156** and **157** and the related amine **158** did not result in observable incorporation into

Phenylalanine

Tyrosine

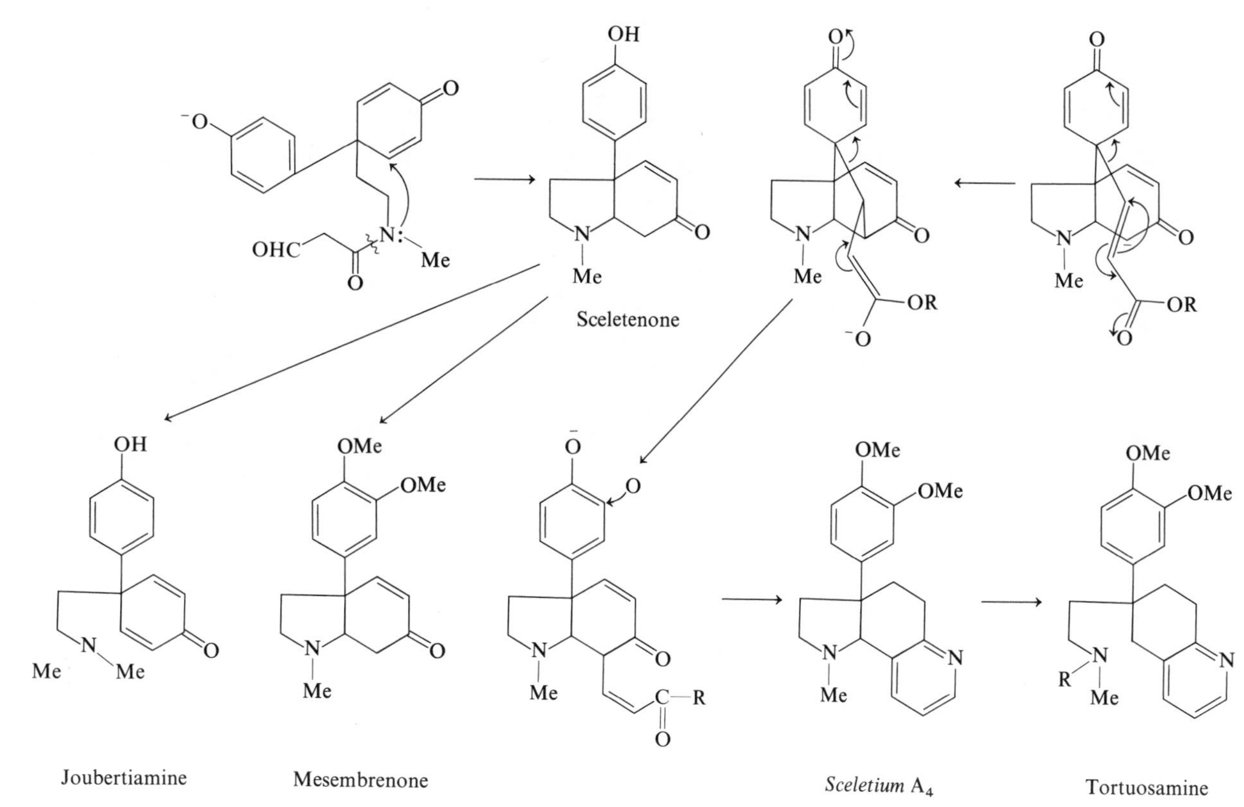

SCHEME 45. *Postulated scheme involving $Ar—C_3—N—C_2—C_6$ precursors for the biogenesis of Sceletium alkaloids.*

the alkaloids (*70*). The extreme insolubility of these amides in aqueous systems necessitated the use of high concentrations (>20%) of Tween which was subsequently shown at these levels to markedly reduce the incorporation of established precursors such as tyrosine (*45*). Consequently, these negative results should be interpreted with caution.

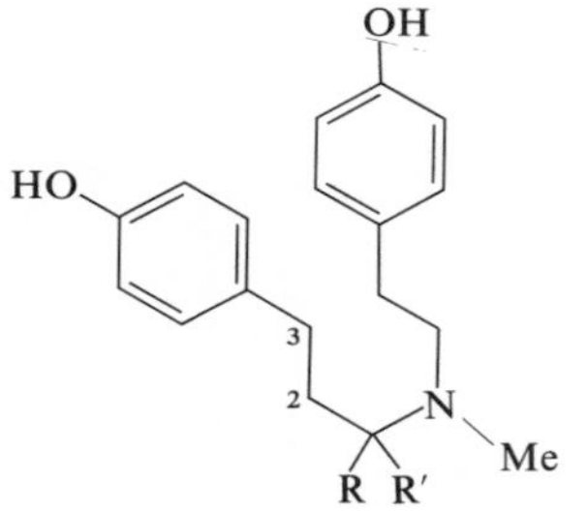

156 R = R′ = O, 2,3-double bond
167 R = R′ = O
158 R = R′ = H

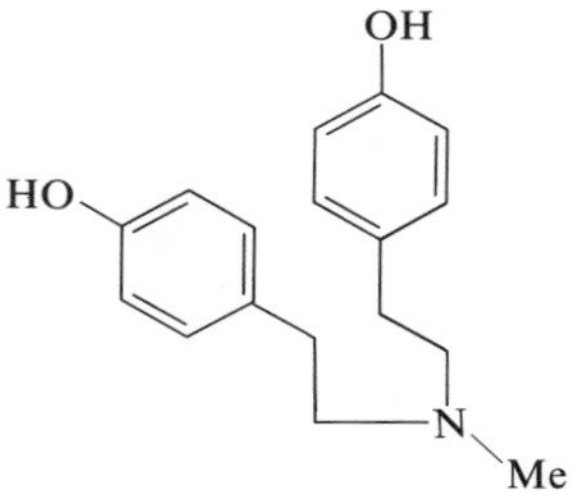

159

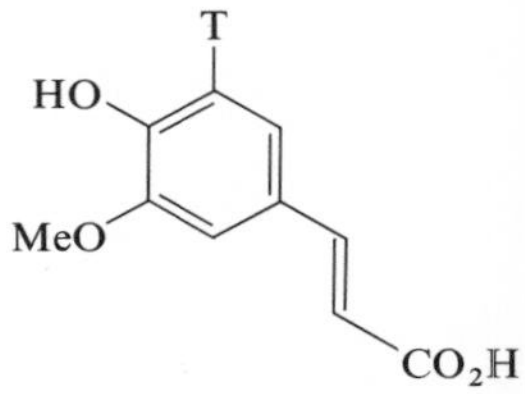

161

Since Ar—C_2—N—C_2—C_6 units are involved in the biosynthesis of many alkaloid families (*76*), the amide **159** was examined as a possible precursor to the *Sceletium* bases in *S. strictum*. Again, no incorporation of radioactivity occurred on feeding a suspension of this compound in 20% Tween solution. Therefore, given the same reservation regarding the effect of high Tween concentrations on this result, it would appear that further studies will be required to determine the fate of *N*-methyltyramine and 4′-hydroxydihydrocinnamic acid in their biosynthetic conversion to the octahydroindole skeleton of the mesembrine alkaloids.

At this stage it would seem that a modified scheme (see Scheme 46), in which an intermolecular phenol–oxidative coupling between *N*-methyltyramine and 4′-hydroxycinnamic acid or its dihydro derivative occur to provide an intermediate **160**, must be considered as a viable alternative to an intramolecular phenol–oxidative process. The intermediate **160** could serve to provide all the structural types by simple modification of the routes given in Scheme 45. Although an intermolecular oxidative coupling of two dissimilar fragments would appear to be without direct precedent in alkaloid biosynthesis, there is no experimental evidence to the contrary which indicates that the 4-hydroxycinnamyl species and the *N*-methyltyramine encounter each other prior to their linkage by an intermolecular phenol–oxidative coupling process.

Utilization of the 3-carbon side chain of the phenylalanine–cinnamic acid intermediates for the construction of the pyridine ring in the A_4 alkaloid, as indicated in Schemes 45 and 46, is easily ascertained in principle by an

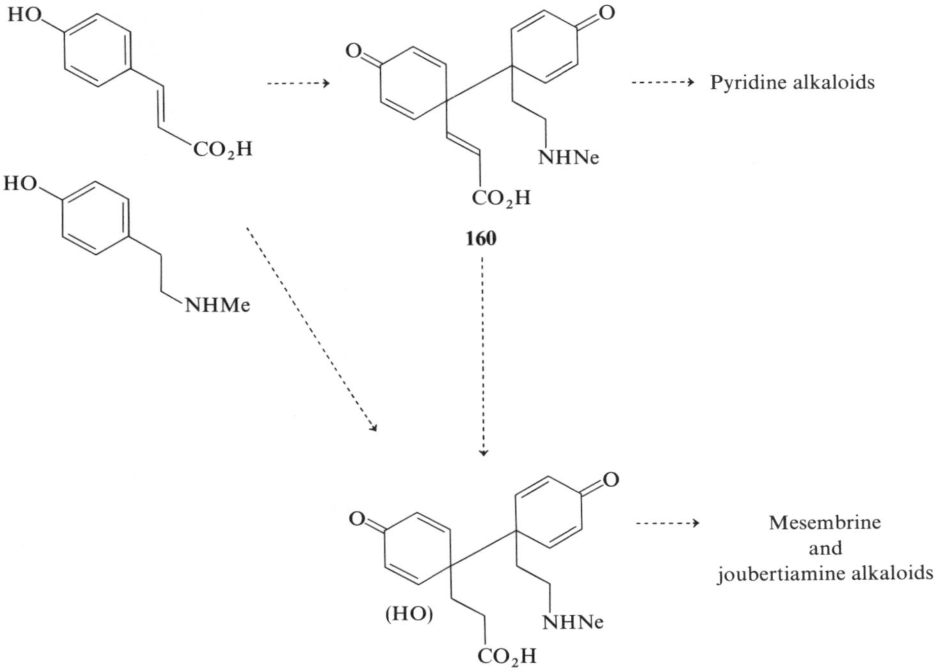

HO
CO_2H
HO
NHMe
O
O
NHNe
CO_2H
160
Pyridine alkaloids
O
O
(HO)
NHNe
CO_2H
Mesembrine
and
joubertiamine alkaloids

SCHEME 46. *Intermolecular oxidative coupling process as a biogenetic alternative.*

experiment with an appropriate double-labeled phenylalanine. Unfortunately, the A_4 base is not produced in *S. strictum*, and *S. namaquense*, in which A_4 is a minor alkaloid, has not been available for biosynthetic work outside South Africa. It is hoped that these simple but crucial experiments will be solved by workers in that country.

F. Late Stages in the Biosynthesis of Mesembrine Bases

The discovery of the monooxyaryl octahydroindole base sceletenone and the existence of the monooxyaryl bases of the joubertiamine series had preceded the results which defined the role of 4′-hydroxycinnamic acid derivatives as precursors to the 3′,4′-dioxygenated bases of the mesembrine alkaloids. The implication of the structure patterns and the biosynthetic results which demonstrated the efficient incorporation of 4′-hydroxycinnamic acid suggested that the 3′-oxygen function of the 3′,4′-dioxyaryl bases might be introduced at a late stage in the pathway. Confirmation of this idea was obtained by the demonstration that sceletenone (**30**) containing tritium labels at the 3′(5′) positions was converted to mesembrenol by *Sceletium strictum*. The high efficiency of this conversion (>2.0%) indicates that hydroxylation of the aromatic ring at the 3′ position of sceletenone represents a major pathway for the biosynthesis of the 3,4′-dioxyaryl bases of the mesembrine group.

In view of the result, it would appear that the low (0.06%) but specific incorporation (*74*) of the double-labeled ferulic acid (**161**) into mesembrenol must represent either aberrant metabolism or a minor pathway in the biosynthetic processes of *S. strictum*.

Some of the other steps involved in the late stages of biosynthesis have been defined (*74*) as summarized in Scheme 47. 4′-*O*-Demethylmesembrenone (**26**) and mesembrenone (**8**), each labeled at the H-5 position with tritium, were converted to the incredible extent of 58% and 62%, respectively, into mesembrenol by three-month-old *S. strictum* plants, whereas somewhat lower incorporations of **8** were observed into the companion alkaloids mesembrine and mesembranol in the same experiment. The absence of labeling in the phenolic base **26** from the feeding experiment with 5-^{3}H-labeled mesembrenones indicated that O-demethylation processes were not significant in the formation of the phenolic bases in *S. strictum*.

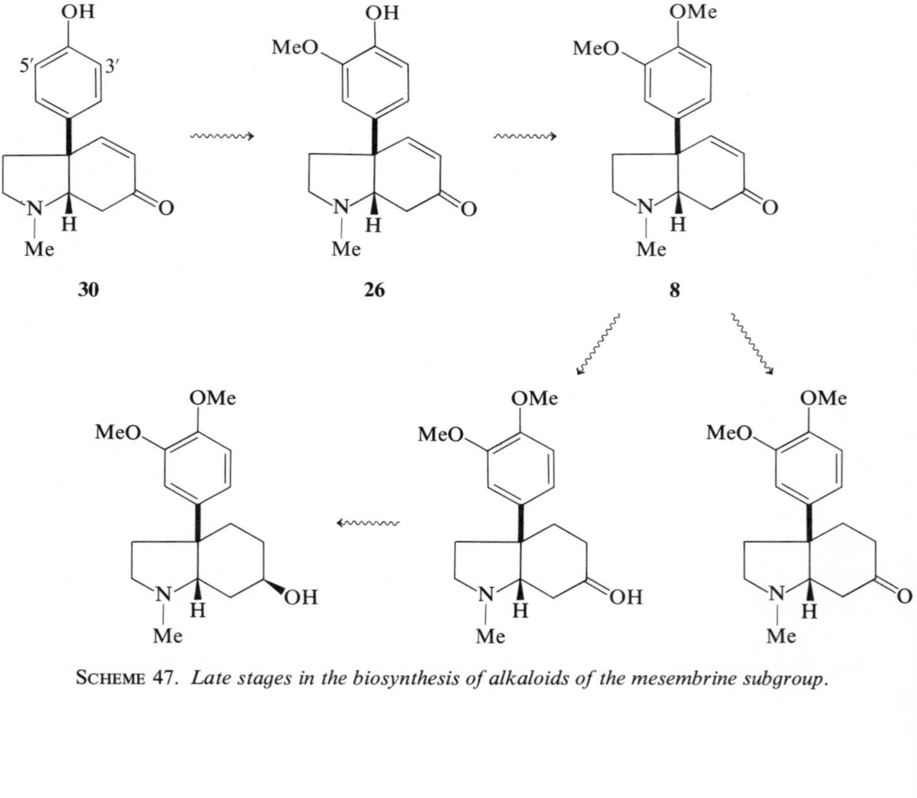

SCHEME 47. *Late stages in the biosynthesis of alkaloids of the mesembrine subgroup.*

It is clear that further work remains to be accomplished before the biosynthesis of these relatively simple alkaloids is fully understood.

REFERENCES

1. O. Dapper, "Umstandliche und Eigentliche Beschriebung von Africa," Amsterdam, 1670.
2. P. Kolben, "The Present State of the Cape of Good Hope" (G. Medley, transl.), 2nd ed., Vol. 1, p. 212. Innys & Manby, London, 1738.
3. L. Lewin, "Phantastica, Narcotic and Stimulating Drugs," p. 225. Dutton, New York, 1964.
4. C. F. Juritz, *Rep. Jt. Meet. Br. Assoc. S. Afr. Assoc. Adv. Sci.* **1**, 216 (1905).
5. E. M. Holmes, *Pharm. J. Trans.* **23**, 810 (1874).
6. J. Watt and M. Breyer-Brandwijk, "The Medicinal and Poisonous Plants of Southern and Eastern Africa," 2nd ed., p. 4. Livingstone, Edinburgh, 1962.
7. J. Meiring, *Trans. South Afr. Philos. Soc.* **9**, 48 (1896).
8. E. Zwicky, Ph. D. Thesis, ETH, Zurich, 1914; see also C. Hartwich and E. Zwicky, *Apoth. Z.* **29**, 961 (1914).
9. H. Jacobsen, "A Handbook of Succulent Plants," Vol. 3, p. 1257. Blandford Press, London, 1960.
10. C. Hartwich and E. Zwicky, *Apoth. Z.* **29**, 949 (1914).
11. C. Remington and G. C. S. Roets, *Onderstepoort J. Vet. Sci. Anim. Ind.* **9**, 187 (1937).
12. K. Bodendorf and W. Krieger, *Arch. Pharm. Ber. Dtsch. Pharm. Ges.* **290**, 441 (1957).
13. A. Popelak, E. Haack, G. Lettenbauer, and H. Spingler, *Naturwissenschaften* **47**, 156 (1960).
14. A. Popelak, G. Lettenbauer, E. Haack, and H. Spingler, *Naturwissenschaften* **47**, 231 (1960).
15. A. Popelak and G. Lettenbauer, *in* "The Alkaloids" (R. H. F. Manske, ed.), Vol. IX, p. 467. Academic Press, New York, 1967.
16. P. W. Jeffs, R. L. Hawks, and D. S. Farrier, *J. Am. Chem. Soc.* **91**, 3831 (1969).
17. F. O. Snyckers, F. Strelow, and A. Weichers, *Chem. Commun.* 1467 (1971).
18. H. W. Whitlock, Jr. and G. L. Smith, *J. Am. Chem. Soc.* **89**, 3600 (1967).
19. E. Smith, N. Hosansky, M. Shamma, and J. B. Moss, *Chem. Ind.* (*London*) 402 (1961).
20. C. Djerassi and W. Klyne, *Proc. Natl. Acad. Sci. U.S.A.* **48**, 1093 (1968).
21. P. Coggin, D. S. Farrier, P. W. Jeffs, and A. T. McPhail, *J. Chem. Soc. B* 1267 (1970).
22. T. M. Capps, K. D. Hargrave, P. W. Jeffs, and A. T. McPhail, *J.C.S. Perkin II* 1098 (1977).
23. J. Hudec, *Chem. Commun.* 829 (1970).
24. K. D. Hargrave, Ph.D. Thesis, Duke Univ., Durham, North Carolina, 1977.
25. M. M. A. Hassan and A. F. Casey, *Org. Magn. Reson.* **1**, 389 (1969): A. J. Jones and M. M. A. Hassan, *J. Org. Chem.* **37**, 2332 (1972).
26. P. W. Jeffs, G. Ahmann, H. F. Campbell, D. S. Farrer, G. Ganguli, and R. L. Hawks, *J. Org. Chem.* **35**, 3512 (1970).
27. P. W. Jeffs, T. Capps, D. B. Johnson, J. M. Karle, N. H. Martin, and B. Rauchman, *J. Org. Chem.* **39**, 2703 (1974).
28. P. E. J. Kruger and R. R. Arndt, *J. S. Afr. Chem. Inst.* **24**, 235 (1971).
29. A. Abou-Donia, P. W. Jeffs, A. T. McPhail, and R. W. Miller, *J.C.S. Chem Commun.* 1078 (1978).
30. J. M. Karle, *Acta Crystallogr., Ser. B* **33**, 185 (1977).
31. R. R. Arndt and P. E. J. Kruger, *Tetrahedron Lett.* 3237 (1970).
32. A. Wiechers, personal communication, cited in R. V. Stevens and J. T. Lai, *J. Org. Chem.* **37**, 2138 (1972).
33. K. Psotta, F. Strelow, and A. Wiechers, *J.C.S. Perkin II* 1063 (1979).
34. T. M. Capps, Ph.D. Thesis, Duke Univ., Durham, North Carolina, 1977
35. P. W. Jeffs, P. A. Luhan, A. T. McPhail, and N. H. Martin, *Chem. Commun.* 1466 (1971).

37. D. B. McClean, *in* "The Alkaloids" (R. H. F. Manske, ed.), Vol. X, p. 305. Academic Press, New York, 1968.
38. P. A. Luhan and A. T. McPhail, *J.C.S. Perkin II* 2006 (1972).
39. F. O. Snyckers, F. Strelow, and A. Wiechers, *Chem. Commun.* 1467 (1971).
40. R. V. Stevens, P. M. Lesko, and R. Lapalme, *J. Org. Chem.* **40**, 3495 (1975).
41. M. Shamma and H. R. Rodriguez, *Tetrahedron Lett.* 4847 (1965); *Tetrahedron* **24**, 6583 (1968).
42. T. J. Curphey and H. L. Kim, *Tetrahedron Lett.* 1441 (1968).
43. R. V. Stevens and M. P. Wentland, *J. Am. Chem. Soc.* **40**, 5580 (1968).
44. S. L. Keely and F. C. Tahk, *J. Am. Chem. Soc.* **90**, 5584 (1968).
45. J. M. Karle, Ph.D. Thesis, Duke Univ., Durham, North Carolina, 1976.
46. R. V. Stevens, *in* "The Total Synthesis of Natural Products" (J. Ap Simmon, ed.), Vol. 3, p. 439. Wiley (Interscience), New York, 1977.
47. R. V. Stevens and J. T. Lai, *J. Org. Chem.* **37**, 2138 (1972).
48. T. Ohishi and H. Kugita, *Chem. Pharm. Bull.* **18**, 299 (1970).
49. T. Ohishi and H. Kugita, *Chem. Pharm. Bull.* **18**, 291 (1970).
50. S. F. Martin, T. A. Puckette, and J. A. Colapret, *J. Org. Chem.* **44**, 3391 (1979).
51. P. W. Jeffs and G. Molina, *J.C.S. Chem. Commun.* 3 (1973).
52. N. A. Cortese, Ph.D. Thesis, Duke Univ., Durham, North Carolina, 1976.
53. P. W. Jeffs and Y. Wolfram, 1980, unpublished observations.
54. J. P. B. A. Wijnberg and W. N. Speckamp, *Tetrahedron* **34**, 2579 (1978).
55. G. Otani and S. Yamada, *Chem. Pharm. Bull.* **21**, 2130 (1973).
56. G. Otani and S. Yamada, *Chem. Pharm. Bull.* **21**, 2125 (1973).
57. M. F. Strauss and A. Wiechers, *Tetrahedron* **34**, 127 (1978).
58. H. Muxfeldt, R. S. Schneider, and J. Mooberry, *J. Am. Chem. Soc.* **88**, 3670 (1966).
59. K. Psotta and A. Wiechers, *Tetrahedron* **35**, 255 (1979).
60. C. P. Forbes, G. L. Wenteler, and A. Wiechers, *Tetrahedron* **34**, 487 (1978).
61. O. Diels and M. Paquin, *Ber Dtsch. Chem. Ges.* **46**, 2000 (1913); O. Diels and E. Fischer, *Ber Dtsch. Chem. Ges.* **47**, 2043 (1914).
62. G. W. Kenner and J. P. Stedman, *J. Chem. Soc.* 2089 (1952); R. Huisgen and F. Jacob, *Justus Liebigs Ann. Chem.* **590**, 37 (1954).
63. K. W. Bentley and D. G. Hardy, *J. Am. Chem. Soc.* **89**, 3281 (1967).
64. P. W. Jeffs, H. F. Campbell, and R. L. Hawks, *Chem. Commun.* 1338 (1971).
65. R. L. Hawks, Ph.D. Thesis, Duke Univ., Durham, North Carolina, 1970.
66. G. Opitz and W. Mertz, *Justus Liebigs Ann. Chem.* **652**, 158 (1962); F. Bohlman and O. Schmidt, *Chem. Ber.* **97**, 1354 (1966).
67. P. W. Jeffs, W. C. Archie, R. L. Hawks, and D. S. Farrier, *J. Am. Chem. Soc.* **93**, 3752 (1971).
68. P. W. Jeffs, H. F. Campbell, D. S. Farrier, G. Ganguli, N. H. Martin, and G. Molina, *Phytochemistry* **13**, 933 (1974).
69. D. H. Barton, G. W. Kirby, J. B. Taylor, and G. M. Thomas, *J. Chem. Soc.* 4545 (1963).
70. K. R. Hanson and E. A. Havir, *in* "The Enzymes (P. D. Boyer, ed.), Vol. 7, p. 75. Academic Press, New York, 1972.
71. N. H. Martin, Ph.D. Thesis, Duke Univ., Durham, North Carolina, 1972.
72. P. W. Jeffs, D. B. Johnson, N. H. Martin, and B. S. Rauckman, *J.C.S. Chem Commun.* 82 (1976).
73. G. W. Kirby and I. T. Bruce, *J.C.S. Chem Commun.* 207 (1968).
74. P. W. Jeffs, J. M. Karle, and N. H. Martin, *Phytochemistry* **17**, 719 (1978).
75. A. R. Battersby, R. B. Herbert, E. McDonald, R. Ramage, and J. H. Clements, *J.C.S. Perkin I* 1741 (1972).
76. D. R. Dalton, *in* "Studies in Organic Chemistry" (P. G. Gassman, ed.), Vol. 7, p. 328. Dekker, New York, 1979.

——CHAPTER 2——

SOLANUM STEROID ALKALOIDS

HELMUT RIPPERGER AND KLAUS SCHREIBER

Institute of Plant Biochemistry of the Academy of Sciences of the GDR, DDR-402 Halle/Saale, German Democratic Republic

I. Introduction

The chemistry and biochemistry of *Solanum* steroid alkaloids as well as the occurrence of these natural products in the plant kingdom have been reviewed completely up to 1966 by Prelog and Jeger in Volumes III and VII as well as by Schreiber in Volume X of this treatise (*1–3*). The present chapter deals with work reported since that time; results of earlier investigations are mentioned briefly only if necessary. The present review includes a tabulated survey of the occurrence of *Solanum* glycoalkaloids and alkamines in plants (Table I), two additional surveys of the well-characterized alkaloid glycosides (Table II) and alkamines (Table III) that have been isolated from plants listed in Table I, and tables of physical constants compiling all the glycosides (Table VIII), sugar-free alkamines (Tables IX–XVI), and their

THE ALKALOIDS, VOL XIX

ISBN 0-12-469519-1

derivatives (together more than 500 compounds) described in the literature only since 1967. However, the two chapters on *Solanum* alkaloids published in Volume X (*3*) and in the present volume, supplemented to a minor extent by the corresponding reviews by Prelog and Jeger (*1*, *2*), constitute as complete a survey as possible of all the well-characterized glycoalkaloids, alkamines, and their derivatives obtained by synthesis or degradation so far known (together more than 1100 compounds) as well as of the distribution of these alkaloids in the plant kingdom.

The chemical work done in the last decade on *Solanum* steroid alkaloids has still been particularly stimulated by the early statements of Sato *et al.* (*4–6*) as well as of Kuhn *et al.* (*7*), who nearly 30 years ago announced the chemical transformation of the spirosolane alkaloids solasodine and tomatidine into 3β-acetoxypregna-5,16-dien-20-one and its 5,6-dihydro derivative, respectively. Because these pregnanes are important intermediates in the industrial production of hormonal steroids, the *Solanum* alkaloids, especially solasodine, have been receiving increased interest and significance as starting materials for the commercial manufacture of steroidal drugs. On the basis of some work already summarized (*3*, *8–10*; cf. *369*), much new activity has been directed again to this topic in the last few years. This recent interest in steroidal *Solanum* alkaloids has been initiated by an increasing demand for steroidal raw materials all over the world and coupled with a simultaneous shortage of diosgenin, the most important starting material to date for the steroid industry (*11*), which has sharply increased in price. According to Djerassi (*12*), "solasodine may well become the 'diosgenin of the next decade' although it is unlikely that any of the major steroid manufacturers will want to become dependent again on any single source" (cf. *13–16*).

On the other hand, the report that potatoes (*Solanum tuberosum* L.) infected (blighted) by the fungus *Phytophthora infestans* (Mont.) de Bary were responsible for an abnormal incidence of congenital defects (*17*) has also attracted considerable research interest (*18*). The results of some of these investigations suggest that the harmful effects of potato preparations, whether toxic or teratogenic, can be attributed to their alkaloid content (*19*). Thus, recently a number of *Solanum* alkaloids were tested for teratogenicity. According to the results obtained, 5α,22α*H*,25β*H*-solanidan-3β-ol (demissidine), 5α,22β*H*,25β*H*-solanidan-3β-ol (22-isodemissidine), and the spirosolane alkaloid tomatidine were shown to be inactive. Solasodine was found to possess weak activity, but 5α,22β*H*,25α*H*-solanidan-3β-ol or 22β*H*,25α*H*-solanid-5-en-3β-ol possessed a high teratogenicity (*20–22*).

Solanum steroid alkaloids have been isolated so far from nearly 350 species of the plant families Solanaceae and Liliaceae, in which they generally occur as glycosides. All of the 75 known steroidal alkamines of the *Solanum* type, the structures of which have been established, possess the C_{27}-carbon skeleton of cholestane and belong to one of the following five groups representing different types of structure: the spirosolanes, e.g., solasodine (**1**); the epiminocholestanes, e.g., solacongestidine (**2**); the solanidanes, e.g., solanidine (**3**); the alkaloids with a solanocapsine skeleton, e.g., solanocapsine (**4**); the 3-aminospirostanes, e.g., jurubidine (**5**).

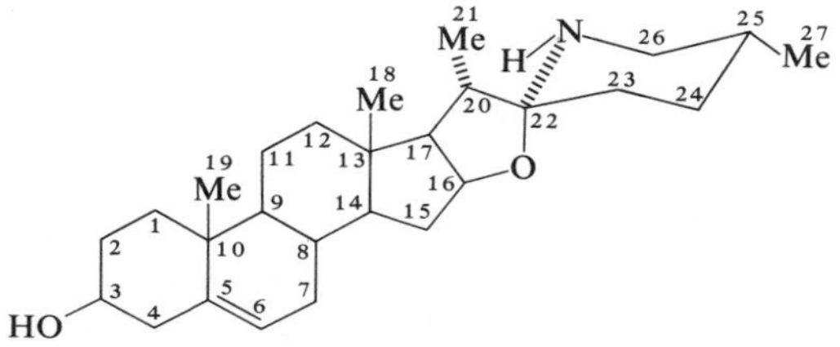

1 Solasodine

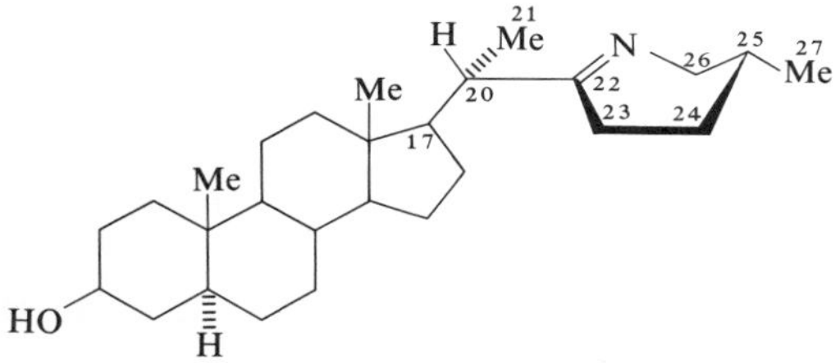

2 Solacongestidine

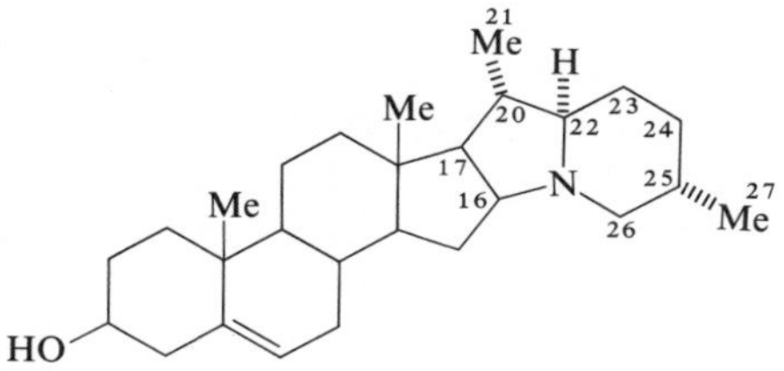

3 Solanidine

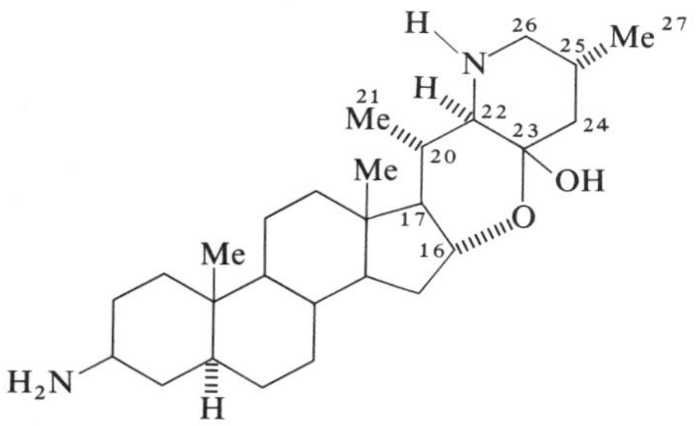

4 Solanocapsine

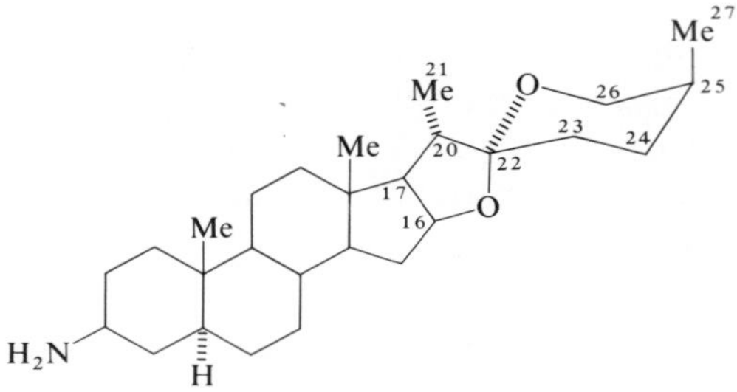

5 Jurubidine

Thus, the C_{27}-steroid alkaloids with C-nor-D-homo ring system, that is, with a jervane or cevane skeleton, with 18-nor-17β-methylcholestane skeleton (*23–27*) or other alterations of the C_{27}-carbon skeleton of cholestane, found in Liliaceae but not yet in Solanaceae, are not included. They are described exhaustively in other volumes of this series by Kupuchan and By (*28*) as well as by Tomko and Votický (*29*). On the other hand, some *Veratrum* bases, for instance, teinemine and baikeine, belong to the *Solanum*-type alkaloids and are mentioned in this chapter.

The nomenclature used in this review is the same as that used in the former chapter on *Solanum* alkaloids (*3*) and follows the IUPAC–IUB recommendations for steroid nomenclature (*30*), with the exception of the "-anine" designation of steroid alkaloids.

The most significant chemical progress in the last decade has been the isolation and structural elucidation of an increasing number of new steroidal alkamines and glycosides the structures of which, however, still belong to one of the known types **1**–**5**. Extensive work has also been directed to a number of new partial and total syntheses as well as to the degradation of the alkamines, particularly to pregnane derivatives. A number of review articles have been published in the last years (e.g., *31–38*).

Besides the steroid alkaloids only rarely have there been found alkaloids of other structural types in *Solanum* species and related genera (*3*). Recently, solamine [4,4′-bis(dimethylaminobutyl)amine] and/or some of its derivatives have been isolated from species of *Solanum* and the related *Cyphomandra*. Thus, *Solanum tripartitum* Dunal contains the solamine derivatives solapalmitine and solapalmitenine (*41*, *42*), *Cyphomandra betacea* Sendtn. solamine and its derivative solacaproine (*43*), and *Solanum carolinense* L. solamine and its carbamate (solaurethine) (*44*). In addition, the last-mentioned species was shown to contain cuscohygrine and anabasine (*44*). The occurrence of cuscohygrine and solamine-derived amides has also been reported for a number of other *Solanum* and *Cyphomandra* species (*45*).

II. Occurrence of Glycoalkaloids and Alkamines

Table I surveys the distribution of alkamines with nonaltered C_{27}-cholestane skeletons and their glycosides according to the literature since 1967 and represents a supplement of the corresponding compilation (*3*).

TABLE I

OCCURRENCE OF *SOLANUM* GLYCOALKALOIDS AND ALKAMINES[a]

Plant species	Alkaloid (aglycone)	Reference
Solanum species (Solanaceae)		
S. acaule Bitt.	Demissine (demissidine), tomatine (tomatidine)	(*46*)
S. aculeatissimum Jacq.	Solasodine	(*47*)
	Solasonine, solamargine (solasodine)	(*48, 49*)
S. aculeatum Jacq.	(Solasodine, 25-isosolafloridine)	(*380*)
S. aethiopicum L.	(Solasodine)	(*50*)
S. ajanhuiri Juz. et Buk.	α-Solanine, α-, β-chaconine (solanidine)	(*46*)
S. albicaule Kotschy	Solamargine (solasodine)	(*372*)
S. arundo Mattei	Solasonine, solamargine (solasodine), α-solanine (solanidine)	(*51*)
S. asymmetriphyllum Specht	Solasonine, solamargine (solasodine)	(*52*)
S. aviculare Forst.	(Solasodine, tomatidenol, solanaviol)	(*53*)
	Solaradixine (solasodine)	(*54*)
	Solashabinine, solaradinine (solasodine)	(*55*)
S. bahamense L.	(Solasodine)	(*380*)
S. brownii Dun.	(Solasodine)	(*52*)
S. burbankii Bitt.	Solasonine, solamargine (solasodine)	(*56*)
S. callium C. T. White ex R. J. Henderson	(25-Isosolafloridine, solacallinidine)	(*58*)
S. campanulatum R. Br.	(Solasodine)	(*52*)
S. capsiciforme (Domin) Baylis	Solasonine, solamargine (solasodine)	(*59*)
	(Solasodine)	(*40*)
S. chacoense Bitt.	Commersonine, demissine (demissidine)	(*60*)
S. chippendalli Symon	(Solasodine)	(*39*)
S. cinereum R. Br.	(Solasodine)	(*52*)
S. commersonii Dun.	Commersonine, demissine (demissidine)	(*60*)
S. congestiflorum Dun.	Solacongestine, α-, β-solacongestinine (solacongestidine)	(*61*)
	(Solacongestidine, 23-oxo-solacongestidine, "24-oxosolacongestidine," solafloridine)	(*62*)

(*Continued*)

TABLE I (*Continued*)

Plant species	Alkaloid (aglycone)	Reference
S. cookii Symon	(Solasodine)	(*52*)
S. cunninghamii Benth.	Solasonine, solamargine (solasodine)	(*52*)
S. curtilobum Juz. et Buk.	α-Solanine, α-chaconine (solanidine), demissine (demissidine), solamarines (tomatidenol)	(*46*)
S. cyananthum Dun.	Solasodine	(*63*)
S. dasyphyllum Schum. et Thonn.	(Tomatidenol)	(*64*)
	Soladulcamarine, α-, β-, γ-solamarine (tomatidenol)	(*65*)
S. dianthophorum Dun.	(Solasodine)	(*52*)
S. dioicum W. V. Fitz	Solasonine, solamargine (solasodine)	(*52*)
S. dimorphospinum C. T. White	(Tomatidine)	(*52*)
S. diplacanthum Dammar	*cf.* S. arundo Mattei	(*51*)
S. diversiflorum F. v. Muell.	Solasonine, solamargine (solasodine)	(*52*)
S. dubium L.	α-Solanine (solanidine), solamargine, β-solamargine (solasodine), solasodine	(*381*)
S. dulcamara L.	(Tomatidine)	(*66*)
	(Tomatidine, tomatidenol, soladulcidine, solasodine, 15α-hydroxytomatidine, -tomatidenol, -soladulcidine, -solasodine)	(*40*)
S. dulcamara var. *villosissimum* Desv.	(Tomatidenol, solasodine)	(*67*)
S. dunalianum Gaudich.	Soladunalinidine	(*68*)
	(Soladunalinidine, tomatidine)	(*52*)
S. eardleyae Symon	(Solasodine)	(*52*)
S. eburneum Symon	(Solasodine)	(*39*)
S. ecuadorense Bitt.	Solaphyllidine, deacetylsolaphyllidine	(*69, 70*)
S. elaeagnifolium Cav.	Solamargine (solasodine)	(*71, 72*)
	Solamargine, solasurine (solasodine)	(*73*)
S. erianthum D. Don ("*S. eriantum*")	(Solasodine)	(*74*)
"*S. eryanthum*"	(Solasodine, tomatidenol, pimpinellidine)	(*75*)
S. euacanthum Phil.	Solamargine (solasodine)	(*71, 89*)
S. giganteum Jacq.	Solanogantine	(*76*)
	Solanogantamine, isosolanogantamine	(*363*)
S. globiferum Dun.	(Solasodine, tomatidenol)	(*77*)
S. grabrielae Domin	(Solasodine)	(*52*)
S. grandiflorum Ruiz et Pav.	(Solasodine)	(*382*)
S. hainanense Hance	Solasodenone (solasodine)	(*78, 79*)
S. havanense Jacq.	(Tomatidenol, etioline)	(*80*)
S. hypomalacophyllum Bitt.	Solaphyllidine	(*81*)
	Deacetoxysolaphyllidine	(*82*)
	Solamaladine	(*83*)

(*Continued*)

TABLE I (*Continued*)

Plant species	Alkaloid (aglycone)	Reference
S. incanum L.	(Solasodine)	(*84*)
	Solasonine (solasodine)	(*85*)
	Solamargine (solasodine)	(*86*)
S. indicum L.	Solasonine, solamargine (solasodine)	(*361*)
S. jasminoides Paxt.	Solasonine, solamargine (solasodine)	(*87, 88*)
S. judaicum Bess.	(Solasodine)	(*354*)
S. juvenale Thell.	Solasodine	(*89*)
S. juzepczukii Buk.	α-Solanine, α-chaconine (solanidine), demissine (demissidine), solamarines (tomatidenol)	(*46*)
S. khasianum C. B. Clarke	Solakhasianine (solasodine), solasodine	(*90*)
	Solasonine, solamargine, khasianine (solasodine)	(*383*)
S. kieseritzkii C. A. Mey.	Solasonine, solamargine (solasodine), tomatine (tomatidine)	(*91*)
S. laciniatum Ait.	Solashabanine, solaradinine (solasodine)	(*55, 92*)
S. lasiophyllum Dun.	(Solasodine)	(*52*)
S. leopoldensis Symon	(Solasodine)	(*39*)
S. linearifolium Her.	Solasonine, solamargine (solasodine)	(*59*)
S. lorentzii Bitt.	Solasonine, solamargine (solasodine)	(*71*)
S. lycocarpum St. Hil.	Solasonine, solamargine (solasodine)	(*93*)
S. macrocarpon L.	Solasonine, solamargine (solasodine), α-, β-solamarine (tomatidenol)	(*94*)
S. mammosum L.	Solaradixine (solasodine)	(*54*)
	"Solasodine trioside", solaradixine (solasodine)	(*95*)
S. maritimum Meyen	(Solasodine)	(*354*)
S. mauritianum Scop.	Solasonine, solamargine (solasodine)	(*96*)
	(Solasodine)	(*52*)
S. melanospermum F. v. Muell.	Solasonine, solamargine (solasodine)	(*52*)
S. nummularium S. Moore	(Solasodine)	(*52*)
S. oldfieldii F. v. Muell.	(Solasodine)	(*52*)
"*S. oleraceum*"	Solasonine, solamargine (solasodine)	(*97, 98*)
S. oligacanthum F. v. Muell.	(Solasodine)	(*52*)
S. orbiculatum Dun.	(Solasodine)	(*52*)
S. palinacanthum Dun.	Solasonine, solamargine (solasodine)	(*93*)
S. paniculatum L.	Isojurubidine, isopaniculidine, isojuripidine, 4 glycosides of isojurubidine, isojuripine	(*99, 100*)
S. persicum Willd. (=*S. dulcamara* L. var. *rupestre* (F. W. Schmidt) Kostel.)	Solasonine, solamargine (solasodine)	(*101*)
	Solapersine (solasodine)	(*102*)
S. petrophilum F. v. Muell.	(Solasodine)	(*52*)
S. phlomoides Benth.	Solasonine, solamargine (solasodine)	(*52*)

(*Continued*)

TABLE I (*Continued*)

Plant species	Alkaloid (aglycone)	Reference
S. pinnatum Cav.	Solasonine, solamargine (solasodine)	(*159*)
S. platanifolium Sims	(Solasodine)	(*103*)
	Solasonine, solamargine, solatifoline (solasodine)	(*104*)
S. pseudocapsicum L.	Solacasine	(*105*)
S. pseudomeum L.	Solasonine, solamargine (solasodine)	(*106*)
S. pseudopersicum Pojark.	Solasonine, solamargine, β-solamargine (solasodine), α-soladulcine (soladulcidine)	(*107*)
S. pseudoquina St. Hil.	Solaquidine	(*108*)
S. pugiunculiferum C. T. White	(Solasodine)	(*52*)
"*S. pyretrifolium*"	Solamargine (solasodine)	(*71*)
S. schimperianum Hochst.	Solamargine, β-solamargine (solasodine), β-, γ-solamarine (tomatidenol)	(*197*)
S. seaforthianum Andr.	Solaseaforthine, isosolaseaforthine	(*109*)
	Solanoforthine	(*110*)
S. simile F. v. Muell.	Solasonine, solamargine (solasodine)	(*59*)
S. stenotomum Juz. et Buk.	α-Solanine, α-, β-chaconine (solanidine)	(*46*)
S. surattense Burm. f. (=*S. xanthocarpum* Schrad. et Wendl.)	Solasonine, solamargine, solasurine (solasodine)	(*111*)
S. symonii Hj. Eichler	Solasonine, solamargine (solasodine)	(*59*)
S. tetrathecum F. v. Muell.	(Solasodine)	(*52*)
S. torvum Sw.	Solasonine (solasodine)	(*198*)
	Jurubine (jurubidine)	(*112*)
S. transcaucasicum Pojark.	Solasonine, solamargine, β-solamargine (solasodine)	(*113*)
S. trilobatum L.	Tomatidenol	(*114*)
S. tuberosum L. var. Kennebec	α-, β-Solamarine (tomatidenol), α-solanine, α-chaconine (solanidine)	(*115*)
S. tudununggae Symon	Solasonine, solamargine (solasodine)	(*52*)
S. umbellatum Mill. (=*S. callicarpaefolium* Kth. et Bché.)	(Solasodine, solafloridine)	(*57*)
"*S. unguiculatum*"	Solasonine, solamargine (solasodine, free)	(*116*)
S. verbascifolium L.	(Solasodine, solafloridine, tomatidenol)	(*117*)
	(Solasodine, tomatidine, solaverbascine)	(*118*)
	Tomatine (tomatidine), solasonine, solamargine (solasodine)	(*384*)
S. viarum Dun. (=*S. khasianum* C. B. Clarke var. *chatterjeeanum* Sengupta)	Solasonine, solamargine, solasurine (solasodine)	(*73*; cf. *119*)

(*Continued*)

TABLE I (*Continued*)

Plant species	Alkaloid (aglycone)	Reference
S. xanthocarpum Schrad. et Wendl.	Solamargine, β-solamargine (solasodine)	(*120*)
	Solasonine, solamargine, β-solamargine (solasodine)	(*121*)
	(Solasodine, tomatidenol)	(*122*)
Lycopersicon species (Solanaceae)		
L. esculentum Mill.	*N*-Nitrosotomatidine, tomatidine	(*385*)
L. pruniforme Mill.	Tomatine (tomatidine)	(*123*)
Other Solanaceae		
Cestrum purpureum Stand.	Solasodine, solanidine	(*124*; cf. *125*)
Nicotiana plumbaginifolia Viv.	Solaplumbine (solasodine)	(*126*)
Liliaceae		
Fritillaria camtschatcensis (L.) Ker.	Solanidine	(*127*)
	(Hapepunine)	(*128*)
	(Solanidine, solasodine, tomatidenol, hapepunine, anrakorinine)	(*367*)
	(Camtschatcanidine)	(*368*)
Korolkowia severtzovii Rgl.	Corsevinine	(*129*)
Petilium eduardi (Rgl.) Vved.	Edpetilidinine	(*130*)
(=*Fritillaria eduardii* Rgl.)	Edpetilinine (edpetilidinine)	(*131*)
Petilium raddeana (Rgl.) Vved. (=*Fritillaria raddeana* Rgl.)	Petiline	(*132*)
Rhinopetalum bucharicum	Solanidine, rhinoline (rhinolidine)	(*350*)
Rhinopetalum stenantherum Rgl.	Solanidine	(*133*)
Veratrum album L. ssp. *lobelianum* (Bernh.) Suesseng.	Veralosine (*O*(16)-acetyletioline?)	(*134*)
	γ-Solanine (solanidine)	(*135*)
	Solasodine	(*136*)
	Veramiline	(*137*)
	Veracintine	(*138*)
	20-(2-Methyl-1-pyrrolin-5-yl)-pregn-4-en-3-one	(*139*)
	Glucoveracintine (veracintine)	(*140*)
	Veralosidine	(*141*)
	Veralosinine	(*141, 142*)
	Veralosidinine	(*143*)
	Veralodisine	(*144*)
	Veralodinine (veralodisine)	(*365*)
Veratrum californicum Durand	Rubijervine, isorubijervine	(*145*)
Veratrum grandiflorum (Max.) Loesen	(Etioline)	(*146*)
	(Hakurirodine)	(*147*)
	(Teinemine, isoteinemine)	(*148*)
	Baikeine, baikeidine	(*149*)
	Procevine	(*150*)
	Verazine	(*293*)
	Epirubijervine	(*366*)

[a] Generally reported since 1967; supplement of the corresponding compilation (*3*). In the course of some extensive phytochemical screenings alkaloids and "saponins" have also been detected in a number of hitherto noninvestigated *Solanum* species (cf. *151–158*). Using radioimmunoassay, alkaloidal glycosides probably of the spirosolane type have been detected in about 250 *Solanum* species, 180 of which have not yet been investigated (*371*).

III. Glycoalkaloids

The glycoalkaloids isolated from plants or obtained by partial hydrolysis of native glycosides since 1967 are compiled in Table II.

By the permethylation procedure, the sugar unit of solamargine has been shown to be β-chacotriose, i.e., *O*-α-L-rhamnopyranosyl-(1 → 2_{glu})-*O*-α-L-rhamnopyranosyl-(1 → 4_{glu})-β-D-glucopyranose. The molecular rotation dif-

TABLE II
COMPOSITION OF THE GLYCOALKALOIDS[a]

Glycoalkaloid	Aglycone	Sugars	Reference
Solaplumbine	Solasodine	L-Rhamnose, D-glucose	*(126)*
Solaplumbinine[b]	Solasodine	L-Rhamnose	*(126)*
Solasurine	Solasodine	L-Rhamnose, D-glucose	*(73, 111)*
Solasodine bioside	Solasodine	L-Rhamnose, D-glucose	*(161)*
Solashabanine	Solasodine	L-Rhamnose, 3 D-glucose, D-galactose	*(55, 92)*
Solaradinine	Solasodine	L-Rhamnose, 4 D-glucose, D-galactose	*(55, 92)*
Khasianine	Solasodine	L-Rhamnose, D-glucose	*(383)*
Solakhasianine	Solasodine	L-Rhamnose, D-galactose	*(90)*
γ-Solasonine[b]	Solasodine	D-Galactose	*(160)*
Solapersine	Solasodine	2 D-Xylose, D-glucose, D-galactose	*(102)*
Solatifoline	Solasodine	L-Rhamnose, D-glucose, D-galactose	*(104)*
Solasodine trioside	Solasodine	L-Rhamnose, 2 D-glucose	*(95)*
α-Soladulcine[c]	Soladulcidine	D-Xylose, 2 D-glucose, D-galactose	*(107)*
Solacongestine	Solacongestidine	D-Glucose, D-galactose	*(61)*
α-Solacongestinine	Solacongestidine	D-Rhamnose, D-xylose, D-glucose	*(61)*
β-Solacongestinine	Solacongestidine	D-Xylose, D-glucose	*(61)*
Glucoveracintine	Veracintine	D-Glucose	*(140)*
Commersonine	Demissidine	3 D-Glucose, D-galactose	*(60)*
Glycoside SG 1	Isojurubidine	Not identified	*(100)*
Glycoside SG 2	Isojurubidine	Not identified	*(100)*
Glycoside SG 3	Isojurubidine	Not identified	*(100)*
Glycoside SG 4	Isojurubidine	Not identified	*(100)*
Isojuripine	Isojuripidine	Not identified	*(99, 100)*
Edpetilinine	Edpetilidinine	D-Xylose	*(131)*
Veralosine	*O*(16)-Acetyl-etioline (?)	D-Glucose	*(134)*
Rhinoline	Rhinolidine	D-Glucose	*(350)*
Rhinolinine	Rhinolidine	2 D-Glucose	*(362)*
Veralodinine	Veralodisine	D-Glucose	*(365)*

[a] Isolated from plants since 1967; supplement of the corresponding compilation *(3)*.
[b] Obtained by partial hydrolysis.
[c] Probably identical with soladulcidine tetraoside.

ference between γ-solamargine and solasodine shows that the glucose unit is joined to solasodine by a β linkage (*161*).

The sugar sequence of commersonine, a new demissidine glycoside from *Solanum chacoense* and *S. commersonii*, was determined by analysis of the mono-, di-, and trisaccharides obtained by partial hydrolysis of the glycoalkaloid. The positions of the glycosidic linkages were determined from gas chromatography and mass spectroscopy (GC-MS) of the partially methylated alditol acetates. The configurations of the anomeric carbon atoms have not been established; however, one might expect an all-β arrangement. Thus, the tetrasaccharide moiety of commersonine should have structure **6**, differing from β-lycotetraose by replacement of the terminal D-xylose by D-glucose (*60*).

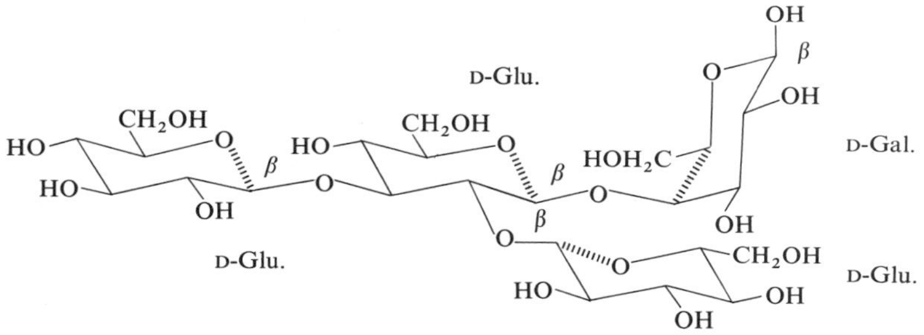

6

Synthetic 2-*O*-α-L-rhamnopyranosyl-D-galactose heptaacetate (*162*) was shown to be not identical with the respective compound isolated from the partial acid hydrolysate of acetylated α-solanine by Kuhn *et al.* (*163*).

A new solasodine glycoside, solaplumbine, has been isolated from aerial parts of *Nicotiana plumbaginifolia*. Acid hydrolysis gave solasodine and the sugars D-glucose and L-rhamnose in a 1:1 molar ratio. Enzymic hydrolysis of solaplumbine with takadiastase afforded a second glycoside, solaplumbinine, and D-glucose, which was interpreted to mean that solaplumbine is a β-glucoside of solaplumbinine and that the latter glycoside is a L-rhamnosylsolasodine. Periodate oxidation was carried out with both solaplumbine and solaplumbinine. In each case, 1 mole of formic acid was generated. In solaplumbine it obviously derived from the glucose part where three vicinal hydroxy groups were free, while in case of solaplumbinine its formation suggested that three vicinal hydroxy groups of L-rhamnose were also free and therefore that the latter was linked to solasodine through either position 1 or position 4. Permethylsolaplumbine yielded 2,3,4,6-tetra-*O*-methyl-D-glucose and 2,3-di-*O*-methyl-L-rhamnose (*126*). Therefore, solaplumbine seems to be the $O(3)$-[O-β-D-glucopyranosyl-$(1\rightarrow 4_{rha})$-α-L-rhamnopyranosyl]solasodine; however, further investigation is necessary to confirm its structure convincingly. This glycoside, which possesses anticancer activity, is the first steroidal alkaloid to be isolated from a *Nicotiana* species.

After natural fermentation of the ripe fruit of *Solanum marginatum*, a novel L-rhamnosyl-D-glucosylsolasodine has been isolated which was shown to be not identical with β-solamargine (*161*).

The root of *Solanum laciniatum* and *S. aviculare* (*54*) as well as the fruit of *S. mammosum* (*54*, *95*) contain solaradixine, the sugar moiety of which consists of D-galactose, D-glucose, and L-rhamnose in the molar ratio 1:2:1. Partial hydrolysis with 0.01 *N* sulfuric acid yielded solasonine. Hydrolysis of permethyl solaradixine furnished 2,3,4-tri-*O*-methyl-L-rhamnose, 2,3,4,6-tetra-*O*-methyl-D-glucose, 4,6-di-*O*-methyl-D-galactose, and 3,4,6-tri-*O*-methyl-D-glucose. Assuming a β linkage of the glucose unit bound to solasonine, the sugar moiety of solaradixine has structure **7** (*54*).

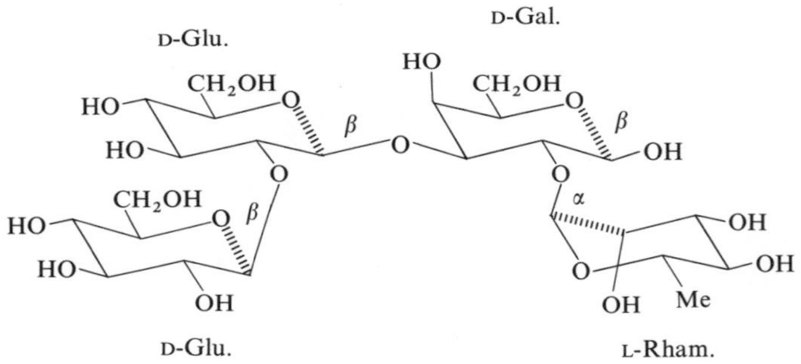

7

In addition to solaradixine, the fruit of *S. mammosum* contains two other solasodine glycosides, the main product of which was shown to be a trioside consisting of 2 mol of D-glucose and 1 mol of L-rhamnose (*95*).

Solaradixine was converted to solasonine by *Aspergillus japonicus* in good yield (*164*). The solasodine glycosides solashabanine and solaradinine have been isolated from *S. laciniatum* and *S. aviculare*. Solashabanine contains 1 mol of D-galactose, 1 mol of L-rhamnose, and 3 mol of D-glucose, whereas solaradinine possesses 1 mol of D-galactose, 1 mol of L-rhamnose, and 4 mol of D-glucose. Enzymic cleavage of solashabanine with β-glucosidase gave solasonine, whereas solaradinine gave solaradixine with loss of two D-glucose moieties in each case (*55*).

Khasianine has been isolated from berries of *Solanum khasianum*. By application of CMR spectroscopy the structure has been elucidated as *O*-α-L-rhamnopyranosyl-(1 → 4_{glu})-*O*(3)-β-D-glucopyranosylsolasodine (*383*).

Glucosyl- and diglucosylsolanidine are synthetized by potato tuber tissue slices or cell suspension cultures that have been incubated with solanidine (*386*). Enzyme preparations from sprouts and dormant tubers of the potato, *Solanum tuberosum*, are able to hydrolyze α-solanine, α-, β_1-, β_2-, and γ-chaconine to solanidine and the respective sugars. The enzymes of tubers, unlike those of sprouts, hydrolyze α-chaconine in a stepwise, but α-solanine in a nonstepwise manner. Contrary to the literature (*3*), not only the enzymes of tubers but those of sprouts are capable of hydrolyzing γ-chaconine (*165*; cf. *166*, *167*).

Five glycosides of the two new (25*R*)-3β-amino-5α,22α*O*-spirostanes, isojurubidine and its 6α-hydroxy derivative isojuripidine, have been isolated from *Solanum paniculatum* (*99*, *100*), the structures of which may be of the same type as that of jurubine, the (25*S*)-3β-amino-5α-furostane-22α,26-diol-*O*(26)-β-D-glucopyranoside from the same species and from *Solanum torvum* (*112*, *168*, *169*; cf. *3*), but their structures have not yet been elucidated.

Some *Solanum* glycoalkaloids have been characterized by GC-MS analysis of their permethyl derivatives (*60*, *170*; cf. *370*) and determined in living plants and herbarium specimens by use of a radioimmunoassay (*371*). Like digitonin and α-tomatine, the steroidal glycoalkaloid mixture from potato (α-solanine and α-chaconine) is able to complex with 3β-hydroxysterols *in vitro* (*370*) which can be used for the quantitative analysis of these alkaloids (*379*).

IV. Alkamines

The alkamines isolated directly from plant material or obtained by acid or enzymatic hydrolysis of naturally occurring glycosides between 1967 and 1979 are summarized in Table III. Physical methods are of increasing importance in structure elucidation. Recent papers have dealt with IR spectra (*173*), ^{1}H NMR (*174*, *175*), ^{13}C NMR (*176–178*, *357*, *358*), EPR (*179*), ORD (*180–183*), and MS (*184*, *351*, *352*) of *Solanum* alkamines. Some X-ray analyses have been carried out since 1967 (*58*, *81*, *149*, *185–189*, *359*). Tables IV–VI survey the characteristic features of ^{1}H NMR, ^{13}C NMR, and MS of *Solanum* alkaloids. Silver nitrate-containing adsorption layers were shown to be useful in TLC separation of 5α-saturated and 5-unsaturated alkamines (*194*, *195*). High-pressure liquid chromatography has been applied successfully for the separation of steroidal *Solanum* and *Veratrum* alkamines (*373*).

TABLE III
ALKAMINES WITH CHOLESTANE SKELETON[a]

Alkamine	Isolation	Reference
Spirosolanes		
Soladunalinidine	Direct	(*68, 356*)
Solasodenone	Direct	(*78*)
Solanaviol	—	(*53*)
Epiminocholestanes		
Solaseaforthine	Direct	(*109*)
Isosolaseaforthine	Direct	(*109*)
Etioline	Hydrolysis	(*80, 146*)
25-Isosolafloridine	Direct or hydrolysis	(*58, 355*)
Hakurirodine	Hydrolysis	(*147*)
Veramiline	Direct	(*137*)
Isoteinemine	Hydrolysis	(*148*)
Hapepunine	Hydrolysis	(*128*)
Anrakorinine	Hydrolysis	(*367*)
Teinemine	Hydrolysis	(*148*)
Baikeine	Direct	(*149*)
Baikeidine	Direct	(*149*)
Solacallinidine	Hydrolysis	(*58, 355*)
Solacongestidine	Hydrolysis	(*62*)
23-Oxosolacongestidine	Hydrolysis	(*62*)
Solafloridine	Hydrolysis	(*62*)
Solaverbascine	Hydrolysis	(*118*)
Deacetoxysolaphyllidine	—	(*82*)
Deacetylsolaphyllidine	Direct	(*69*)
Solaphyllidine	Direct	(*69*)
Solaquidine	Direct	(*108*)
"24-Oxosolacongestidine"	Hydrolysis	(*62*)
Solamaladine	—	(*83*)
Veracintine	Direct	(*138, 139*)
20-(2-Methyl-1-pyrrolin-5-yl)-pregn-4-en-3-one	Direct	(*139*)
Solanidanes		
Solanogantine	Direct (?)	(*76*)
Solanogantamine	—	(*363*)
Isosolanogantamine	—	(*363*)
Procevine	—	(*150*)
Epirubijervine	Hydrolysis	(*366*)
Camtschatcanidine	Hydrolysis	(*368*)
Solanocapsine group		
Solanoforthine	Direct	(*110*)
Solacasine	Direct	(*105*)

(*Continued*)

TABLE III (*Continued*)

Alkamine	Isolation	Reference
3-Aminospirostanes		
Isojurubidine	Direct or hydrolysis	(*100*)
Isojuripidine	Direct or hydrolysis	(*99*, *100*)
Isopaniculidine	Direct	(*100*)
Alkamines with unknown structure		
Corsevinine ($C_{27}H_{41}NO_3$)[b]	Direct	(*129*)
Veralosidine ($C_{27}H_{43}NO_2$)[c]	Direct	(*141*)
Petiline ($C_{27}H_{43}NO_2$)[d]	Direct	(*132*)
Edpetilidinine ($C_{28}H_{47}NO_2$)[e]	Direct	(*130*)
Rhinolidine ($C_{28}H_{47}NO_2$)[f]	Hydrolysis	(*350*)
Veralodisine ($C_{29}H_{43}NO_4$)[g]	Direct	(*144*)
Veralosinine ($C_{29}H_{45}NO_3$)[h]	Direct	(*141*, *142*)
Veralosidinine ($C_{29}H_{45}NO_4$)[i]	Direct	(*143*)

[a] Directly isolated from plants or obtained after hydrolysis of the respective glycosides since 1967; supplement of the corresponding compilation (*3*).

[b] Isolated from *Korolkowia severtzovii*; insufficient data for structural determination.

[c] Isolated from *Veratrum album* subsp. *lobelianum*; containing two hydroxy groups, a methylpiperideine moiety, and a 5-double bond; possibly identical with etioline (identical ^1H NMR, melting point, and melting point of triacetate, but the optical rotation does not correspond to etioline (cf. *142*). By CD the (25*S*)-configuration was shown (*353*).

[d] Isolated from *Petilium raddeana*; containing one hydroxy and one ketone group as well as a methylpiperideine moiety; the ^1H-NMR data do not unequivocally prove the position of the functional groups (cf. *171*).

[e] Isolated from *Petilium eduardi*; insufficient data for structural determination.

[f] Isolated from *Rhinopetalum bucharicum*; insufficient data for structural determination.

[g] Isolated from *Veratrum album* subsp. *lobelianum*; containing one acetoxy, one hydroxy, and one ketone group; the authors (*144*) assumed a similarity with tomatillidine, for which, however, a new structural formula was proposed later (*172*); Huang–Minlon reduction of deacetylveralodisine led to veralosidine (*144*).

[h] Isolated from *Veratrum album* subsp. *lobelianum*; probably *O*(16)-acetyletioline (cf. *142*). By CD the (25*S*)-configuration was demonstrated (*353*).

[i] Isolated from *Veratrum album* subsp. *lobelianum*; containing one acetoxy and two hydroxy groups as well as a methylpiperideine moiety; the ^1H-NMR arguments concerning the positions of functional groups are not convincing (cf. *143*). By CD the (25*S*)-configuration was shown (*353*).

TABLE IV

^{1}H-CHEMICAL SHIFTS (δ) OF THE METHYL PROTONS OF *SOLANUM* ALKAMINES ($CDCl_3$, TMS)

Alkamine or its derivative	C-18	C-19	C-21	C-27	Reference
Tomatidine (**9**)	0.83	0.83	0.97	0.85	(*190*)
Soladunalinidine (**8**)	0.82	0.82	0.96	0.85	(*68*)
Soladulcidine (**161**)	0.83	0.79	0.93	0.84	(*190*)
25-Isosolafloridine (**22**)	0.69	0.80	1.10	0.91	(*58*)
Solacallinidine (**23**)	0.69	0.78	1.10	0.91	(*58*)
Veramiline (**29**)	0.68	1.00	0.86	1.03	(*137*)
Acetyldemissidine (**275**)	0.82	0.82	0.91	0.83	(*168, 191*)
Solanoforthine (**77**)	0.79	1.00	0.92[a]	0.87[a]	(*110*)
N,*N'*-Diacetylsolanocapsine (*N*,*N'*-diacetyl **4**)	0.79	0.79	—	—	(*192*)
N-Acetyljurubidine (*N*-acetyl **5**)	0.74	0.79	1.08	0.99	(*168*)

[a] May be reversed.

TABLE V

^{13}C-CHEMICAL SHIFTS (δ) OF *SOLANUM* ALKAMINES ($CDCl_3$, TMS)

C atom	Tomatidine (**9**) (*176*)	Soladulcidine (**161**) (*176*)	25-Isosolafloridine (**22**) (*358*)	Demissidine (**278**) (*176*)	Solanocapsine (**4**) (*358, 380*)	Jurubidine (**5**) (*358, 387*)
1	37.0	37.0	37.0	37.1	37.5	37.7
2	31.5	31.5	31.4	31.6	32.5	32.3
3	71.0	71.1	71.0	71.3	51.1	50.9
4	38.2	38.2	38.2	38.3	39.3	40.1
5	44.9	44.9	44.9	45.0	45.7	45.5
6	28.6	28.6	28.7	28.8	28.7	28.6
7	32.3	32.3	32.0	32.3	31.9	31.7
8	35.0	35.2	35.2[a]	35.4	35.0	35.2
9	54.4	54.4	54.3	54.6	55.0	54.5
10	35.5	35.6	35.5	35.6	35.7	35.6
11	21.1	21.1	21.1	21.1	20.5	21.0
12	40.2	40.1	40.4	40.2	39.3	37.6
13	40.9	41.0	44.2	40.6	41.8	40.6
14	55.8	56.3	53.3	57.4	55.0	56.4
15	32.6	32.1	35.0[a]	33.5	28.4	30.9
16	78.5	80.0	76.7	69.0	74.4	80.9
17	62.0	62.6	63.7	63.3	60.7	62.1
18	16.9	16.5	14.0	17.1	13.7	16.5
19	12.3	12.4	12.4	12.4	12.4	12.3
20	43.0	41.6	44.7	36.7	33.1	42.2
21	15.8	15.0	18.9	18.3	15.1	14.3
22	99.3	98.3	177.1	74.7	68.9	109.7
23	26.6	33.3	29.7	29.3	96.1	26.0
24	28.6	29.6	28.0[b]	31.1[a]	46.2	25.8
25	31.0	30.3	27.4[b]	31.3[a]	30.0	27.1
26	50.2	46.9	56.1	60.2	55.0	65.1
27	19.3	19.1	19.2	19.5	18.7	16.1

[a,b] May be reversed.

TABLE VI

DIAGNOSTIC MS FRAGMENTS OF *SOLANUM* ALKAMINES

Alkamine type	Fragments (*m/e*)	Reference
Spirosolanes	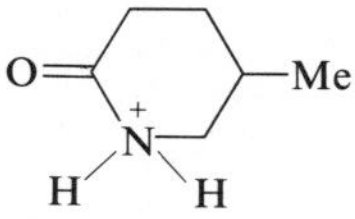 (114)	(*193*)
	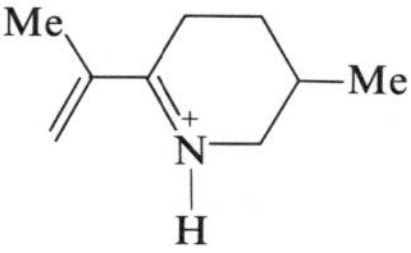 (138)	
22(*N*)-Unsaturated 22,26-epimino-cholestanes	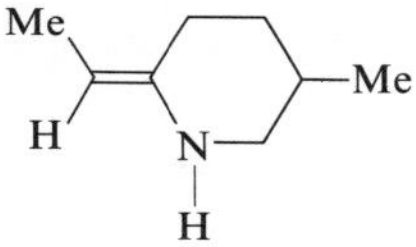 (125)	(*146*)
Saturated 22,26-epimino-cholestanes	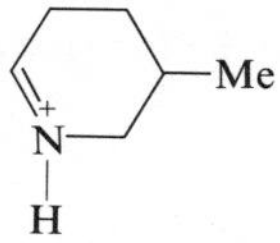 (98)	(*137*)
Solanidanes	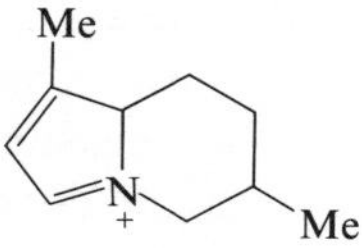 (150)	(*193*)
	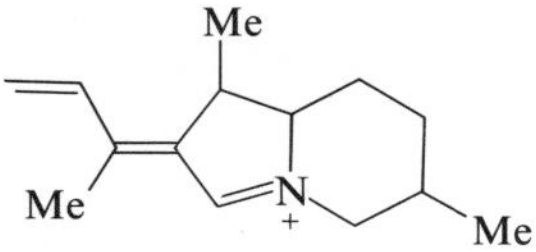 (204)	
Solanocapsine	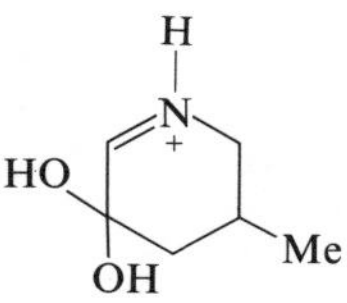 (130)	(*110*)
	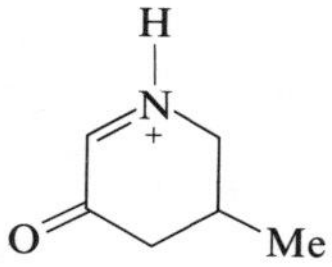 (112)	

(*Continued*)

TABLE VI (*Continued*)

Alkamine type	Fragments (m/e)	Reference
3-Aminospirostanes	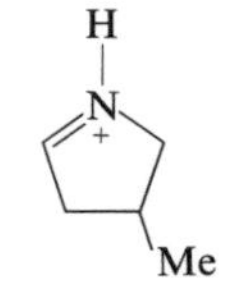 (84), 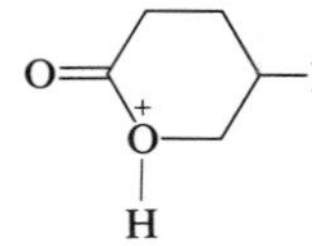 (115), 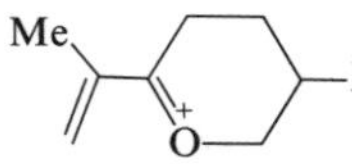 (139)	(*168*)

A. Structure Elucidation

1. The Spirosolanes

a. Soladunalinidine. Soladunalinidine has been obtained from *Solanum dunalianum*. The structure was proved to be **8** (3-deoxy-3β-aminotomatidine) by IR, ^{1}H NMR, and MS. The MS shows a molecular ion m/e 414 and fragmentation peaks at m/e 138 and 114, consistent with a spirosolane structure (see Table VI). The ^{13}C-NMR spectrum corresponds to that of tomatidine (**9**) except for the chemical shifts of the ring A carbon atoms. The

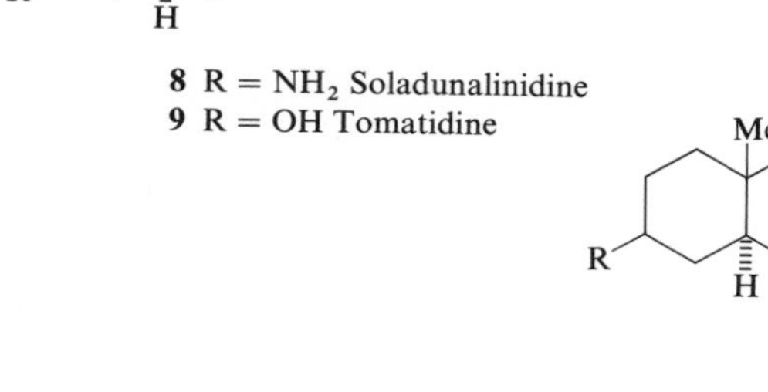

8 R = NH_2 Soladunalinidine
9 R = OH Tomatidine

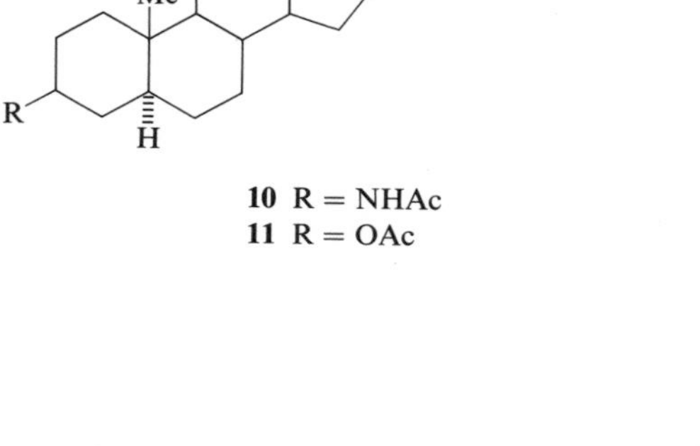

10 R = NHAc
11 R = OAc

latter are comparable with those of 3β-amino-5α-cholestane. The positive sign of the CD curve of the *N*-salicylidene derivative at 316 and 253 nm (dioxane) confirms the 3β-amino group. Verification of the structure was achieved by a chemical correlation between soladunalinidine (**8**) and tomatidine (**9**). After diacetylsoladunalinidine (**10**) had been nitrosated with sodium nitrite in acetic acid, diacetyltomatidine (**11**) was isolated from the reaction mixture (*68*, *356*).

b. Solasodenone. Solasodenone was isolated from *Solanum hainanense*. The IR spectrum shows the presence of a spiroaminoketal system (883, 913, 965, 978 cm^{-1}) as well as an α,β-unsaturated ketone. The UV data also indicate an enone chromophore. The MS fragmentation pattern (*m*/*e* 411, 138, 114) is in agreement with a spirosolane skeleton (see Table VI). Solasodenone was identified as **12** by comparison (*78*) with (25*R*)-22α*N*-spirosol-4-en-3-one (*196*) available by Oppenauer oxidation of solasodine (**1**).

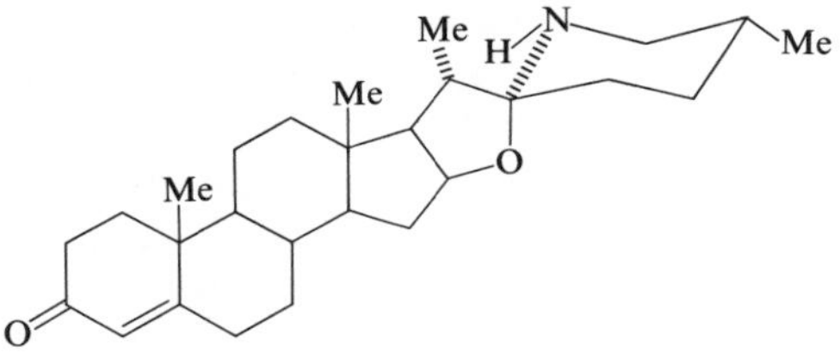

12 Solasodenone

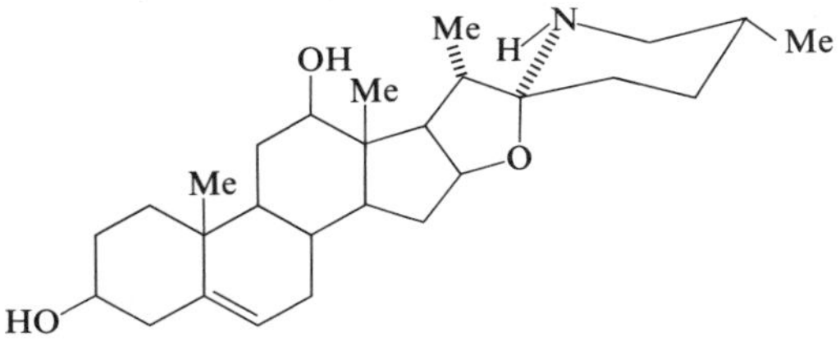

13 Solanaviol

c. Solanaviol. Solanaviol has been obtained from *Solanum aviculare*. The MS contains a molecular ion *m*/*e* 429 and peaks at *m*/*e* 138 and 114, consistent with a spirosolane structure (see Table VI). The signals for C-20, C-23, and C-26 in the ^{13}C-NMR spectrum are nearly the same as in solasodine (**1**) and differ from those of tomatidine (**9**). The resonance of C-18 is shifted upfield (Δδ − 6 ppm) because of γ-gauche interaction with a β-equatorial hydroxy group at C-12, and the signals of C-11 and C-13 are shifted downfield as a result of β effect with the hydroxy group. Structure **13** (12β-hydroxysolasodine) was proved by degradation to 3β,12β-diacetoxy-5α-pregn-16-en-20-one and by conversion to solasodine (**1**) (*53*).

2. The Epiminocholestanes

a. Solaseaforthine and Isosolaseaforthine. The spectral data of both alkaloids isolated from *Solanum seaforthianum* showed them to be stereoisomers (**14**, **15**). The MS of the *N*,*N*-dimethyl derivatives display M^+ at *m*/*e* 454 and fragments of *m*/*e* 84 and 110 characteristic of 3-dimethylamino steroids. The 3β-amino-5α-stereochemistry is evident from the CD spectra of the salicylidene derivatives of **14** and **15**. The singlets for one aromatic methyl (δ 2.2 ppm) and two aromatic protons (δ 6.84 and 7.76 ppm) in the ^{1}H-NMR spectra of the *N*,*N*-dimethyl derivatives coupled with the diagnostic peak at *m*/*e* 137 (**16**) in the MS of the alkaloids indicated the presence of a 3-hydroxy-5-methylpyridine moiety. The UV spectra of **14** and **15** in neutral and alkaline media are in good agreement with that of 3-hydroxypyridine. The presence of a phenolic OH group was further corroborated by reaction of the *N*,*N*-dimethyl derivatives with diazomethane to afford the methyl ethers **17** and **18**. Jones oxidation of the latter furnished ketones both of which exhibit IR absorption at 1730 cm^{-1} for a five-membered ring ketone. Strong peaks at *m*/*e* 276 and 191 in their MS could only be rationalized with the keto group localized at C-16 (bond fission C-13/C-17 and C-15/C-16). Sodium borohydride reduction of the ketones afforded alcohols different from the original ones (**17** and **18**). As this reagent is expected to attack the carbonyl group from the α side, the α-configuration could be assigned to the

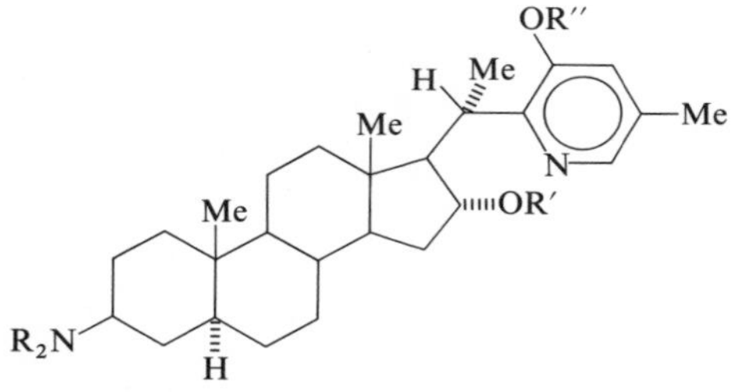

14 R, R′, R″ = H Solaseaforthine
17 R, R″ = Me, R′ = H
19 R, R″ = Me, R′ = Ac

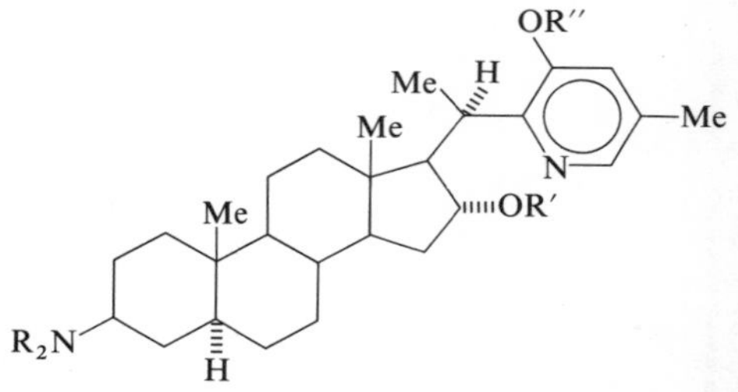

15 R, R′, R″ = H Isosolaseaforthine
18 R, R″ = Me, R′ = H
20 R, R″ = Me, R′ = Ac

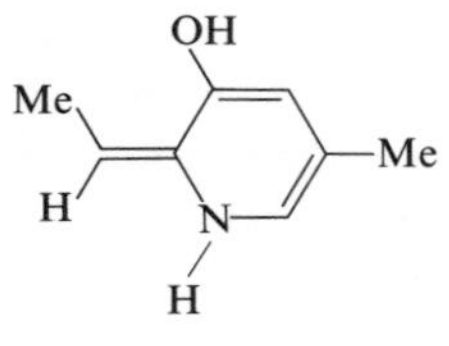

16

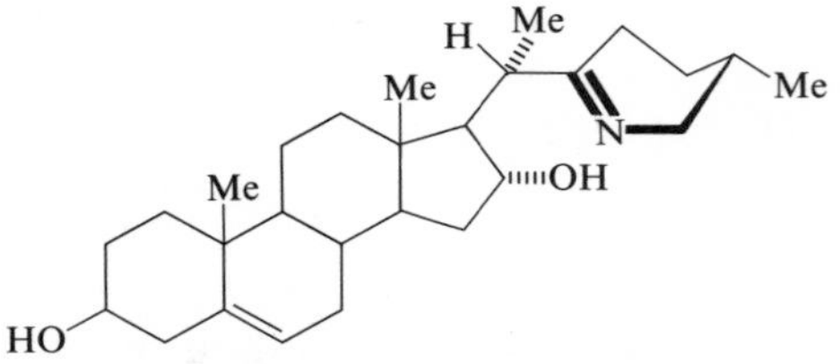

21 Etioline

alcoholic function of solaseaforthine and isosolaseaforthine. That **14** and **15** are epimeric at C-20 was demonstrated by the mirror-image CD spectra of **19** and **20**. The signs of the Cotton effects of **19** agree well with those reported for (20*S*)-20-pyridylpregnane; **20** must, therefore, have the (20*R*)-configuration (*109*).

b. Etioline. Leaves of budding *Veratrum grandiflorum* gave a glycoside mixture from which etioline was obtained after acid hydrolysis (*146*). Recently, etioline was isolated from *Solanum havanense* (*80*). The molecular formula $C_{27}H_{43}NO_2$ was determined by elemental analysis and MS. Presence of a double bond was revealed by the ^{1}H-NMR spectrum (δ 5.36 ppm). Etioline was converted to an α,β-unsaturated ketone by Oppenauer oxidation. An IR absorption band of etioline at 1660 cm^{-1} is characteristic of a C═N group; a UV absorption maximum is observed at 238 nm. The base peak *m*/*e* 125 in the MS is consistent with structure **21** (16α-hydroxyverazine) (see Table VI). Etioline formed an *N*,*O*,*O*′-triacetate. Chromic acid oxidation of etioline yielded a diketone showing absorption at 1745 cm^{-1} (five-membered ring ketone) and 1715 cm^{-1} (six-membered ring ketone). The α-orientation was assigned to the hydroxy group at C-16 because etioline failed to cyclize to a spirosolane when refluxed in alcoholic KOH (*146*). The ^{1}H-NMR signals of all methyl groups correspond to those of the 5α,6-dihydro compound, 25-isosolafloridine, if the 5-double bond is taken into account (*80*).

c. 25-Isosolafloridine and Solacallinidine. Two new steroidal alkaloids, 25-isosolafloridine and solacallinidine, have been obtained by hydrolysis of the crude glycoalkaloid extracted from *Solanum callium*. The spectroscopic properties of 25-isosolafloridine (UV, IR, MS, ^{1}H NMR and ^{13}C NMR) indicate that it is a 22,26-epimino-5α-cholest-22(*N*)-ene. Structure **22** (5α,6-dihydroetioline) was determined by X-ray diffraction measurements on the hydrochloride. Acetylation gave an *N*,*O*,*O*′-triacetyl derivative (*58*, *355*). 25-Isosolafloridine had already been synthesized some years before its isolation from plant material (see Section IV,B,2). The MS of solacallinidine, $C_{27}H_{46}N_2O$, shows the parent ion at *m*/*e* 414 and a prominent peak at *m*/*e* 125 (see Table VI). The chemical shifts in the ^{13}C-NMR spectrum correspond closely with those of 25-isosolafloridine (**22**) except for those in ring A. The chemical shift differences between 3β-hydroxy- and 3β-amino-5α-cholestane parallel those between 25-isosolafloridine and solacallinidine. This finding indicated that solacallinidine (**23**) has the same structure as 25-isosolafloridine except that a 3β-amino group replaces the 3β-hydroxy group. The CD of **23** (negative Cotton effect at 242 nm in dioxane) is consistent with the (25*S*)-configuration (*58*, *355*).

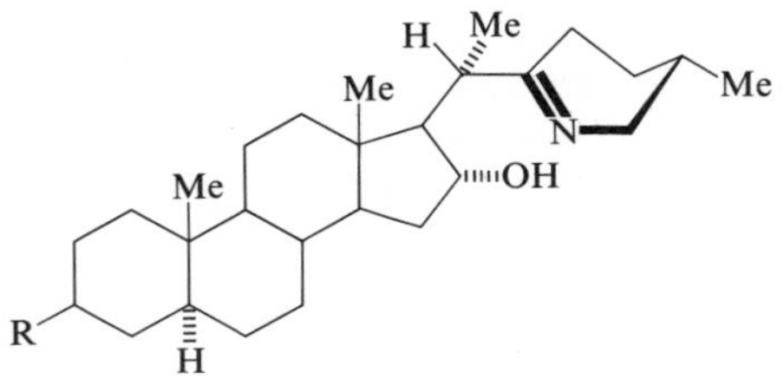

22 R = OH 25-Isosolafloridine
23 R = NH_2 Solacallinidine

d. Hakurirodine. The rhizomes of dormant *Veratrum grandiflorum* accumulate glycosides, from which hakurirodine was isolated after acid hydrolysis. The spectroscopic properties of the alkaloid and its derivatives (IR, UV, ^1H NMR, MS) are consistent with the 22,26-epiminocholestane structure **24** (12α-hydroxyetioline). Hydrogenation of hakurirodine gave a mixture of the 22-epimers **25** and **26**. After chromic acid oxidation of **25** followed by hydrogenation, 5α,22α*H*,25β*H*-solanidane-3,12-dione (**27**) was isolated, also available from rubijervine (**28**) by hydrogenation and sub-

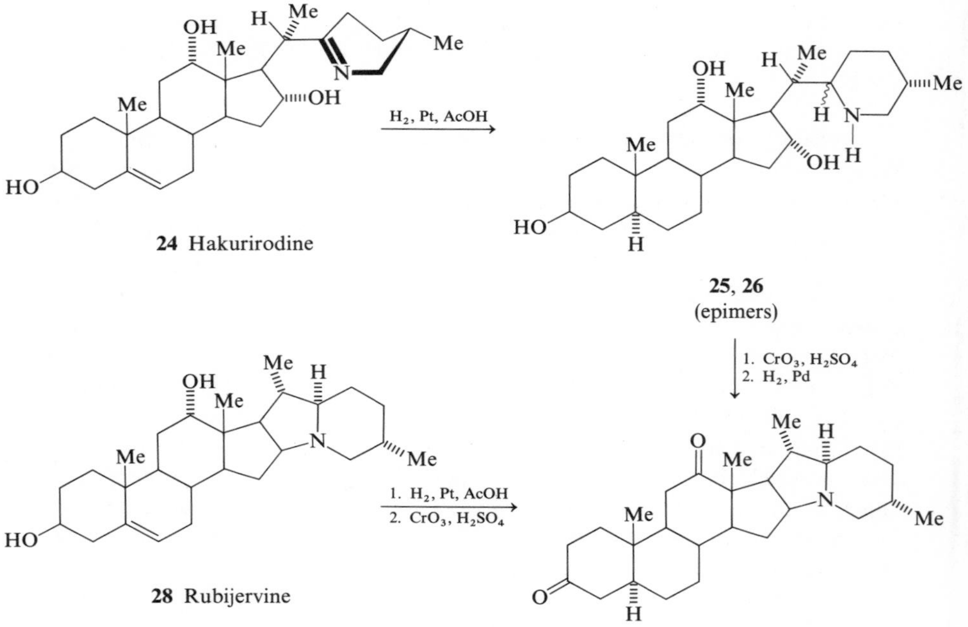

24 Hakurirodine

H_2, Pt, AcOH

25, 26
(epimers)

1. CrO_3, H_2SO_4
2. H_2, Pd

28 Rubijervine

1. H_2, Pt, AcOH
2. CrO_3, H_2SO_4

27

sequent oxidation. On the basis of ^{1}H-NMR arguments, an α-orientation was assigned to the hydroxy groups of hakurirodine (**24**) at C-12 and C-16 (*147*).

e. Veramiline. Veramiline was isolated from the aerial parts of *Veratrum album* subsp. *lobelianum*. Structure **29** [(22*S*)-22,*N*-dihydroverazine] was assigned to this alkaloid by spectroscopy (IR, ^{1}H NMR, MS). The MS reveals a base peak at *m*/*e* 98 indicating the presence of a methylated piperidine ring (see Table VI). The dihydro derivative was identified with synthetic (22*S*, 25*S*)-22,26-epimino-5α-cholestan-3β-ol (*137*). Compound **29** had already been synthesized from tomatidenol in 1967 (*199*, *200*).

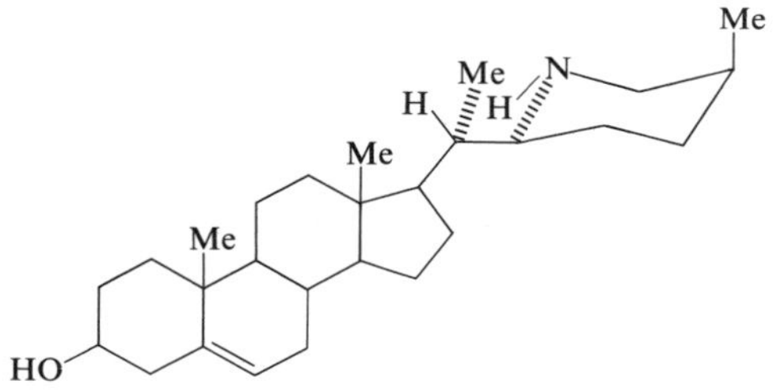

29 Veramiline

f. Teinemine and Isoteinemine. From *Veratrum grandiflorum* two new alkaloids, named teinemine and isoteinemine after the Ainu name "teine" for *Veratrum*, were obtained after HCl hydrolysis. Spectroscopy (IR, ^{1}H NMR, MS) indicated teinemine to be a 22,26-epiminocholestane alkaloid. It was obtained as the main product of hydrogenation of etioline (**21**). Dihydroteinemine (**30**) was converted to 5α,22α*H*,25β*H*-solanidan-3-one (**31**) by oxidation with the Kiliani reagent and subsequent reduction. From these results, teinemine is (22*R*,25*S*)-22,26-epiminocholest-5-ene-3β,16α-diol [(22*R*)-22,*N*-dihydroetioline; see structure **32**]. Isoteinemine is identical with the second hydrogenation product of etioline (**21**). Therefore, the structure was postulated as (22*S*,25*S*)-22,26-epiminocholest-5-ene-3β,16α-diol [16α-hydroxyveramiline, structure **33**] (*148*).

g. Hapepunine and Anrakorinine. A new alkaloid named hapepunine after the Ainu name for the original plant, "hapepui," was isolated from *Fritillaria camtschatcensis*. From spectroscopic investigations an *N*-methyl-22,26-epiminocholest-5-ene-3β,16β-diol structure was expected. The ^{1}H-NMR spectrum displays a methyl signal at δ 2.30 ppm for an *N*-methyl

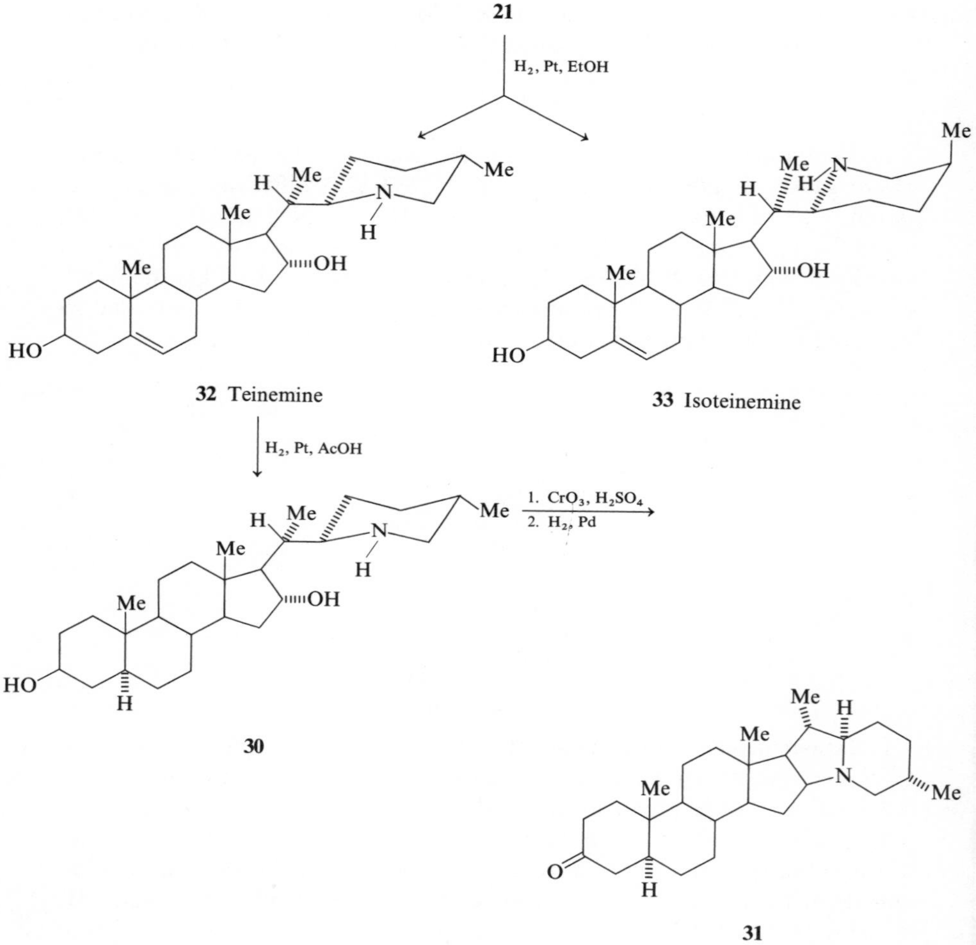

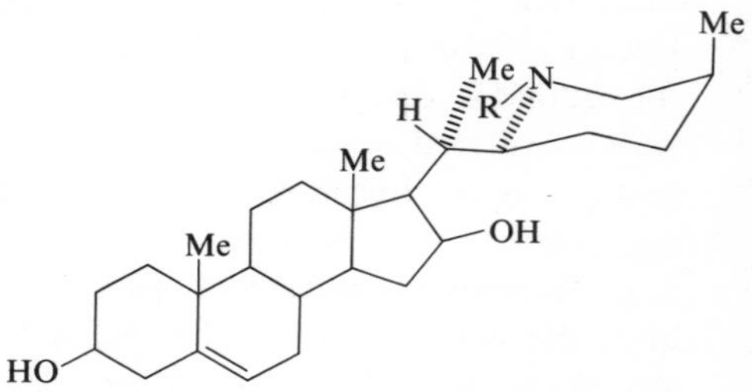

21

H_2, Pt, EtOH

H Me Me N H Me OH Me HO

32 Teinemine

Me Me N H H Me OH Me HO

33 Isoteinemine

H_2, Pt, AcOH

H Me Me N H Me OH Me HO H

30

1. CrO_3, H_2SO_4
2. H_2, Pd

Me H Me N Me Me O H

31

Me Me N R H Me OH Me HO

34 R = Me Hapepunine
35 R = H

group. The MS of hapepunine reveals the base peak at *m/e* 112 (dimethylpiperidyl residue, see Table VI). Hapepunine (**34**) was obtained by methylation of (22*S*,25*S*)-22,26-epiminocholest-5-ene-3β,16β-diol [(22*S*)-22,*N*-dihydrotomatidenol, 16β-hydroxyveramiline, structure **35**] (*3*) with methyl iodide. The (25*R*)-isomer of hapepunine was also synthesized (*128*, *367*). Recently, 18-hydroxyhapepunine (anrakorinine) has been isolated from *F. camtschatcensis* (*367*).

h. Baikeine and Baikeidine. Baikeine and baikeidine, named after the Japanese term for the plant, "baikeiso," have been isolated from *Veratrum grandiflorum*. The ^{1}H-NMR spectrum of baikeine exhibits four methyl signals. This, when coupled with its molecular formula, suggested that the alkaloid had a cholestane carbon skeleton. X-Ray crystallographic analysis of *N*-ethylbaikeine hydrobromide established structure **36** of baikeine (12α-hydroxyteinemine). Most probably, baikeidine is *O*(3)-acetylbaikeine. It could not be isolated in pure form. On acetylation in methanol, **36** afforded *N*-acetylbaikeine and in pyridine tetraacetylbaikeine. *N*-Ethylbaikeine, available by lithium aluminum hydride (LAH) reduction of *N*-acetylbaikeine, yielded a triacetyl derivative and the 4-unsaturated 3,12-dione (*149*).

i. Solacongestidine, 23-Oxosolacongestidine, "24-Oxosolacongestidine," and Solafloridine. These alkaloids were isolated from *Solanum congestiflorum*. Although 23-oxo- and "24-oxosolacongestidine" probably exist in the plant per se, there appears to be some augmentation during the work-up of solacongestidine (*62*). Solacongestidine (**2**) possesses the formula $C_{27}H_{45}NO$ with two tertiary and two secondary methyl groups according to the ^{1}H-NMR spectrum, a pattern typical of the *Solanum*-type steroidal alkaloids. IR bands show the presence of a hydroxy group (3597 and 3333 cm^{-1}) and a C=N moiety (1653 cm^{-1}). The MS displays a prominent peak at *m/e* 125 (see Table VI). Dehydrogenation of *O*-acetylsolacongestidine (**37**) by palladium produced the β-picolyl derivative **38**. It was found (*62*) that solacongestidine was identical with a synthetic specimen of (25*R*)-22,26-epimino-5α-cholest-22(*N*)-en-3β-ol (*201*) prepared in an unambiguous manner. Wolff–Kishner reduction of 23-oxosolacongestidine (**39**) afforded **2**, which can be oxidized with manganese dioxide or selenium dioxide to a mixture of **39** and "24-oxosolacongestidine." An interesting feature of the 23-oxo compound **39** is that it readily undergoes aromatization into **38** by brief reflux in acetic anhydride. The spectral data show 23-oxosolacongestidine to be **39** (*62*). When dihydrotomatillidine (**53**; see Section IV,A,2,m) was treated with HCl, "24-oxosolacongestidine" (**40**) was isolated. Hence these compounds appear to be stereoisomeric at C-20 (*172*). The ^{1}H-NMR spectrum of solafloridine indicates the presence of an extra hydroxy group,

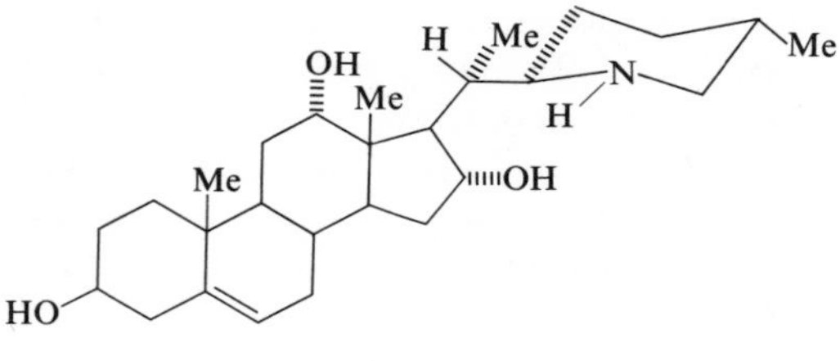

36 Baikeine

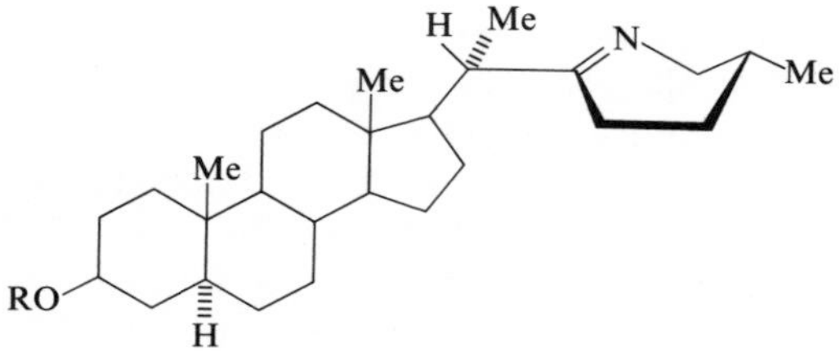

2 R = H Solacongestidine
37 R = Ac

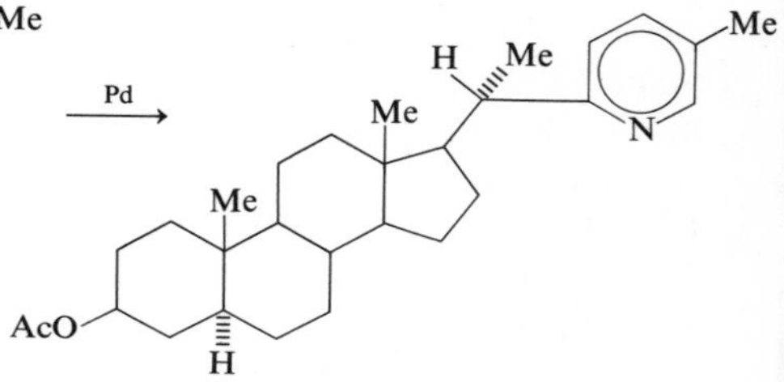

Pd →

38

SeO_2 or MnO_2

Ac_2O

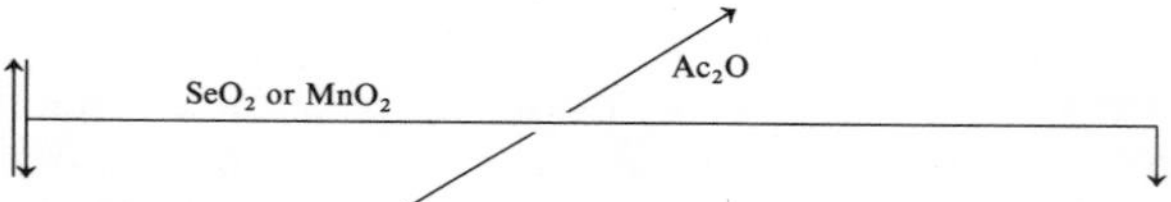

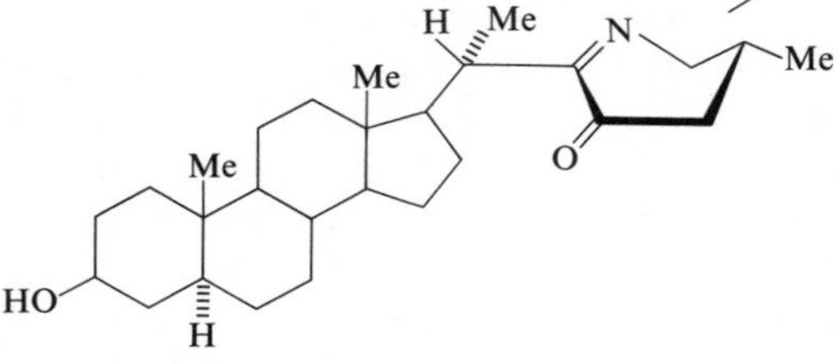

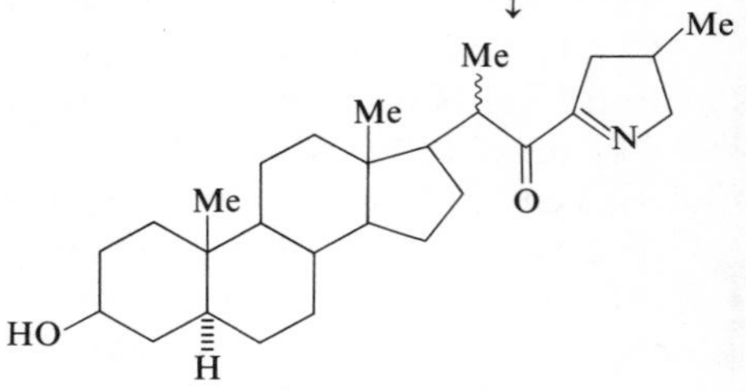

39 23-Oxosolacongestidine

40 "24-Oxosolacongestidine"

when compared with solacongestidine (**2**). Solafloridine (**41**, 16α-hydroxysolacongestidine) and its 22,*N*-dihydro derivative **42** were found to be identical (*62*) with synthetic specimens (*202*). Oxidation of **41** with chromic acid afforded a diketone. Dihydrosolafloridine (**42**) was converted into 5α,22β*H*,25α*H*-solanidan-3-one (**43**) (*62*).

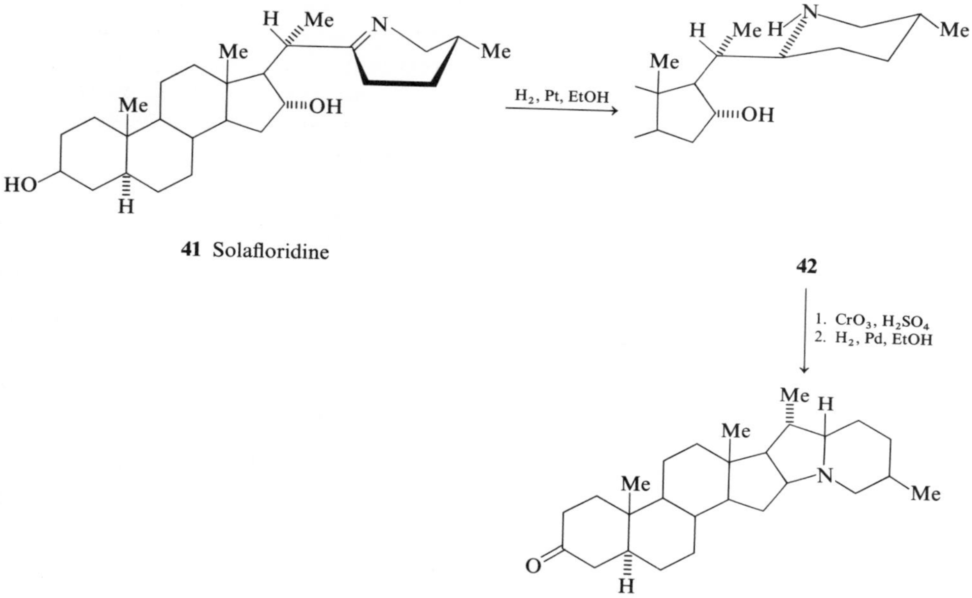

41 Solafloridine

H_2, Pt, EtOH

42

1. CrO_3, H_2SO_4
2. H_2, Pd, EtOH

43

j. Solaverbascine. This alkaloid has been obtained from *Solanum verbascifolium* and identified as (22*S*,25*R*)-22,26-epiminocholest-5-ene-3β, 16β-diol [(22*S*)-22,*N*-dihydrosolasodine, structure **44**] (*118*). It is obtainable by sodium borohydride reduction of solasodine (**1**) (*203*).

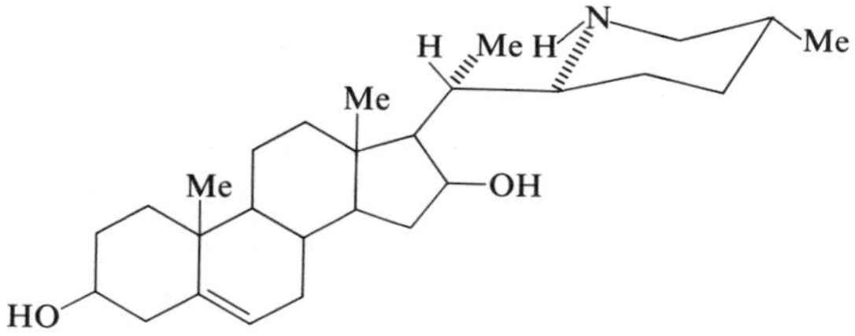

44 Solaverbascine

k. Solaphyllidine, Deacetylsolaphyllidine, and Deacetoxysolaphyllidine. Structure **45** has been assigned to solaphyllidine, the most abundant alkaloid found in *Solanum hypomalacophyllum*, on the basis of an X-ray diffraction analysis of a single crystal which contained no heavy atoms. The absolute configuration was assumed to be the same as in cholesterol. Chemical and spectroscopic data of unexceptional character supported the structure elucidation (*81*). Later, **45** and its deacetyl derivative **46** were isolated from *S. ecuadorense* (*69, 70*). Recently, deacetoxysolaphyllidine (**47**) has been obtained from *S. hypomalacophyllum* (*82*).

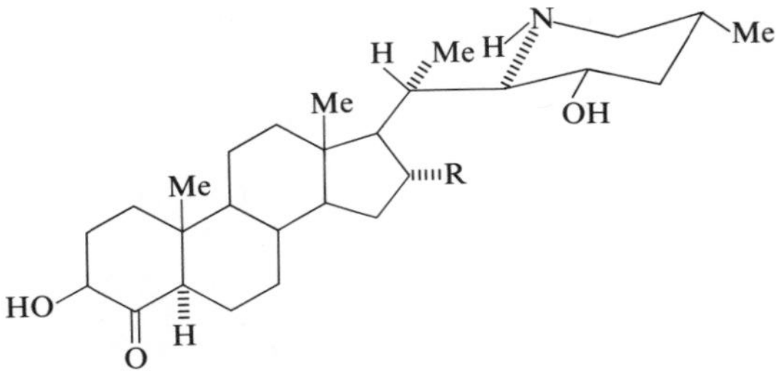

45 R = OAc Solaphyllidine
46 R = OH Deacetylsolaphyllidine
47 R = H Deacetoxysolaphyllidine

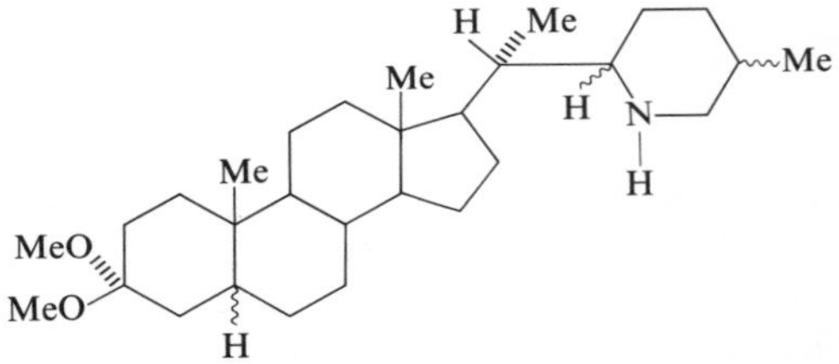

48 Solaquidine

l. Solaquidine. From green berries of *Solanum pseudoquina*, solaquidine was isolated. The ^{1}H-NMR spectrum indicates two methoxy groups (δ 3.10 and 3.15 ppm). The $M^+ - 1$ in the MS is found at *m/e* 444; the base peak at *m/e* 98 shows the presence of a methylpiperidine side chain (see Table VI). Two abundant fragments, *m/e* 101 and 127, indicate that both methoxy groups are located at C-3. A 3,3-dimethoxy-22,26-epiminocholestane structure (**48**) was assumed for solaquidine. The absolute configurations at C-5, C-22, and C-25 remain undetermined (*108*).

m. Tomatillidine. It was found that during thin-layer chromatography (TLC) (silica gel) 23-oxosolacongestidine (**39**) was converted to dihydrotomatillidine (available from tomatillidine by catalytic hydrogenation).

Sodium borohydride reduction of dihydrotomatillidine was carried out, followed by acetylation to produce three triacetates. By ^{1}H-NMR spectroscopy of the latter, structures **49–51** were identified for the triacetates and hence structures **52** and **53** for tomatillidine and dihydrotomatillidine, respectively. The signals assigned to 22-H in compounds **49–51** appear at δ 5.24 ppm (quartet, $J = 8$ and 1 Hz), 5.36 ppm (doublet, $J = 4$ Hz), and 5.46 ppm (quartet, $J = 4$ and 2 Hz). More complicated signals should have been expected for 24-H in the alternative structures **54** derived from the old structural proposal. Decoupling experiments with compound **51** provided strong support for the new structure. Irradiation at the center of the 22-H quartet sharpened the 23-H signals. Conversely, irradiation at the 23-H signal changed the 22-H quartet to a doublet. Tomatillidine (**52**) was synthesized starting from *O*-acetylsolasodine (**55**). Compound **55** was reduced to the epiminocholestane **56**. Its benzyloxycarbonyl derivative (**57**) was oxidized to 16-oxo compound **58**, the thioketal (**59**) of which was treated with Raney nickel to give **60**. The chlorination of **60** and dehydrochlorination of the *N*-chloro compound afforded the 22(*N*)-unsaturated 22,26-epimino compound **61**, which was oxidized with manganese dioxide in chloroform to the 23-ketone **62**. The latter on TLC (SiO_2) was converted to an isomeric compound, which was identified with tomatillidine. During the transformation from **55** to **52** no change of the configuration is expected, except at C-20, suggesting a (25*R*)-configuration for **52** (*172*).

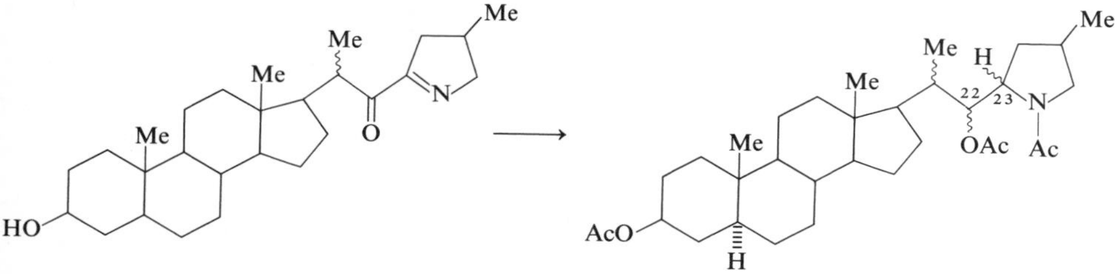

52 Δ^5 Tomatillidine
53 5αH Dihydrotomatillidine

49, 50, 51

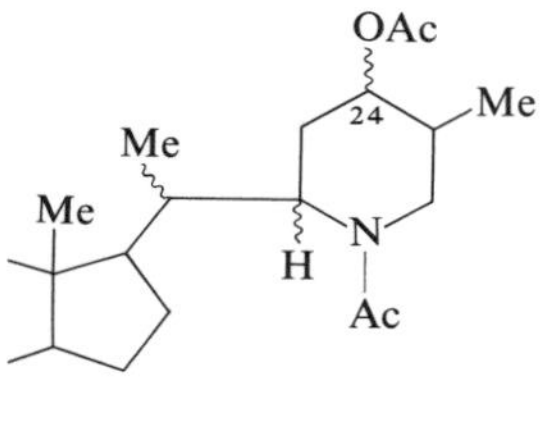

54

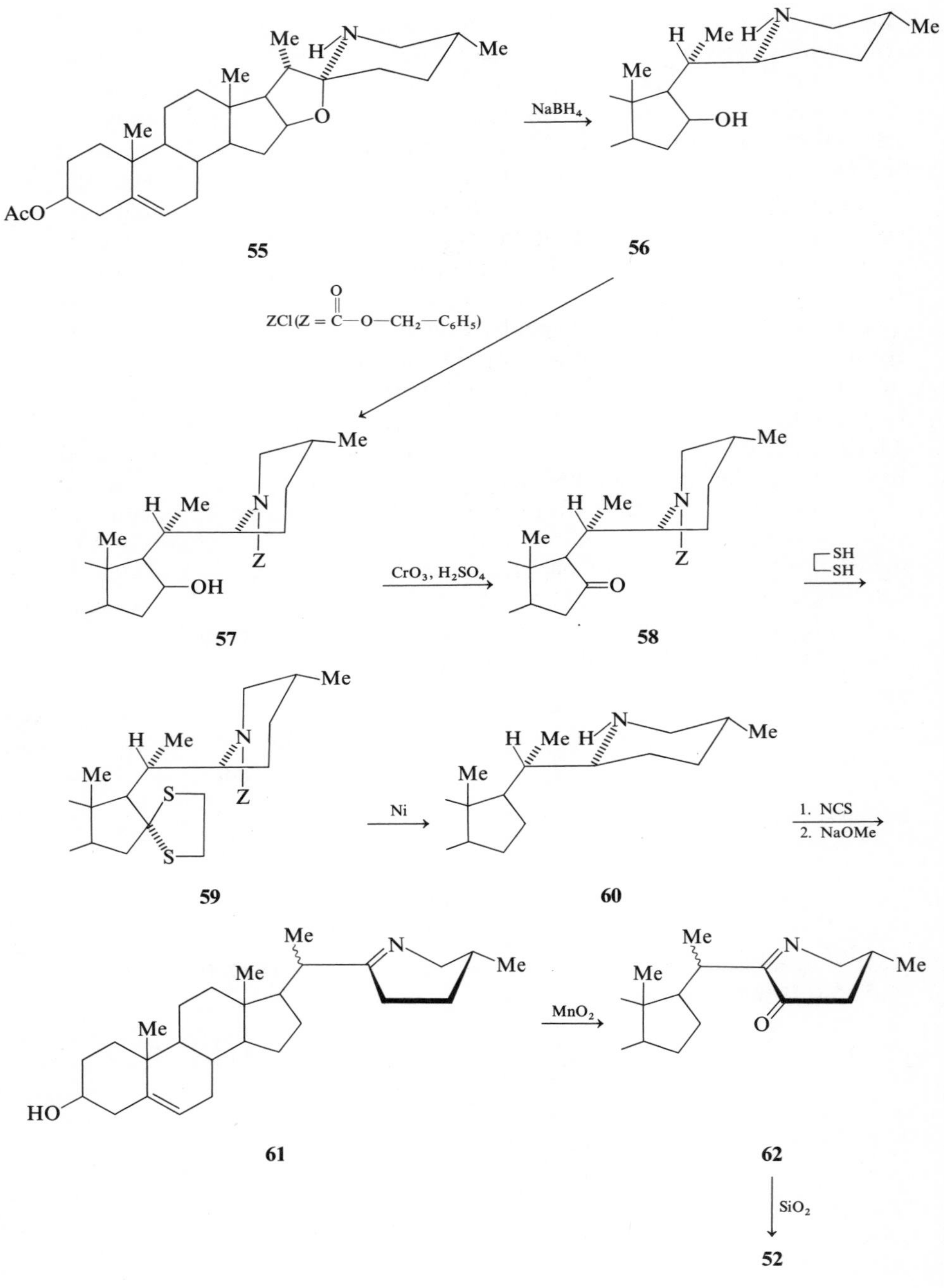

Me, H, N, Me, Me, O, Me, AcO

55

$\xrightarrow{NaBH_4}$

H, Me, H, N, Me, Me, OH

56

$ZCl(Z = \overset{O}{\overset{||}{C}}—O—CH_2—C_6H_5)$

Me, H, Me, N, Me, Z, OH

57

$\xrightarrow{CrO_3,\ H_2SO_4}$

Me, H, Me, N, Me, Z, O

58

$\xrightarrow{HS-CH_2-CH_2-SH}$

Me, H, Me, N, Me, Z, S, S

59

$\xrightarrow{Ni}$

H, Me, H, N, Me, Me

60

$\xrightarrow[2.\ NaOMe]{1.\ NCS}$

Me, N, Me, Me, Me, HO

61

$\xrightarrow{MnO_2}$

Me, N, Me, Me, O

62

$\downarrow SiO_2$

52

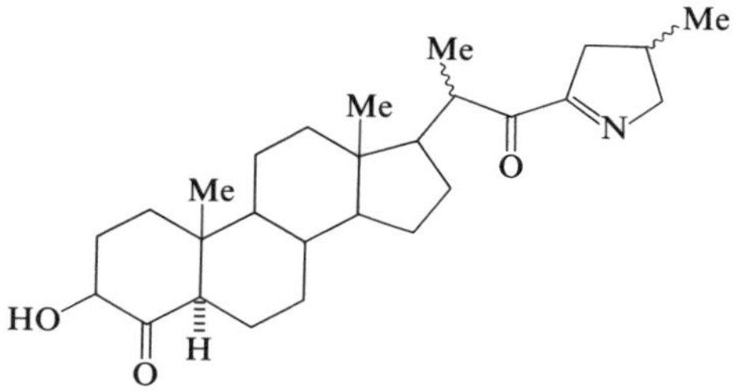

63 Solamaladine

n. Solamaladine. Solamaladine (**63**) was isolated from green fruit of *Solanum hypomalacophyllum*. By high-resolution MS the molecular formula was shown to be $C_{27}H_{41}NO_3$. The base peak at *m/e* 140 ($C_8H_{14}NO$) as well as further fragments at *m/e* 111 and 110 indicated a tomatillidine-type structure. The ^{1}H-NMR spectrum displays signals at similar positions as in the tomatillidine spectrum (e.g., at δ 0.69 ppm for methyl C-18, 0.98 and 1.06 ppm for the secondary methyl groups) except for the signal of the methyl group C-19 (δ 0.69 ppm), which corresponds to the value of solaphyllidine (**45**). Solamaladine shows IR absorption at 1710 (4-keto group) and 1683 cm^{-1} (tomatillidine 1672 cm^{-1}). By reaction with acetic anhydride in pyridine, solamaladine (**63**) and tomatillidine (**52**) formed an *O*-acetyl derivative at room temperature. By reaction with boiling acetic anhydride both compounds gave *N,O*-diacetates with 23 double bonds. The UV absorption of both *N,O*-diacetyl derivatives is similar (bands at 222 and 273 nm). All these observations are consistent with a tomatillidine-like structure (*83*). In 1976 a revised structural proposal for tomatillidine was published by Kusano *et al.* (*172*). Therefore, the given formula **63** is different from that in Usubillaga's original publication (*83*).

o. Veracintine and 20-(2-Methyl-1-pyrrolin-5-yl)pregn-4-en-3-one. Both alkaloids have been obtained from *Veratrum album* subsp. *lobelianum*. In naming veracintine, Professor Tomko considered his daughter Hyacinte. High-resolution MS of veracintine (**64**) shows a parent peak in accordance with the molecular formula $C_{26}H_{41}NO$. The base peak at *m/e* 82 is formed as a result of the C-20–C-22 bond fission. The IR band of medium intensity at 1655 cm^{-1} is characteristic of a C=N double bond. The ^{1}H-NMR spectrum displays a singlet (3 protons) at δ 2.1 ppm corresponding to a methyl group in the neighborhood of a double bond. These and further spectroscopic results show veracintine to have structure **64**. Compound **64** forms a 25,*N*-dihydro and a tetrahydro derivative after hydrogenation over platinum in ethanol or acetic acid (*138*). By analogous spectroscopic investigations, the structure 20-(2-methyl-1-pyrrolin-5-yl)pregn-4-en-3-one

(**65**) was proved for another alkaloid from *V. album* subsp. *lobelianum*. The perhydro derivatives of veracintine (**64**) and alkaloid **65** are identical. Oppenauer oxidation of veracintine gave **65** (*139*).

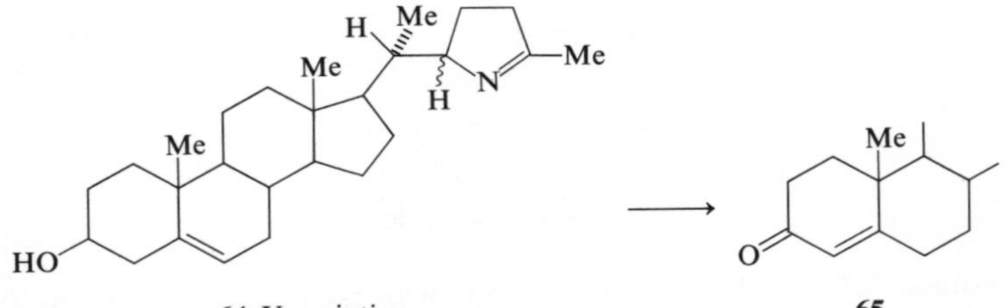

64 Veracintine

65

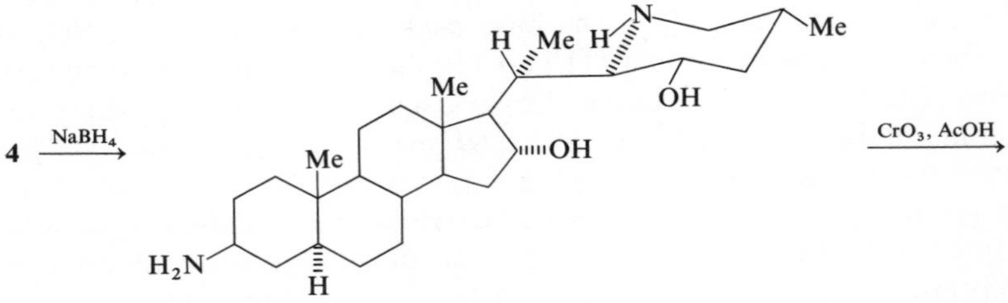

4 $\xrightarrow{NaBH_4}$ ... $\xrightarrow{CrO_3,\ AcOH}$

67

$\xrightarrow{NaBH_4}$

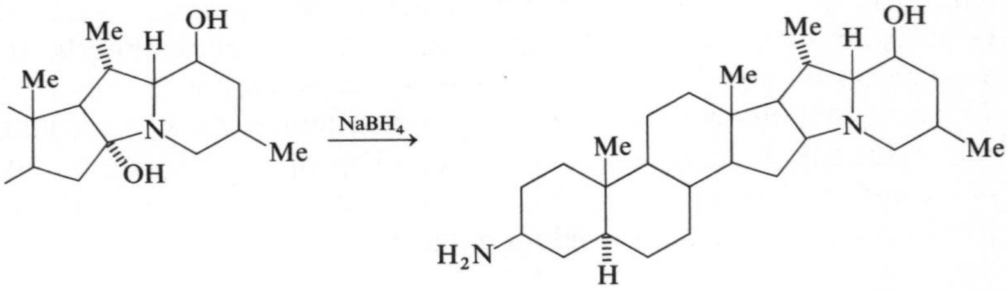

68

66 Solanogantine

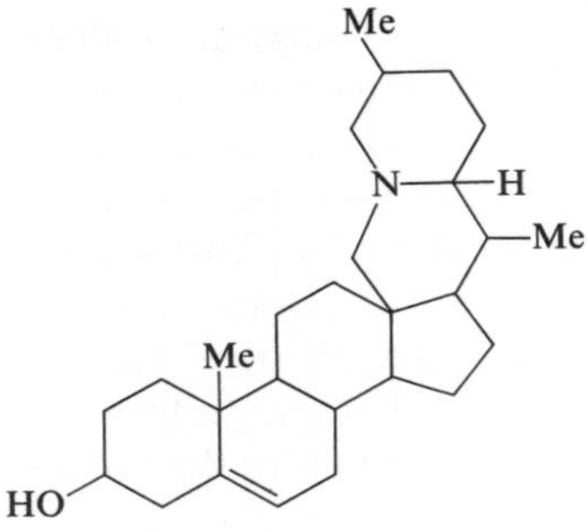

69 Procevine

3. The Solanidanes

a. Solanogantine, Solanogantamine, and Isosolanogantamine. Solanogantine was isolated from the leaves of *Solanum giganteum*. The alkaloid has the molecular formula $C_{27}H_{46}N_2O$ (M^+, *m/e* 414). The MS shows fragments for 3-aminosteroids (*m/e* 56, 82) and for a solanidane skeleton with a hydroxy group at ring E or F (*m/e* 166, *m/e* 150 + oxygen; see Table VI). Structure **66** was proved by its synthesis from solanocapsine (**4**). Reduction of **4** afforded the diol **67**, the 23-hydroxy group of which was assigned an equatorial orientation from mechanistic considerations. Oxidation led to the carbinolamine **68**, which was reduced to solanogantine (**66**). Reduction of **68** is expected to occur from the less hindered α side (*76*). Solanogantine (**66**) is the first naturally occurring solanidane with both 3β-amino function and 22β*H*,25α*H*-configuration. All four of the 3-amino-22α*H*,25β*H*-solanidanes which are stereoisomeric at C-3 and C-5 have been synthesized before (*204*, *205*). Because of the facile mercuric acetate dehydrogenation of **66** and its *N*,*O*-diacetate as well as the high-resolution ^{1}H-NMR spectra of the *N*,*N*-dimethyl and the *N*,*O*-diacetyl derivatives of **66**, the configuration of the indolizidine moiety (E/F ring junction) was shown to be trans, putting the nitrogen nonbonding electrons into an α position (*76*). This configuration was considered to hold for all the 22β*H*-solanidanes (*22*), in contrast to former suggestions (*206*). Recently, solanogantamine and isosolanogantamine (3β- and 3α-amino-5α,22α*H*-solanidan-23β-ol with unknown stereochemistry at C-25) have been isolated from *S. giganteum* (*363*).

b. Procevine. This alkaloid was isolated from the aerial part of *Veratrum grandiflorum* that had been cultivated under a sunlamp for 2 days. The MS shows the parent peak at *m/e* 397 and the base peak at *m/e* 112. Procevine was shown to be identical with (22*R*,25*S*)-*N*(16→18)abeosolanid-5-en-3β-ol (**69**) synthesized earlier from isorubijervine (*150*).

c. Epirubijervine. Recently, the 12-isomer of rubijervine has been isolated from illuminated *Veratrum grandiflorum* (*366*).

d. Camtschatcanidine. This alkaloid has been isolated from *Fritillaria camtschatcensis* and has the structure of 27-hydroxysolanidine (*368*).

4. The Solanocapsine Group

a. Solanocapsine. Synthetic studies (see Section IV,B,4) indicated that the proposed 16β,23-oxido structure **70** (*3*) for solanocapsine was erroneous. *N*,*N*′-diacetylsolanocapsine had been dehydrated and oxidized to the secosteroid **71**, which had given acetyltigogeninlactone (**72**) by acid hydrolysis followed by deamination and acetylation. Surprisingly, during the acid hydrolysis of the secosteroid **71**, inversion of the configuration at C-16 had occurred, as the following results confirm. Lithium aluminum hydride reduction of **71** and subsequent acetylation gave **73**, the hydrolysis of which yielded **74**. The negative molecular rotation difference between **73** and **74** is only consistent with a 16βH configuration for both compounds. In addi-

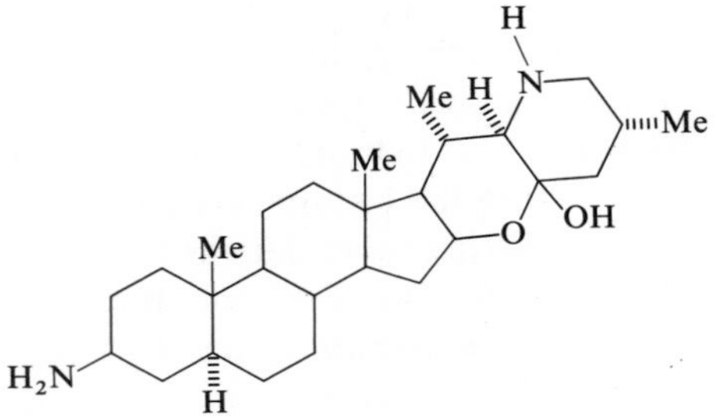

70

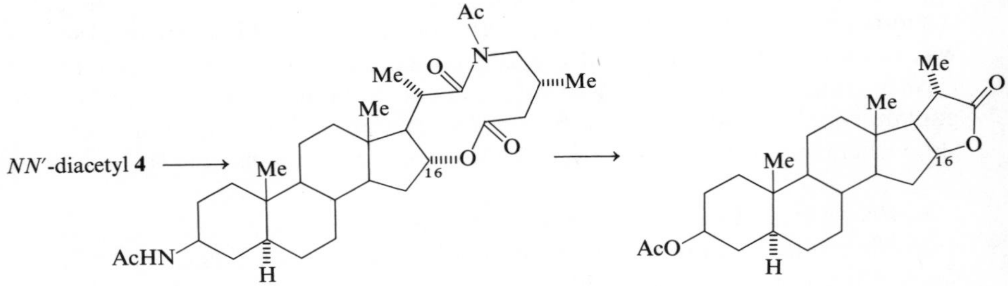

NN'-diacetyl **4** ⟶ **71** ⟶ **72**

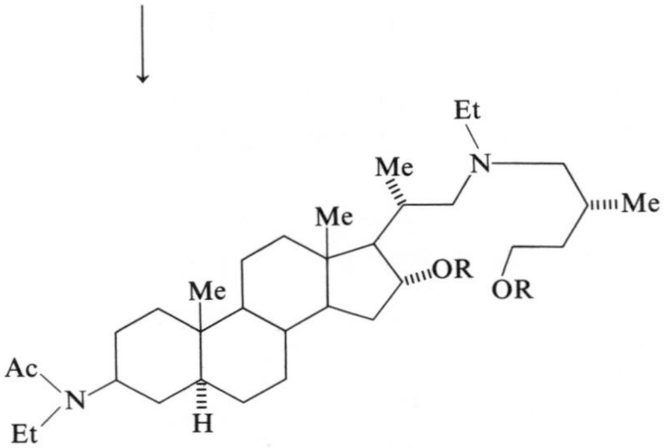

73 R = Ac
74 R = H

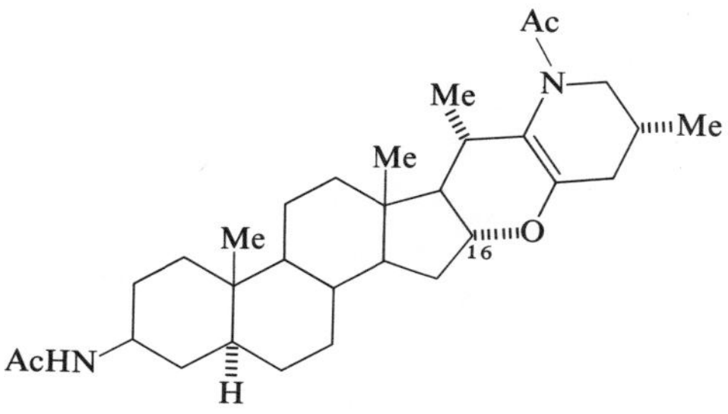

75 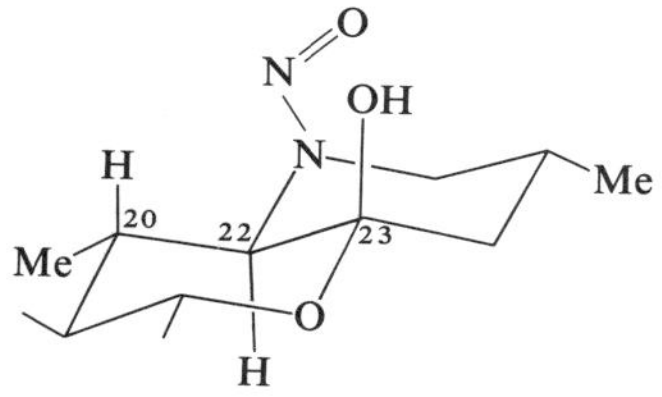 **76**

tion, the ^{1}H-NMR spectrum of the acetyl derivative **75** obtained from solanocapsine proves the β position of the proton at C-16. At δ 4.05 ppm the signals of 16-H can be recognized as the X part of an ABCX system in the form of an octet. The distance between the outer signals should be about 30 Hz according to the torsional angles, which were estimated from the Dreiding model. This value as well as the number of lines correspond to expectations. In contrast, in the case of the 16-epimeric compound, a distance of about 18 Hz was estimated. The configuration at C-22 and C-23 was proved by the ^{1}H-NMR spectrum of *N*-nitroso-3-deamino-3β-hydroxysolanocapsine (**76**). Because of steric reasons, the nitroso group has the anti position. The chemical shift of 22-H at δ 3.38 ppm indicates axial conformation; furthermore, the coupling constant of 11 Hz is only consistent with a diaxial relationship between 20-H and 22-H. This is the case in steric structure **76** (*192*). By X-ray analysis of *N*-(2-bromobenzylidene)solanocapsine, the structure 3β-amino-22,26-epimino-16α,23-epoxy-5α,22αH,25βH-cholestan-23β-ol (**4**) has been confirmed (*187*).

b. Solanoforthine. Solanoforthine was isolated from *Solanum seaforthianum*. Methylation of the alkaloid (**77**) by heating with formaldehyde and formic acid afforded the *N,N,N′*-trimethyl compound **78**, which was hydrogenated over platinum in acetic acid solution containing a few drops of HCl to yield the dihydro derivative, identical with *N,N,N′*-trimethylsolanocapsine (**79**). Treatment of **78** and **79** with sodium borohydride for 12 hr gave the diols **80** and **81**, respectively. Acetylation of solanoforthine furnished the *N,N′*-diacetate and, in addition, the *N,N′O*-triacetate **82** with IR absorption at 1710 cm^{-1} (*110*).

c. Solacasine. Solacasine is the main antibacterial constituent of *Solanum pseudocapsicum*. Reaction of solacasine with sodium borohydride led to the dihydro derivative **83**, also available from solanocapsine (**4**). Therefore, solacasine most probably possesses structure **84** (*105*).

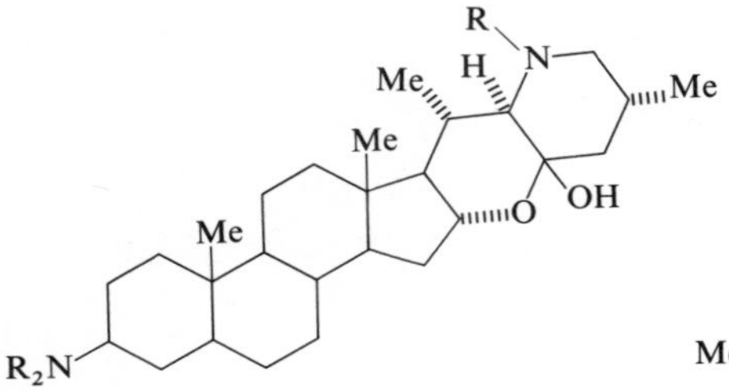

77 R = H, Δ^5 Solanoforthine
78 R = Me, Δ^5
79 R = Me, 5αH

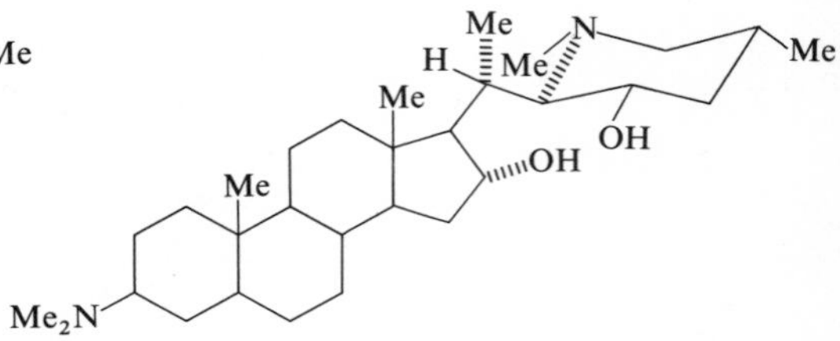

80 Δ^5
81 5αH

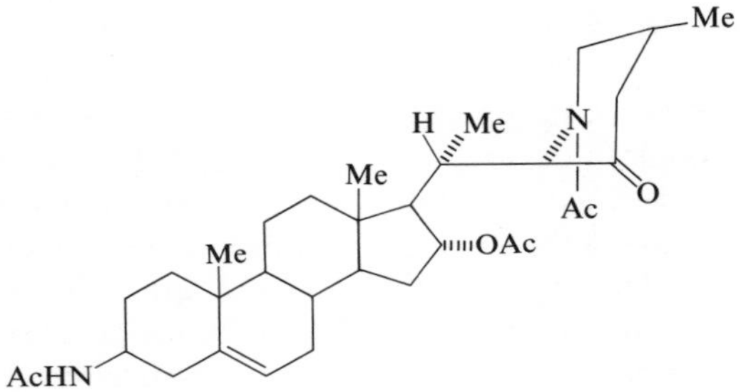

82

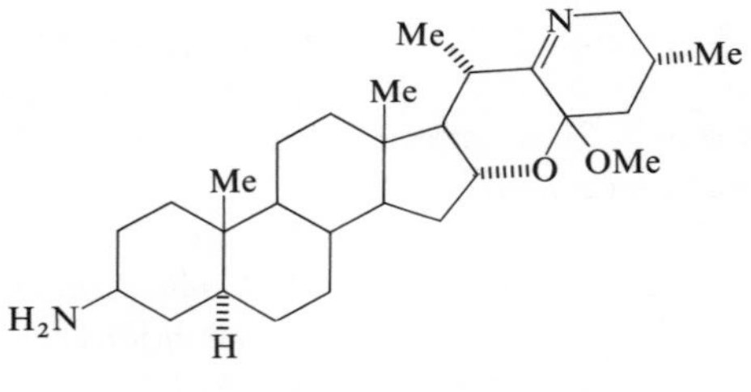

84 Solacasine

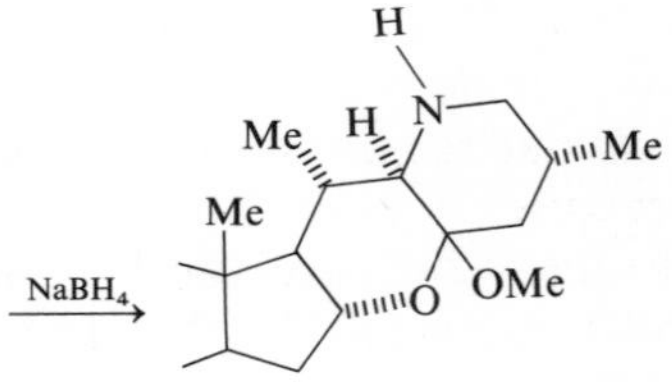

NaBH$_4$

83

H$^+$, MeOH

4

5. The 3-Aminospirostanes

Isojurubidine, Isopaniculidine, and Isojuripidine. Isojurubidine, isopaniculidine, and isojuripidine have been obtained from roots of *Solanum paniculatum*. The molecular formulas of these alkaloids were established by elemental analysis and MS. Their MS fragmentation is consistent with 3-aminospirostane structures (*m/e* 56, 139), the configurations of which at C-25 were shown to be *R* by IR and ^{1}H-NMR measurements (spiroketal bands at

915 cm^{-1} much weaker than those at 895 cm^{-1}; doublets for the methyl groups C-27 at δ 0.73–0.78 ppm). Deamination of isojurubidine carried out with nitrous acid led to tigogenin (**85**) as the main product and the corresponding 3α-epimer, showing isojurubidine to have structure **86**. Isopaniculidine possesses an additional tertiary hydroxy group. Comparison of the ^{1}H-NMR shifts of the methyl groups C-18 and C-19 in acetylisopaniculidine and acetylisojurubidine confirmed the presence of a 9α-hydroxy group (**87**). Isojuripidine contains a hydroxy group which can be acetylated. Instead of a fragment ion at *m/e* 82 of 5-saturated 3-aminosteroids, isojuripidine displays a fragment at *m/e* 98, which is to be expected whenever an amino group in position 3 is associated with a hydroxy function in position 5, 6, or 7. Differences between the ^{1}H-NMR shifts of the methyl groups C-18 and C-19 in acetylisojurubidine and diacetylisojuripidine are consistent with an additional 6α-acetoxy group in the latter compound and hence with structure **88** for isojuripidine. Deamination of isojuripidine afforded chlorogenin (**89**) as the main product and a 3-epimeric by-product (*100*). Isojuripidine was synthesized from diosgenin (**90**), which was converted to (25*R*)-3β-azido-22α*O*-spirost-5-ene (**91**). Reaction of **91** with a solution of diborane in THF, followed by oxidation with alkaline hydrogen peroxide, gave (25*R*)-3β-azido-5α,22α*O*-spirostan-6α-ol (**92**) along with small amounts of its 6β-hydroxy isomer. Both isomers were oxidized to (25*R*)-3β-azido-5α,22α*O*-spirostan-6-one (**93**). The reduction of **92** with LAH, as well as the reduction of **93** with sodium and *n*-propanol, yielded (25*R*)-3β-amino-5α,22α*O*-spirostan-6α-ol (**88**), found to be identical with isojuripidine. Isojuripidine (**88**) was also obtained by the sodium/*n*-propanol reduction of 3,6-bisdehydrochlorogenin-3-oxime (**94**). In addition, the 5α-, 6β-, 7α-, and 7β-hydroxy derivatives of isojurubidine (**86**) have been synthesized (*99*).

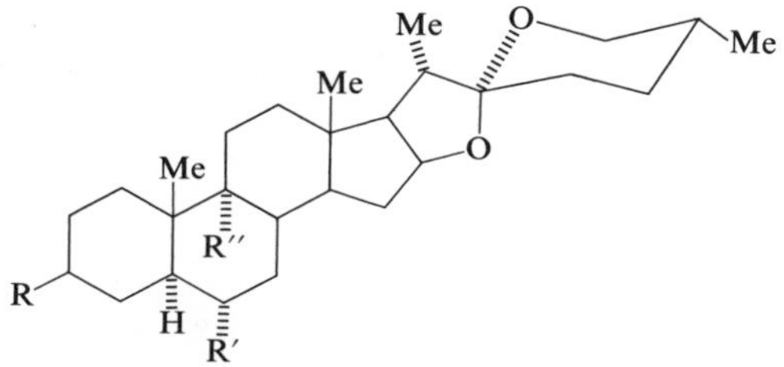

85 R = OH, R′, R″ = H Tigogenin
86 R = NH_2, R′, R″ = H Isojurubidine
87 R = NH_2, R′ = H, R″ = OH Isopaniculidine
88 R = NH_2, R′ = OH, R″ = H Isojuripidine
89 R, R′ = OH, R″ = H Chlorogenin

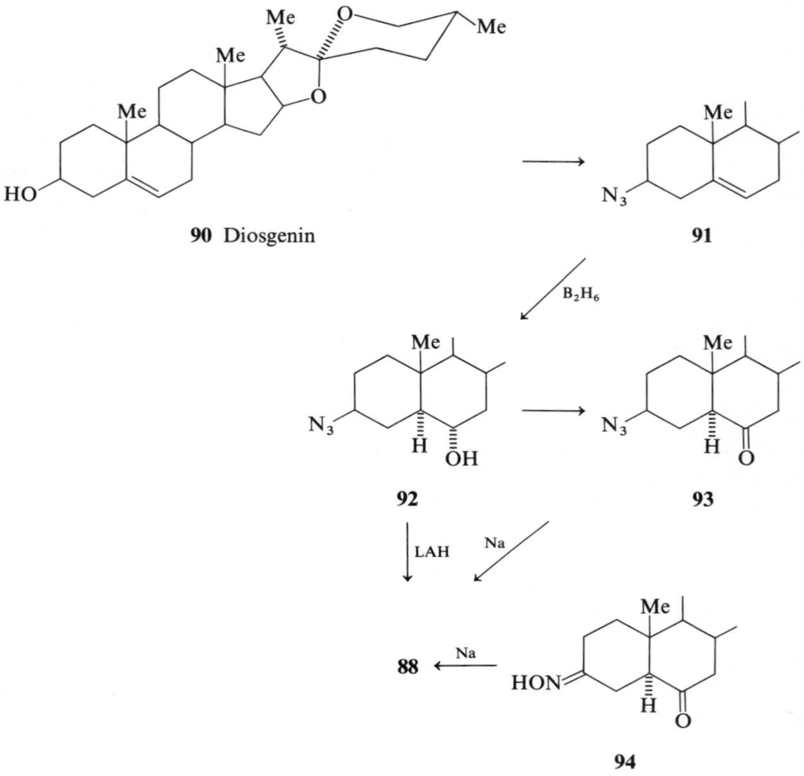

90 Diosgenin

91

B_2H_6

92

93

LAH

Na

88 ← Na

94

B. Syntheses

The syntheses of teinemine, isoteinemine, hapepunine, tomatillidine, solanogantine, and isojuripidine have already been described in this chapter (see Section IV,A).

1. Tomatidenol and Solasodine

Michael addition of the (*S*)-nitroester **95** to the unsaturated ketone **96** gave a mixture of **97** and the (22*S*)-isomer (*207*, *208*). Sodium borohydride reduction of **97** in an acidic medium afforded the nitrodiol **98** in excellent yield. Reduction of the nitro group and subsequent lactam formation (**99**) proceeded smoothly when the nitrodiol **98** was refluxed with zinc and acetic acid. Further reduction of the amide **99** with LAH yielded the piperidine **100**, which was converted into its *N*-chloro derivative. Treatment of this compound with sodium methoxide effected dehydrochlorination which was followed by spontaneous cyclization. The base thus obtained (**101**) was shown to be identical with natural tomatidenol in all respects.

The (22*S*)-isomer of **97**, when carried through the above reaction sequence, afforded the same alkaloid in comparable yield. The structures of the intermediate compounds were confirmed by MS. The base peak in the spectrum of lactam **99** is observed at *m*/*e* 112, which can be rationalized on the basis of a C-20–C-22-bond homolysis.

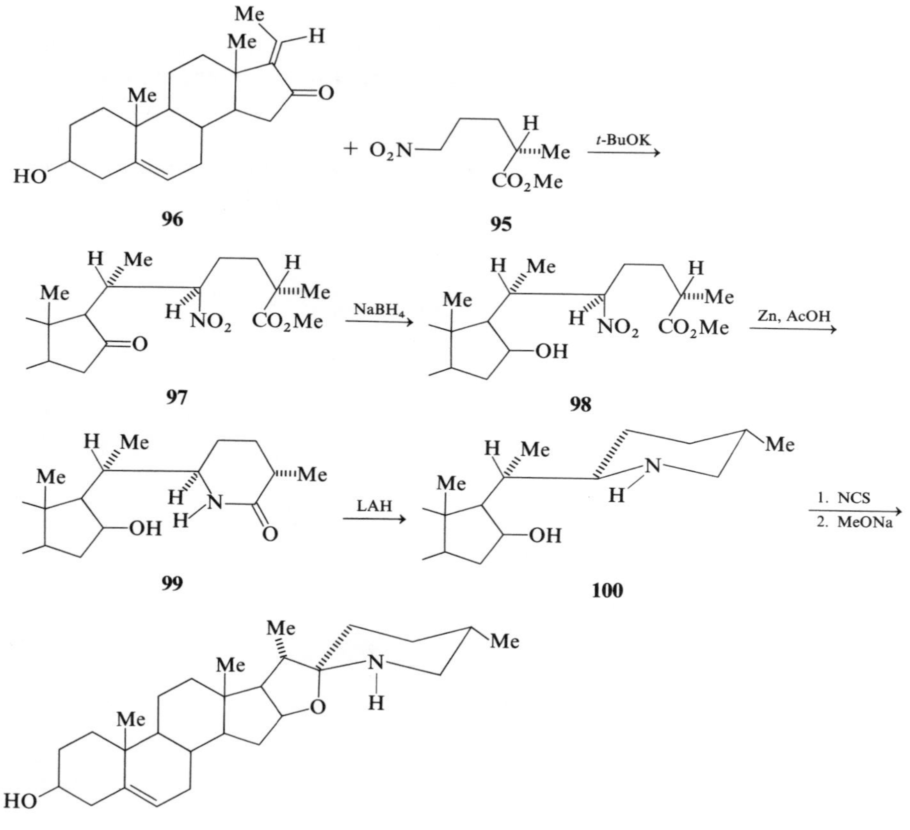

96 + **95** $\xrightarrow{t\text{-BuOK}}$ **97** $\xrightarrow{NaBH_4}$ **98** $\xrightarrow{\text{Zn, AcOH}}$ **99** $\xrightarrow{\text{LAH}}$ **100** $\xrightarrow{\text{1. NCS; 2. MeONa}}$ **101**

101 Tomatidenol

To synthesize solasodine, methyl (*R*)-2-methyl-5-nitropentanoate was added to the unsaturated ketone **96**. The product obtained was again resolved into the 22-isomers. As the separation was very laborious and the asymmetry at C-22 is ultimately eliminated in a subsequent step, the unseparated mixture was carried through the sequence of steps described in the synthesis of tomatidenol. The solasodine thus obtained (**1**) was shown to be identical with a natural sample (*209*, *210*). 15ξ,17α-Dideuterosolasodine has been synthesized from solasodine (**1**) (*357*).

2. Solacongestidine, Solafloridine, 25-Isosolafloridine, and Verazine

Solacongestidine (**2**) had already been synthesized from 3β-acetoxypregn-5-en-20-one (*3*). Two conversions of readily available solasodine (**1**) to this alkaloid were described later. *O*-Acetyldihydrosolasodine (**56**) was hydrogenated to **102**. Oxidation of the benzyloxycarbonyl derivative **103** with the Kiliani reagent gave the 16-ketone **104**. Reaction with ethanedithiol yielded the thioketal **105**. Desulfurization of the thioketal moiety with Raney nickel led to concomitant elimination of the benzyloxycarbonyl function to afford (22*S*)-dihydrosolacongestidine acetate (**106**). Finally, **106** was converted to its *N*-chloro derivative by means of *N*-chlorosuccinimide, the treatment of

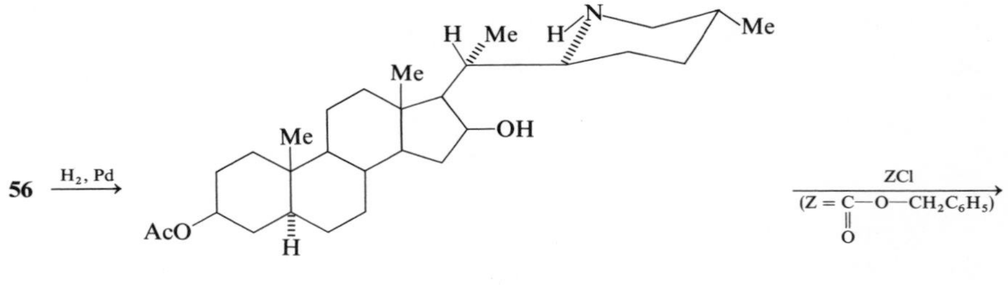

56 $\xrightarrow{H_2, Pd}$ **102** $\xrightarrow[(Z = C(=O)-O-CH_2C_6H_5)]{ZCl}$ **103** $\xrightarrow{CrO_3, H_2SO_4}$ **104** $\xrightarrow{HS-CH_2CH_2-SH}$ **105** $\xrightarrow{Ni}$ **106** $\xrightarrow[2.\ MeONa]{1.\ NCS}$ **2**

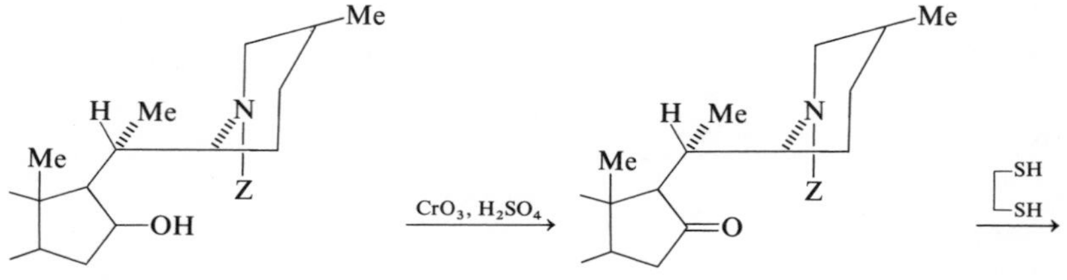

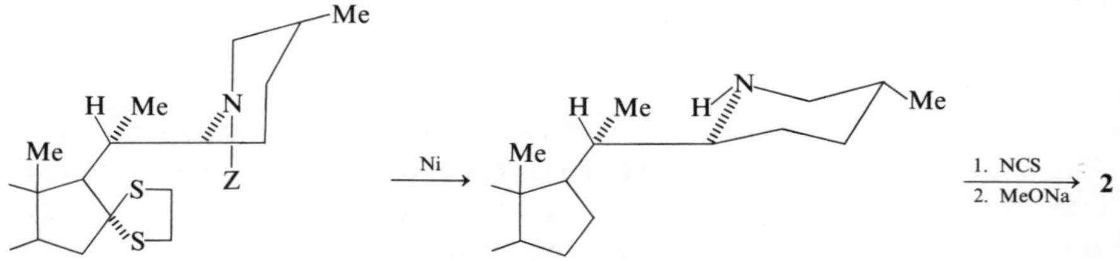

which with sodium methoxide effected dehydrochlorination and hydrolysis of the acetyl function at C-3 to supply solacongestidine (**2**) (*211*).

During a similar sequence of reactions, solasodine (**1**) was reduced to the (22*S*)-dihydro derivative **107**. Acetylation with acetic anhydride in pyridine gave the corresponding *N,O,O'*-triacetate, which was hydrolyzed to the *N*-acetyl derivative **108**. Partial oxidation yielded the 16-ketone **109**, the thioketal **110** of which was treated with Raney nickel to give a mixture of **111** and

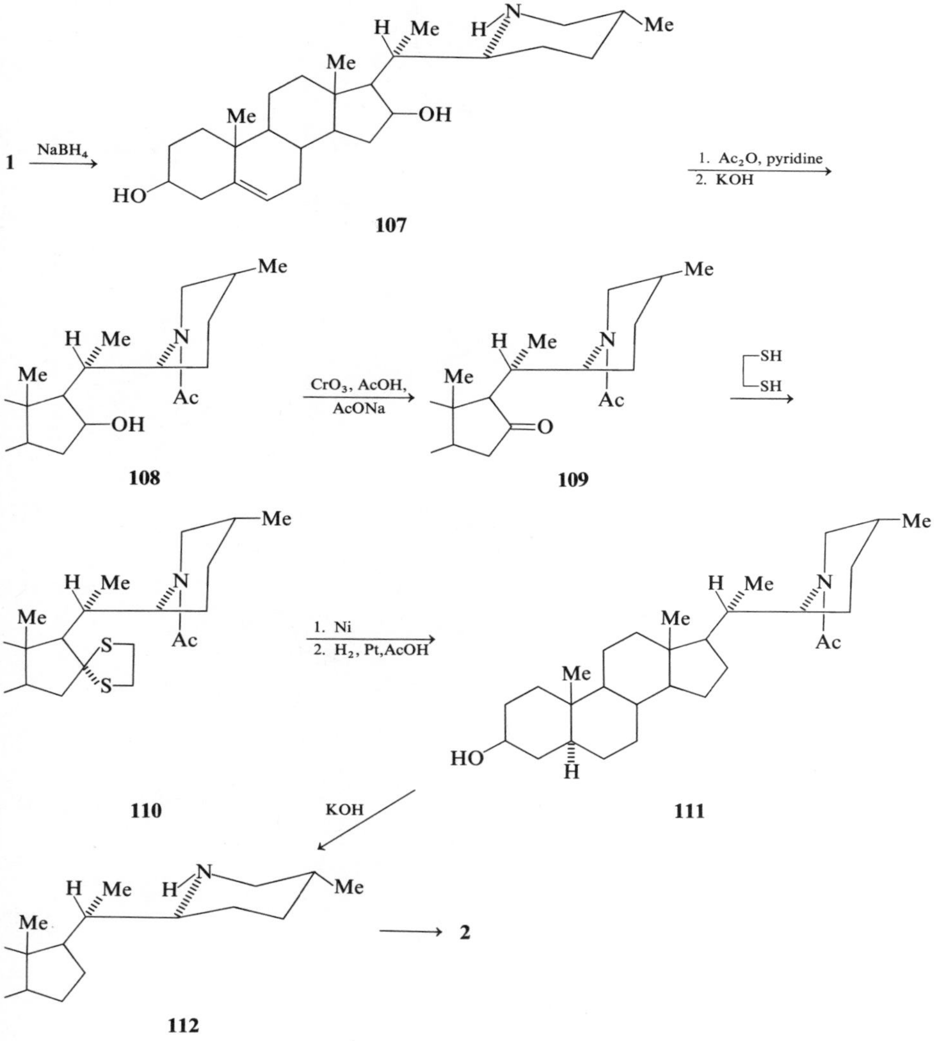

1 $\xrightarrow{NaBH_4}$ **107** $\xrightarrow{1.\ Ac_2O,\ pyridine;\ 2.\ KOH}$

108 $\xrightarrow{CrO_3,\ AcOH,\ AcONa}$ **109** $\xrightarrow{HSCH_2CH_2SH}$

110 $\xrightarrow{1.\ Ni;\ 2.\ H_2,\ Pt,AcOH}$ **111**

KOH

112 ⟶ 2

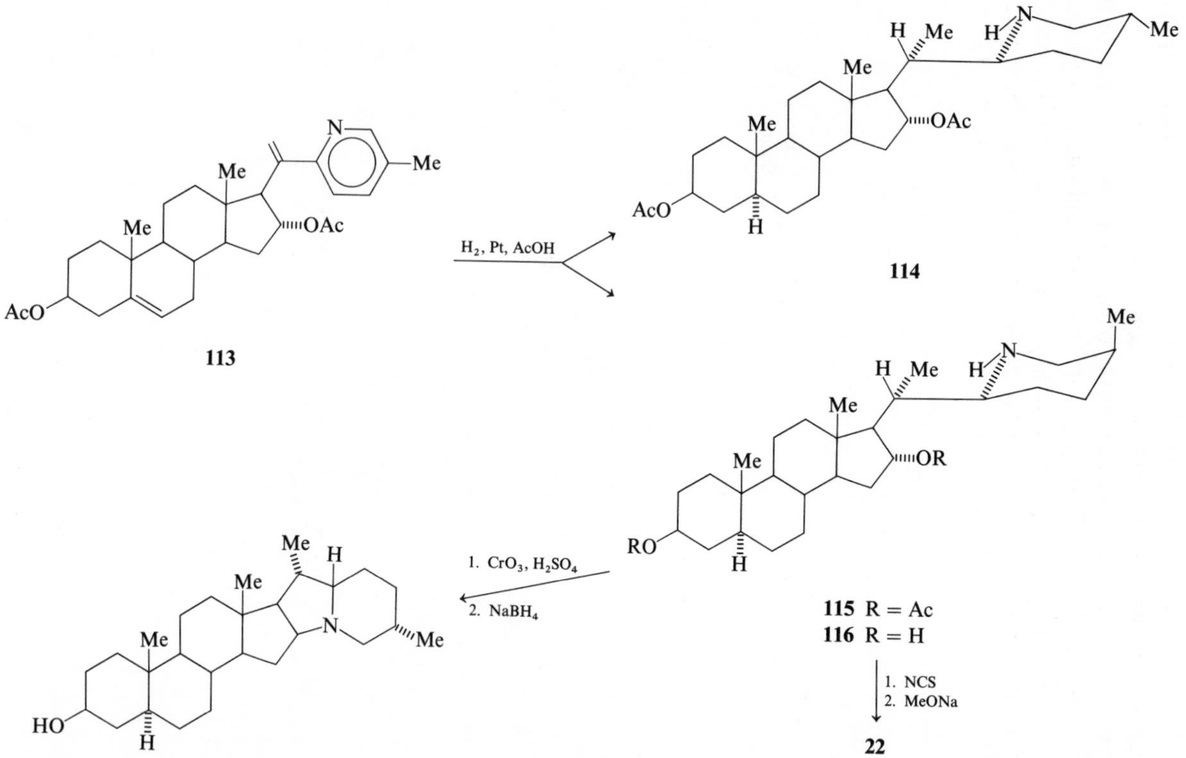

113

H_2, Pt, AcOH

114

115 R = Ac
116 R = H

1. NCS
2. MeONa

22

1. CrO_3, H_2SO_4
2. $NaBH_4$

117

its 5-unsaturated derivative. The mixture was hydrogenated to **111**, which was hydrolyzed by potassium hydroxide in ethanediol to (22*S*)-dihydrosolacongestidine (**112**) (*212*). Deacetylation was also carried out by means of diisobutyl aluminum hydride (*213*). Compound **112** was converted to solacongestidine (**2**) by dehydrochlorination of its *N*-chloro compound (*212*). An improved transformation of ketone **109** into solacongestidine (**2**) was recently described (*358*).

In an analogous manner, the *Veratrum* alkaloid verazine [(25*S*)-22,26-epiminocholesta-5,22(*N*)-dien-3β-ol] has been synthesized from tomatid-5-en-3β-ol (*199*, *200*).

Reaction of the 16-ketone **104** with sodium in 2-propanol afforded the 16α-hydroxy-bearing (22*S*)-dihydrosolafloridine (**42**) in good yield. A somewhat smaller yield was obtained by reduction with lithium in ammonia. In contrast, reaction of **104** with sodium borohydride gave the 16β-hydroxy compound. Finally, (22*S*)-dihydrosolafloridine (**42**) was converted to solafloridine (**41**) by dehydrohalogenation of the *N*-chloro compound (*211*).

Catalytic hydrogenation of the pyridyl steroid **113** gave the piperidyl steroids **114** and **115** in yields of 24% and 9%, respectively. The main product (**114**) was hydrolyzed to the known diol **42**, the minor product (**115**) to the diol **116**. The *N*-chloro derivatives of **115** and **116** display negative Cotton effects and therefore have the (22*S*)-configuration. Dehydrochlorination of the *N*-chloro derivative of **116** led to 25-isosolafloridine (**22**), which shows a negative Cotton effect consistent with a (25*S*)-configuration. To confirm the stereochemistry at C-20, diol **116** was converted to the known 22-isodemissidine **117** (*214*).

The nitroketone **118** (see Section IV,B,1) was treated with hydrogen chloride in ethanedithiol to afford the thioketal **119**. Reduction of **119** with zinc and acetic acid and further treatment with Raney nickel in ethanol furnished lactam **120**. Hydrolysis of **120** in aqueous methanol containing K_2CO_3 gave **121**, which on LAH reduction afforded the amine **122**. On treatment with sodium methoxide, the *N*-chloro derivative of **122** gave verazine (**123**) (*215*).

3. Solanidine, Leptinidine, and Further 23-Hydroxysolanidanes

The nitroester **97** (see Section IV,B,1) was reduced with zinc in acetic acid and the product separated into neutral and basic fractions. The neutral portion on chromatography furnished amide **124** and its acetate **125**. The basic fraction **126** also led to amide **124** via sodium borohydride reduction and cyclization. Treatment of **124** with LAH gave solanidine (**3**) (*207*, *208*).

Acetylation of tomatidenol (**101**) with acetic anhydride–zinc chloride–acetic acid yielded the azomethine **127**, the reaction of which with selenium

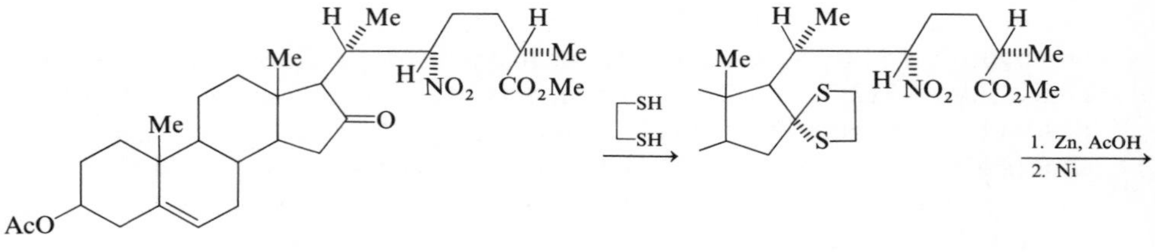

118

119

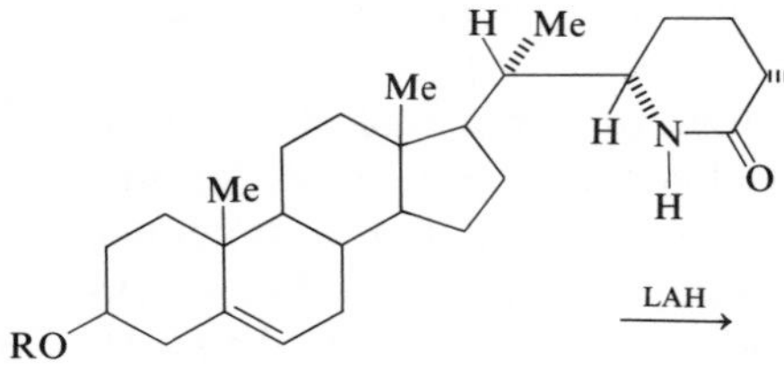

120 R = Ac
121 R = H

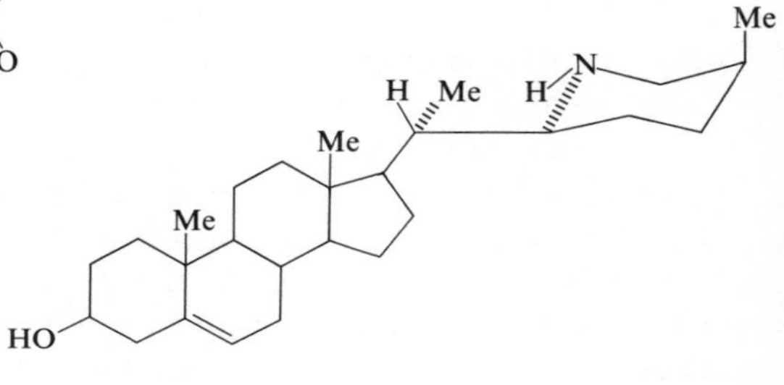

122

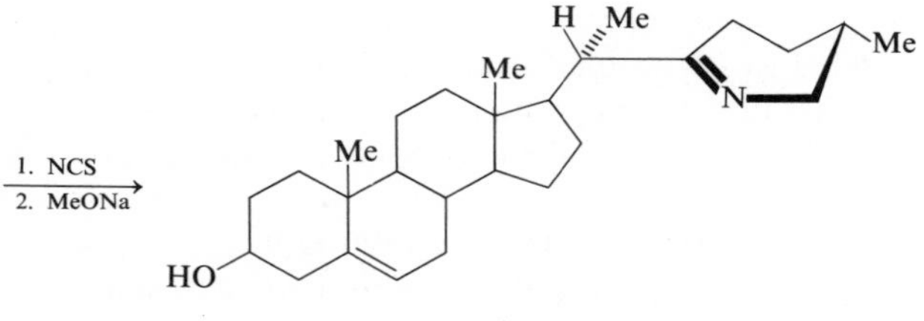

123 Verazine

dioxide gave the 23-ketone **128**, which was reduced with sodium borohydride to supply the piperidinol **129** as the main product. The hydrolysis product **130** was oxidized at C-16 under controlled conditions and the resulting carbinolamine reduced to leptinidine (**131**). Analogously, starting from tomatidine (**9**), four stereoisomeric (25*S*)-3β,16β-diacetoxy-22,26-epimino-5α-cholestan-23-ols (**132**–**135**) were obtained (*216*). Their configurations at C-22 and C-23 were determined by X-ray analysis of (22*S*,23*R*,25*S*)-22,26-epimino-5α-cholestane-3β,16β,23-triol hydrobromide (corresponding to **133**) (*188*) as well as by ^{1}H-NMR and ORD studies. The stated yields refer to the reduction of the 23-ketone. The piperidinols **132**, **133**, **136**, and **137** supplied the stereoisomeric solanidane-3β,23-diols **138** (dihydroleptinidine), **139**, **140** (3-deamino-3β-hydroxysolanogantine), and **141**, respectively (*188*, *216*).

97 $\xrightarrow{\text{Zn, AcOH}}$ 126

126 $\xrightarrow{NaBH_4}$ 124

124 R = H
125 R = Ac

$\xrightarrow{\text{LAH}}$ **3**

101 $\xrightarrow{Ac_2O,\ ZnCl_2}$ **127** $\xrightarrow{SeO_2}$

128 $\xrightarrow{NaBH_4}$ **129** $\xrightarrow{\text{KOH}}$

130 $\xrightarrow{\text{1. } CrO_3\text{, AcOH, AcONa; 2. } NaBH_4}$ **131** Leptinidine

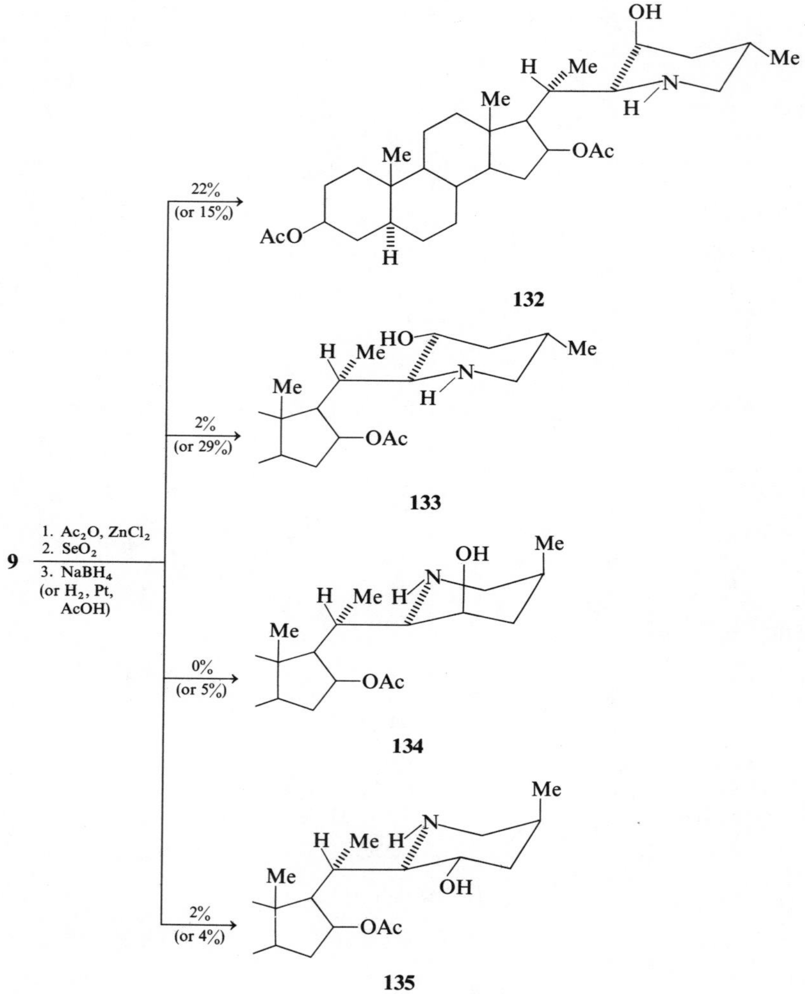

9 → 1. Ac_2O, $ZnCl_2$ 2. SeO_2 3. $NaBH_4$ (or H_2, Pt, AcOH)

22% (or 15%) → **132**

2% (or 29%) → **133**

0% (or 5%) → **134**

2% (or 4%) → **135**

OH Me H Me Me N H Me OAc AcO H

132

H Me HO Me Me N H OAc

133

OH Me H Me H N Me OAc

134

Me H Me H N Me OH OAc

135

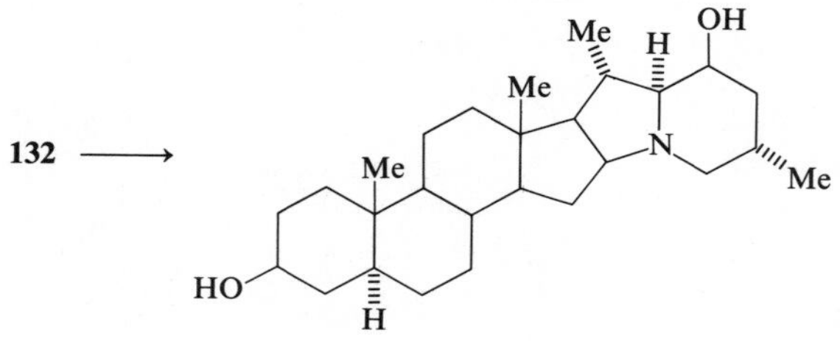

132 ⟶

Me H OH Me N Me Me HO H

138

133 ⟶ **139**

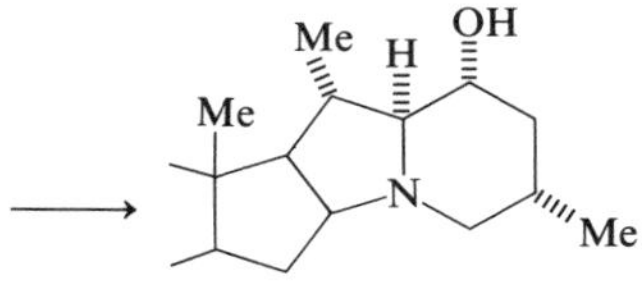

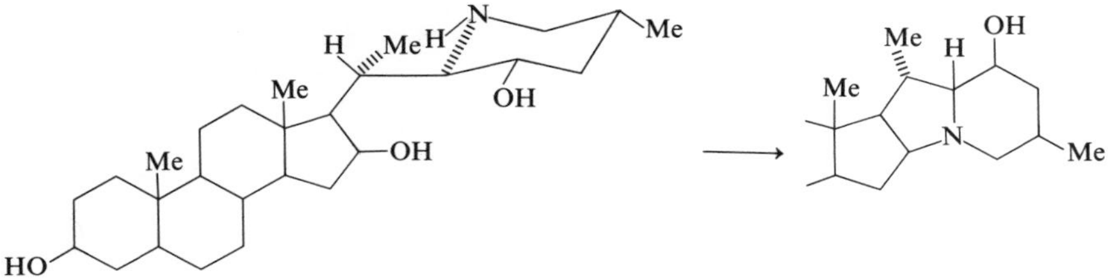

136 5αH
137 Δ^5

⟶ **140** 5αH
141 Δ^5

4. Solanocapsine

Acetylation of solafloridine (**41**) with acetic anhydride–zinc chloride–acetic acid or photolysis of *N*-nitrosodihydrosolafloridine diacetate (**142**) furnished the diacetate **143**. Subsequent oxidation with manganese dioxide in $CHCl_3$ at room temperature yielded ketone **144**, which was reduced with sodium borohydride, without purification, to the piperidinol **145** (yield from **143** 55%). As a by-product of the MnO_2 oxidation of **143**, the pyridyl steroid **146** was isolated.

After an attempt of purify **144** by Al_2O_3 chromatography, a compound was obtained the spectroscopic properties of which correspond to those of "24-oxosolacongestidine" (**40**) (*217*); hence it should have structure **147** (cf. *172*). The stereochemistry of the piperidinol **145** at C-22 and C-23 was demonstrated by its ^{1}H-NMR spectrum. The coupling pattern of the sextet at δ 3.45 ppm is in agreement with an axial conformation of the 23-hydrogen. The doublet at δ 0.83 ppm corresponds to an equatorial 25-methyl group. This stereochemistry is also confirmed by comparison with the spectrum of solaphyllidine (**45**). The most important signals in both spectra have almost identical positions. Reaction of the piperidinol **145** with benzylchloroformate afforded the *N*-benzyloxycarbonyl derivative **148**, which was oxidized by chromium trioxide–sulfuric acid in acetone to the amorphous ketone **149**. Partial alkaline hydrolysis gave the cyclic hemiketal **150**, which served as a relay substance, because it could be synthesized from solanocapsine (**4**).

Reaction of **4** with HNO_2 yielded 3-deamino-3β-hydroxy-*N*-nitroso-solanocapsine (**151**). The *N*-nitroso group was eliminated by treatment with

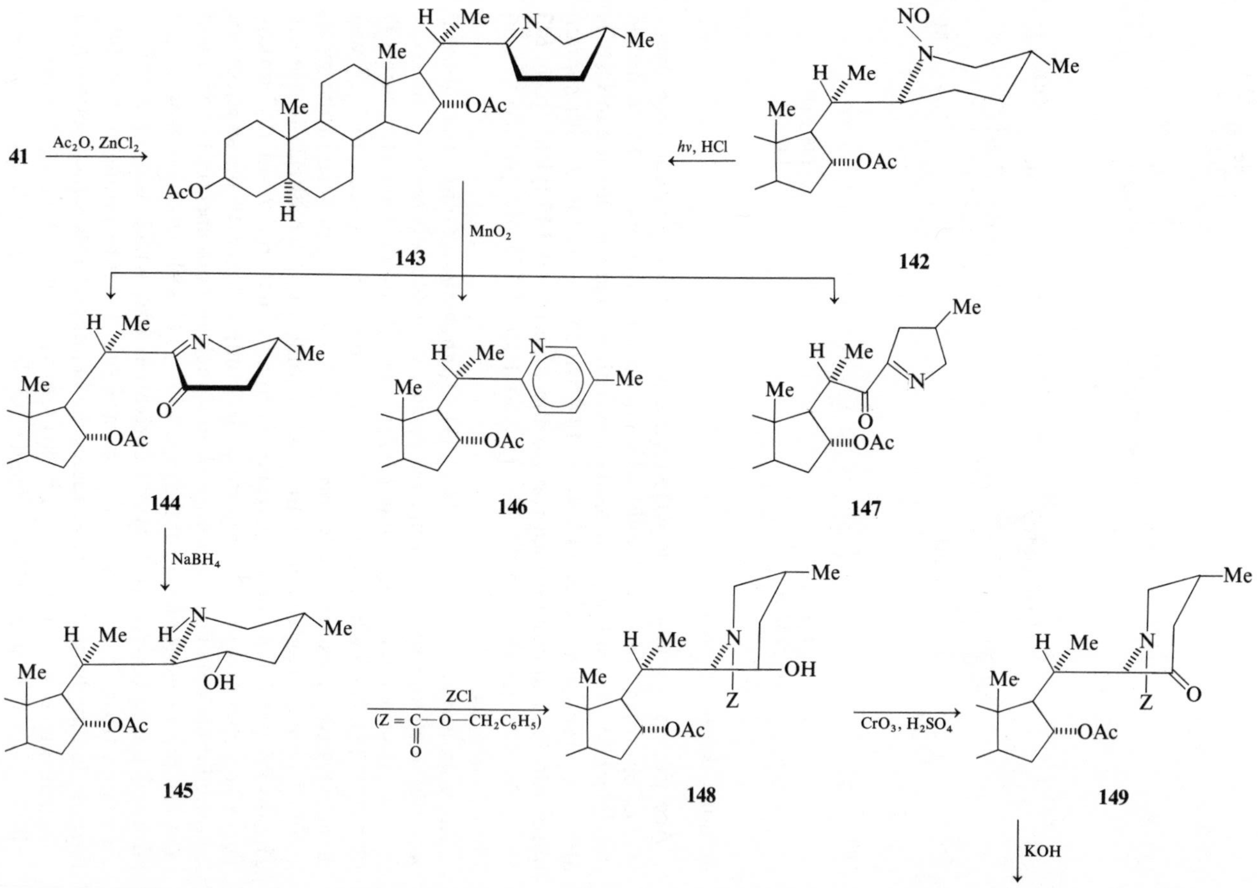

41 $\xrightarrow{Ac_2O,\ ZnCl_2}$ **143** $\xleftarrow{h\nu,\ HCl}$ **142**

143 $\xrightarrow{MnO_2}$ **144**, **146**, **147**

144 $\xrightarrow{NaBH_4}$ **145** $\xrightarrow[(Z = C(=O)-O-CH_2C_6H_5)]{ZCl}$ **148** $\xrightarrow{CrO_3,\ H_2SO_4}$ **149** $\xrightarrow{KOH}$

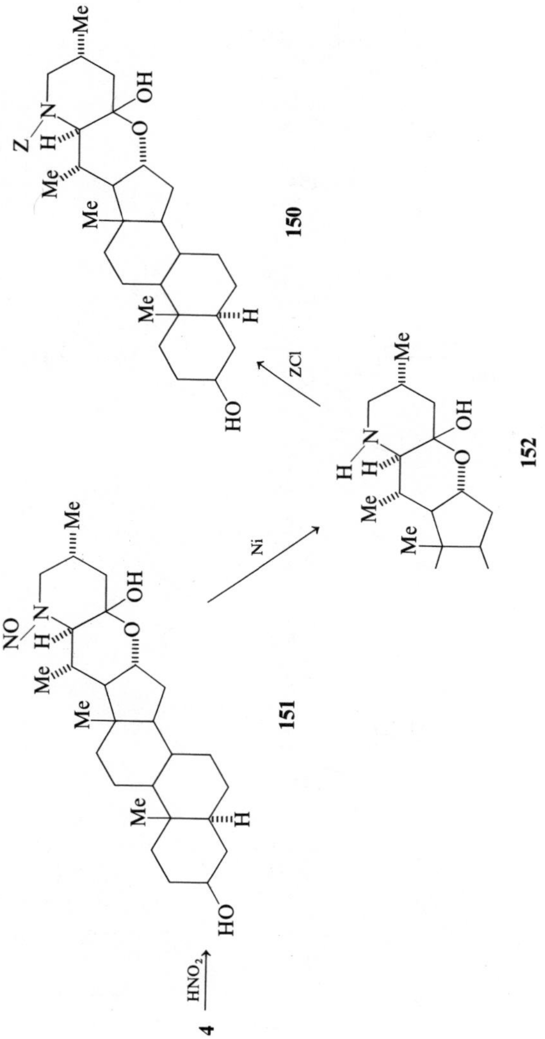

4 $\xrightarrow{HNO_2}$

151

$\xrightarrow{Ni}$

152

$\xrightarrow{ZCl}$

150

150 $\xrightarrow{CrO_3,\ H_2SO_4}$ **154**

150 $\xrightarrow{CrO_3,\ pyridine}$ **153**

153 $\xrightarrow{HBr,\ AcOH}$ **155**

155 R = O
156 R = NOH

156 $\xrightarrow{H_2,\ Pt,\ AcOH}$ **4** + **157**

144 $\xrightarrow{1.\ Zn,\ AcOH;\ 2.\ Ac_2O}$ **158** $\xrightarrow{KOH}$ **159**

alkaline Raney nickel to give 3-deamino-3β-hydroxysolanocapsine (**152**), which was reacted with benzylchloroformate to supply the relay substance **150**, which was oxidized with chromium trioxide in pyridine to the 3-ketone **153**. In contrast, oxidation of **150** with chromium trioxide–sulfuric acid gave the ring-E-opened 3,16,23-triketone **154**. Cleavage of the benzyloxycarbonyl residue of **153** by hydrogen bromide–acetic acid afforded the basic ketone **155**, the oxime (**156**) of which was reduced by catalytic hydrogenation in the presence of platinum in acetic acid to solanocapsine (**4**) and the 3-epimeric amine **157** (*217*, *218*).

The *N*,*O*(3)-diacetyl derivative of compound **152** has been synthesized independently from solafloridine (**41**) by Nagai and Sato (*219*). They obtained the crystalline ketone **144** via **143**. Reduction of the C=N bond in **144** was effected with zinc and acetic acid. Subsequent acetylation gave the 22,26-acetylepimino compound **158**. Treatment of **158** with base and reacetylation afforded *N*-acetyl-3-deamino-3β-acetoxysolanocapsine (**159**) (*219*).

In an analogous sequence of reactions, 16,23-diisosolanocapsine (**160**) was synthesized from soladulcidine (**161**). Catalytic hydrogenation of ketone **162** furnished the piperidinols **163** and **164** in yields of 46% and 17%, respectively, whereas reduction with sodium borohydride gave **163** exclusively. The benzyloxycarbonyl derivatives of **163** and **164** yielded the same 23-ketone, **165**. Cyclization to ketal **166** was effected by hydrogen chloride in aqueous methanol. Compound **166** was oxidized to the 3-ketone, and its oxime **167** was treated with sodium in liquid ammonia and catalytically hydrogenated to furnish 16,23-diisosolanocapsine (**160**) (*192*).

The IR spectra of *N*(22,26)-unsubstituted and *N*(22,26)-acyl derivatives of solanocapsine (**4**) display 23-OH · · · N(22,26) hydrogen bonds ($\Delta\tilde{\nu}$ 80–145 cm^{-1}). However, the spectra of *N*(22,26)-alkyl compounds only show bands of free hydroxy groups, indicating an axial alkyl substituent. *N*(22,26)-Unsubstituted and *N*(22,26)-acyl derivatives of 16,23-diisosolanocapsine (**160**) do not form hydrogen bridges; this observation proves that rings E and F are cis-fused. For thermodynamic reasons the 22β,23β-configuration was excluded (*220*).

5. Modified Spirosolanes

Particularly because of the potential importance of the spirosolane alkaloids as starting materials for the industrial production of hormonal steroids, a number of modified spirosolanes have been synthesized. The degradation of these products (see Section IV,C) should lead to modified pregnanes.

Oppenauer oxidation of solasodine (**1**) gave the corresponding α,β-unsaturated ketone **12**, the catalytic hydrogenation of which over $Pd/CaCO_3$

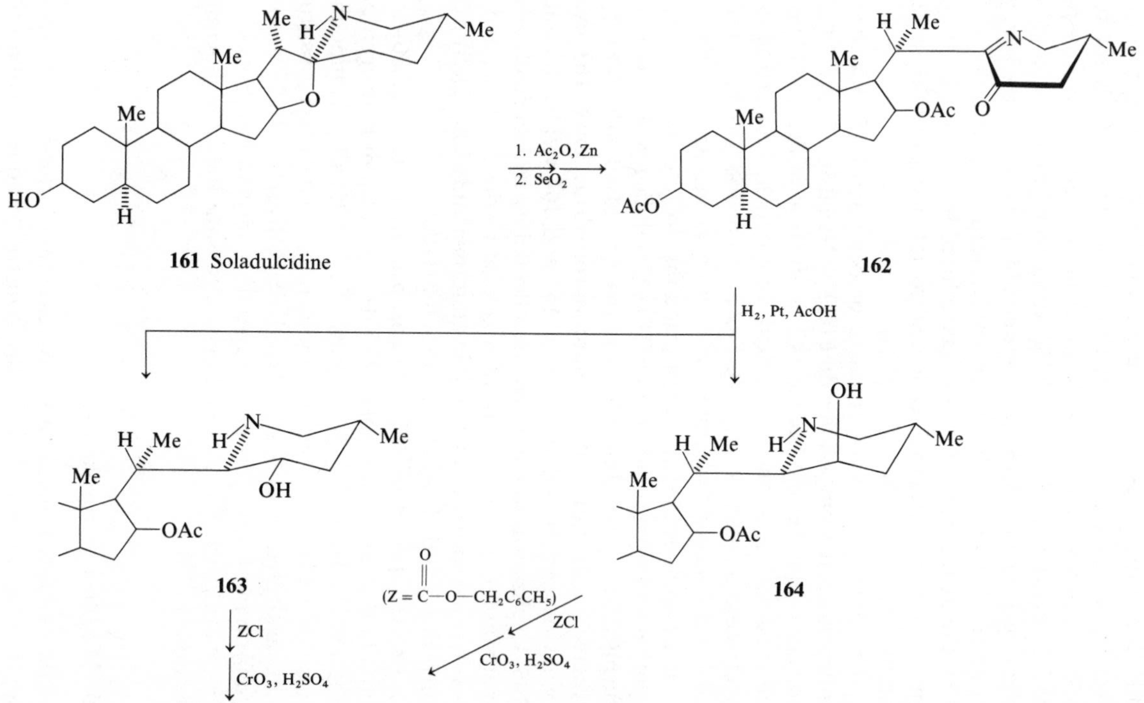

161 Soladulcidine

1. Ac_2O, Zn
2. SeO_2

162

H_2, Pt, AcOH

163

ZCl

CrO_3, H_2SO_4

164

$(Z = \overset{O}{\overset{\|}{C}}—O—CH_2C_6CH_5)$

ZCl

CrO_3, H_2SO_4

165 $\xrightarrow{\text{HCl, MeOH}}$ **166** $\xrightarrow[\text{2. } NH_2OH]{\text{1. } CrO_3\text{, pyridine}}$

167 $\xrightarrow[\text{2. } H_2\text{, Pt, AcOH}]{\text{1. Na, } NH_3}$ **160**

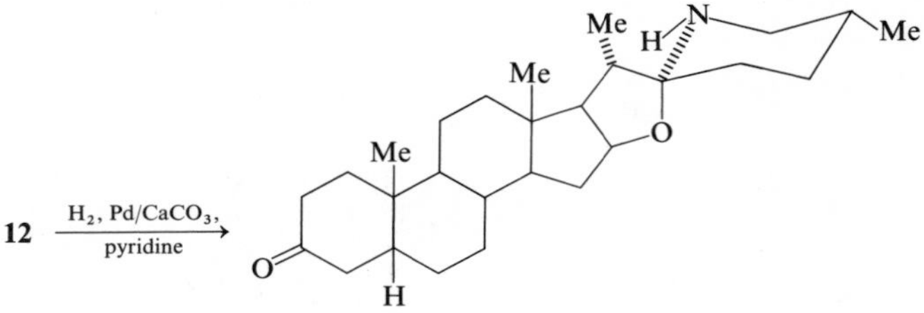

12 $\xrightarrow[\text{pyridine}]{H_2,\ Pd/CaCO_3,}$

168

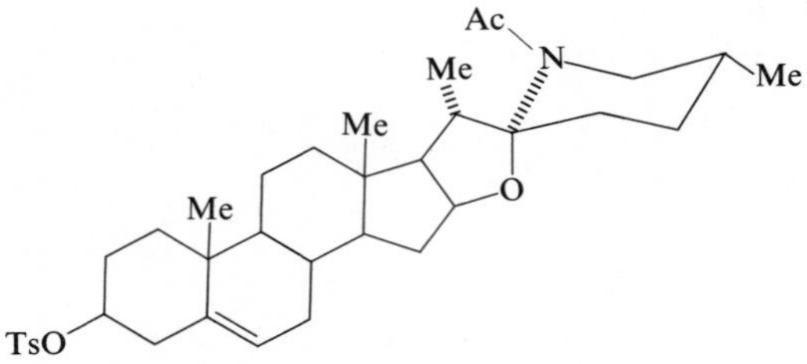

169

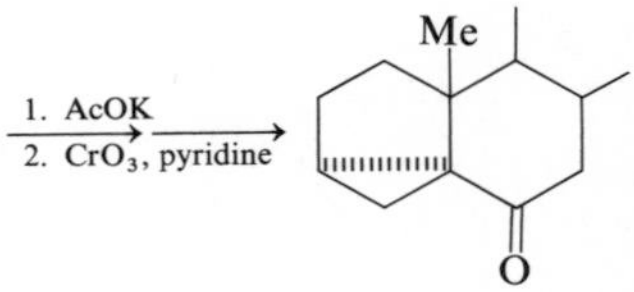

$\xrightarrow[\text{2. } CrO_3\text{, pyridine}]{\text{1. AcOK}}$

170

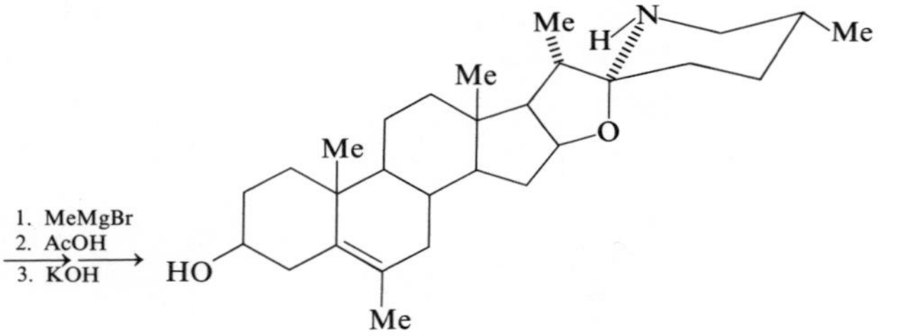

1. MeMgBr
2. AcOH
3. KOH

171

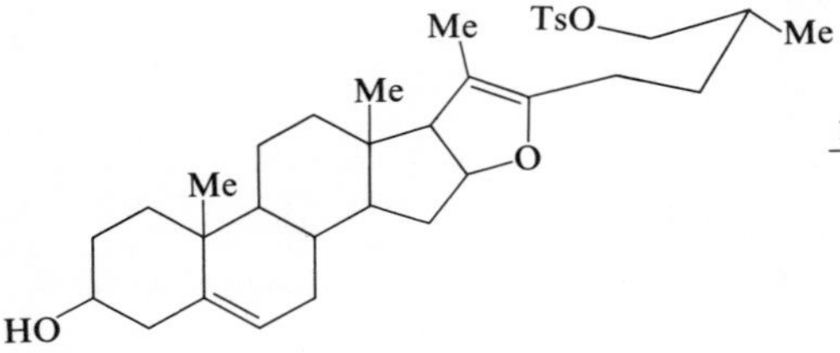

172

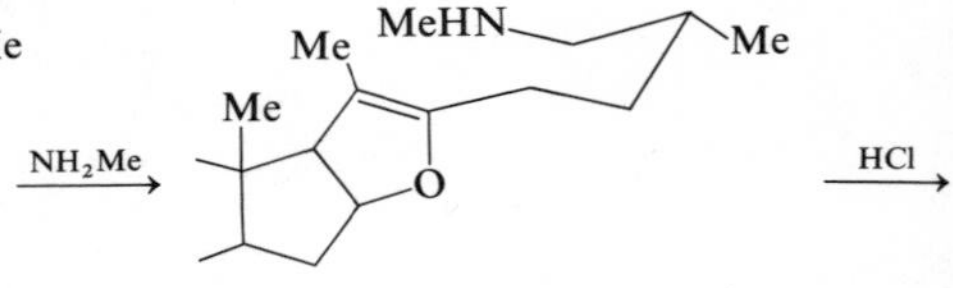

$\xrightarrow{NH_2Me}$

173

$\xrightarrow{HCl}$

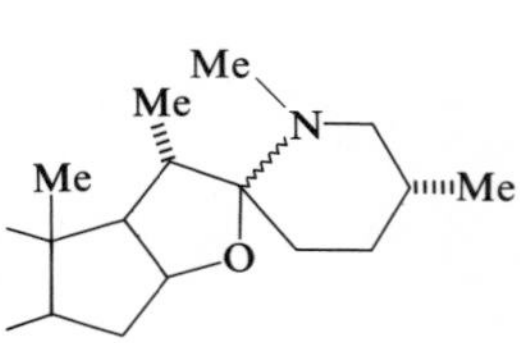

174

in pyridine stereoselectively afforded (25*R*)-5β,22α*N*-spirosolan-3-one (**168**). Hydrogenation of **12** in ethanol yielded a mixture of ketones, epimeric at C-5, whereas hydrogenation in ethanol in presence of Raney nickel furnished a mixture of alcohols, (25*R*)-5α,22α*N*-spirosolan-3α-ol; (25*R*)-5α,22α*N*-spirosolan-3β-ol; and (25*R*)-5β,22α*N*-spirosolan-3α-ol. Ring-E-opened 5β-compounds were also prepared (*196*, *221*). The Wolff–Kishner reduction of solasodanone to solasodane has been described (*364*).

N-Acetyl-*O*-tosylsolasodine (**169**) was rearranged with potassium acetate in boiling aqueous acetone into (25*R*)-*N*-acetyl-3α,5α-cyclo-22α*N*-spirosolan-6β-ol. On oxidation with chromium trioxide in pyridine, the corresponding 6-ketone **170** was obtained. When treated with methylmagnesium bromide in refluxing benzene, **170** afforded a mixture of 6-epimeric (25*R*)-6-methyl-3α,5α-cyclo-22α*N*-spirosolan-6-ols. Cleavage of the *N*-acetyl group had occurred in the course of the reaction. Treatment with glacial acetic acid at reflux temperature gave rise to (25*R*)-3β-acetoxy-6-methyl-22α*N*-spirosol-5-ene, which was further converted to the alcohol **171**. Compounds **170** and **171** were degraded to the corresponding pregnane derivatives (*222–224*).

Pseudodiosgenin *O*(26)-toluenesulfonate (**172**) reacted with methylamine to give the 26-methylaminofurostadienol **173**. Acid-catalyzed cyclization of **173** afforded **174** with an MS indicative of a spiroaminoketal (spirosolane) structure (prominent peaks at *m/e* 128, 152, and 427; see Table VI). Compound **174** was shown to differ from a previously synthesized *N*-methylsolasodine; both compounds seem to differ in stereochemistry (C-22?) (*225*). Isomerism of *N*-methyl- and *N*-formylsolasodine derivatives caused by nitrogen inversion is discussed by Kusano *et al.* (*226*).

The Beckmann rearrangement of the oximes **175**–**177** leading to the lactams **178**–**181**, respectively, was studied. Reduction of **180** and **181** via LAH gave the corresponding azepines, 3- and 4-aza-A-homospirosolane (*227*, *228*). Lactones **182** and **183** were obtained by Baeyer–Villiger reaction of **184** and **185** (*229*). Ozonolysis of solasodine (**1**) and solasodenone (**12**) furnished the ketoacids **186** and **187**, respectively, which were lactonized by heating in acetyl chloride–acetic anhydride to yield **188** and **189** (*230*, *231*). The ketoacid **187** was also obtained by oxidation of solasodenone (**12**) with periodate and permanganate (*232*).

Reaction of solasodenone (**12**) and the saturated ketones **184** and **185** with ethyl formate in the presence of sodium methoxide followed by treatment with hydroxylamine or hydrazine gave isoxazolospirosolanes **193**–**195** or pyrazolospirosolanes **196**–**198**, respectively. Derivatives substituted in the new heterocycle were also prepared (*233*, *234*). (25*R*)-5α,22α*N*-Spirosolan-3-one (**184**) was brominated with bromine in acetic acid. The bromoketone **199**, when reacted with thiourea, provided the 2-aminothiazolospirosolane **200**. Derivatives substituted at the amino group were also synthesized (*235*).

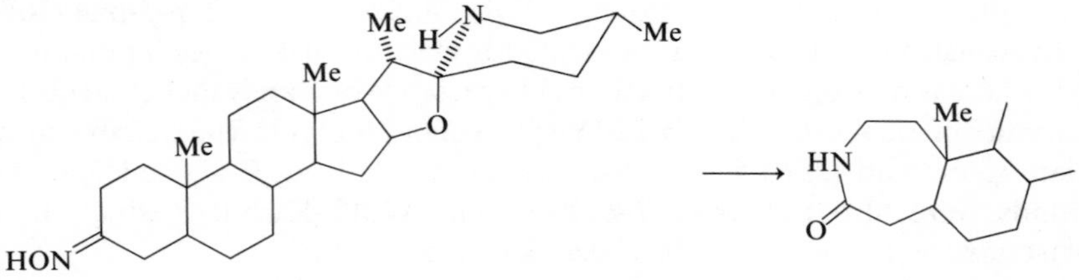

175 Δ^4
176 5βH

178 Δ^{4a}
179 5βH

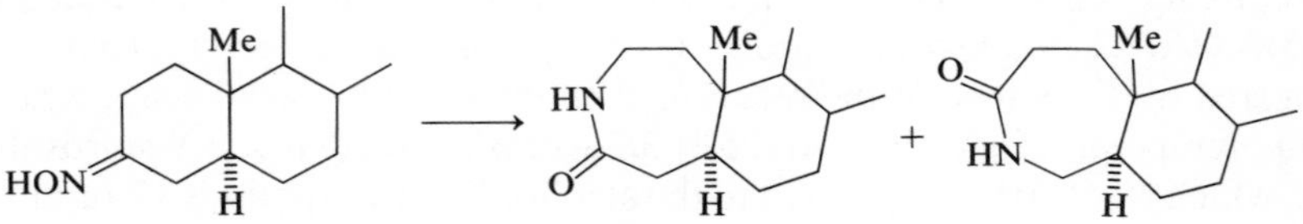

177 **180** **181**

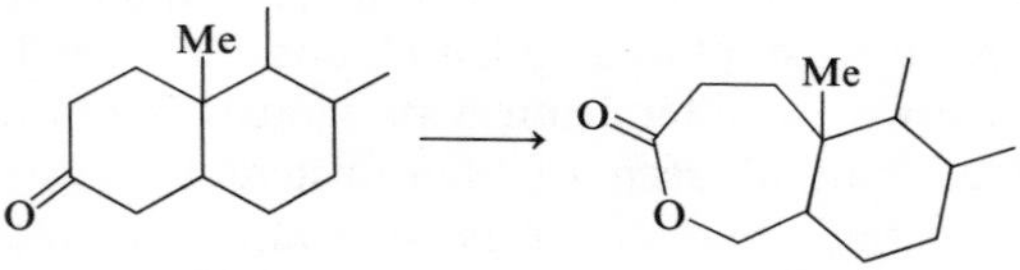

184 5αH
185 5βH

182 5αH
183 5βH

Recently, a synthesis of an E-homo-F-norspirosolane has been described (*236*). By Grignard reaction and reacetylation, 3β,16α-diacetoxy-5α-pregnan-20-one (**201**) was converted to a mixture of the 20-epimers **202**. Refluxing with formic acid gave the aldehyde **203** (mixture). Oxidation of **203** afforded acid **204**, which could be separated from its 20-stereoisomer. Treatment of **204** with $SOCl_2$ gave the acyl chloride which, on reaction with pyrrylmagnesium iodide and subsequent reacetylation, led to the pyrryl derivative **205**. Compound **205**, subjected to catalytic reduction and hydrolysis, simultaneously underwent reduction of the pyrrole nucleus and hydrogenolysis of the carbonyl to a methylene group, yielding **206** as a mixture of the 23-epimers. N-Chlorination followed by treatment with sodium methoxide gave the spiroaminoketal **207**, which is not sterically homogeneous.

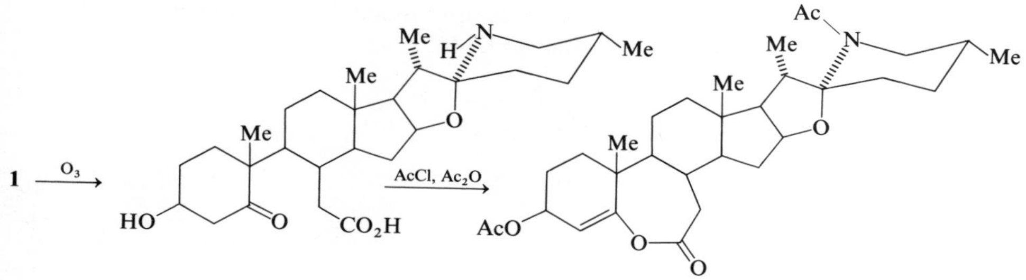

1 $\xrightarrow{O_3}$ **186** $\xrightarrow{AcCl,\ Ac_2O}$ **188**

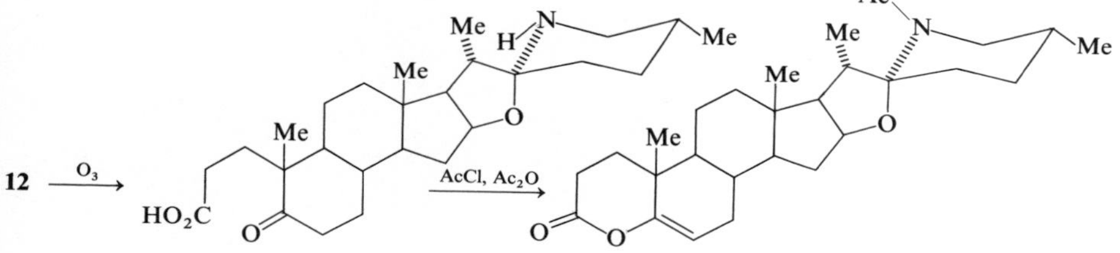

12 $\xrightarrow{O_3}$ **187** $\xrightarrow{AcCl,\ Ac_2O}$ **189**

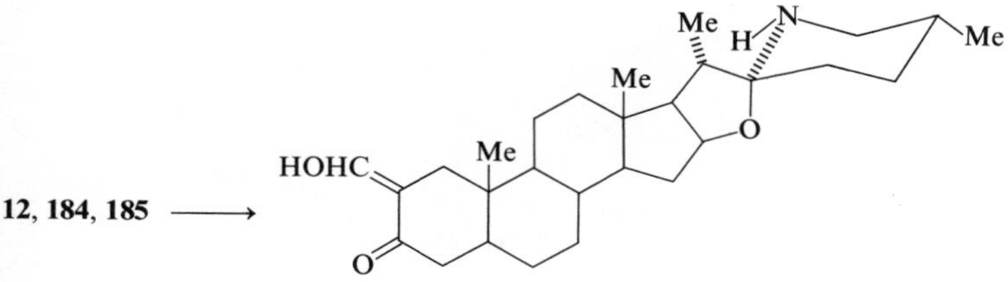

12, **184**, **185** $\longrightarrow$

190 Δ^4
191 5αH
192 5βH

NH_2OH / NH_2NH_2

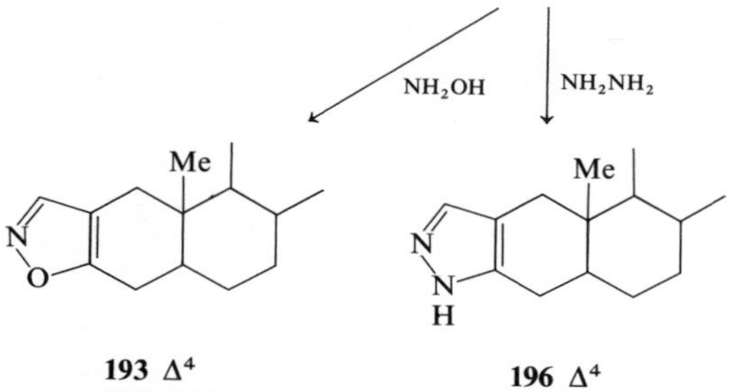

193 Δ^4
194 5αH
195 5βH

196 Δ^4
197 5αH
198 5βH

Cyclization of 16β-hydroxylated 22,26-epiminocholestanes to spirosolanes by manganese dioxide can be achieved (*360*).

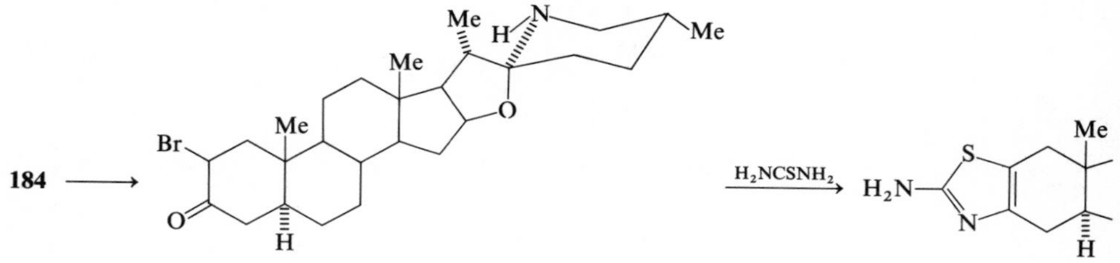

199 **200**

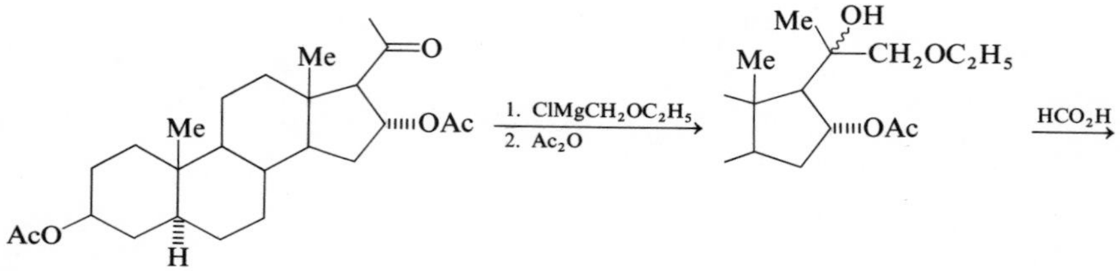

201 **202**

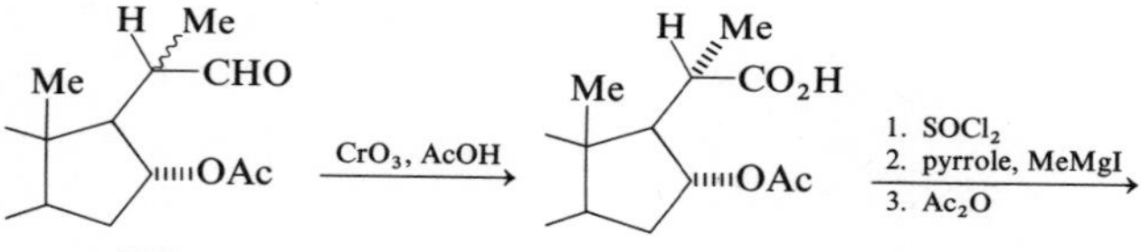

203 **204**

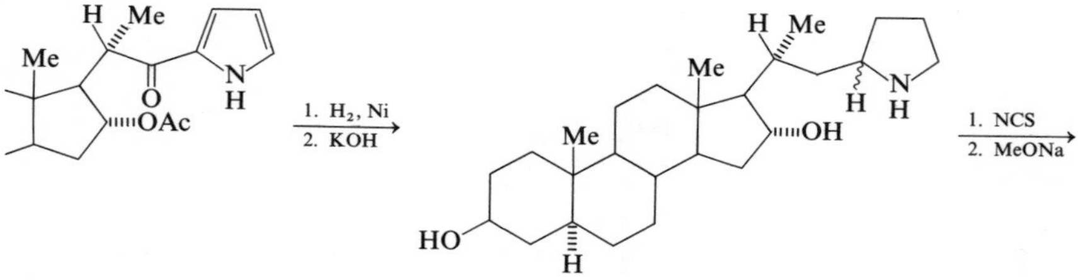

205 **206**

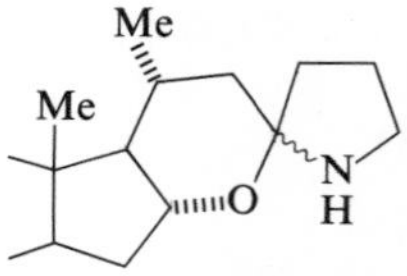

207

6. 22,26-Epiminocholestanes

Spirosolanes are known to be reduced to ring-E-opened 22,26-epiminocholestanes by catalytic hydrogenation or by reaction with LAH with or without $AlCl_3$ (*3*). They also can be easily converted by means of KBH_4 (*237*) or $NaBH_4$ (*203*); e.g., solasodine (**1**) is reduced to **107** (see Section IV, B,2) and tomatidenol (**101**) to a mixture of **208** and **209** (*203*). By catalytic hydrogenation of (20*R*)-3β-acetoxy-20-(5-methyl-3-pyridyl)pregn-5-en-20-ol (**210**), all four theoretically possible stereoisomers (**211**–**214**) were obtained

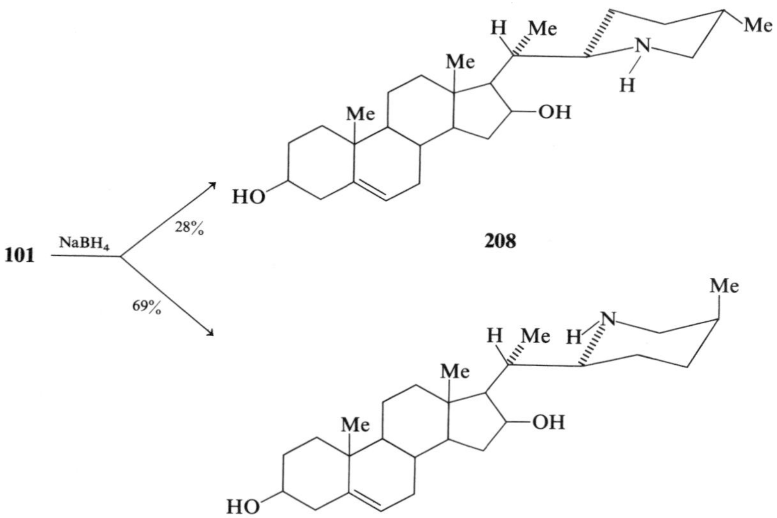

101 —NaBH₄→ 28% **208**; 69% **209**

208

209

(*238*). Hydrogenation of the 16α-acetoxy steroid **215** gave the (22*S*,25*S*)-epiminocholestane **216** as the main product and, as a consequence of partial hydrolysis during column chromatography, the (22*S*,25*R*)-compound **217** with a 16α-hydroxy group as a by-product (*239*). The 16β-acetoxy steroid **218** furnished the (22*R*,25*R*)- and (22*R*,25*S*)-epiminocholestanes **219** and **220**. Triol **221** was isolated after alkaline hydrolysis of the mixture of hydrogenation products (*238*).

The structures of **211** and **216** were assigned by X-ray analysis of the corresponding diol and triol hydroiodide, respectively (*185*, *186*). The stereochemistry of the remaining epiminocholestanes was established by ^{1}H-NMR and ORD measurements (*238*, *239*). 22,26-Epiminocholestenes deuterated at positions 15 and 17 have been synthesized from solasodine (**1**) (*357*).

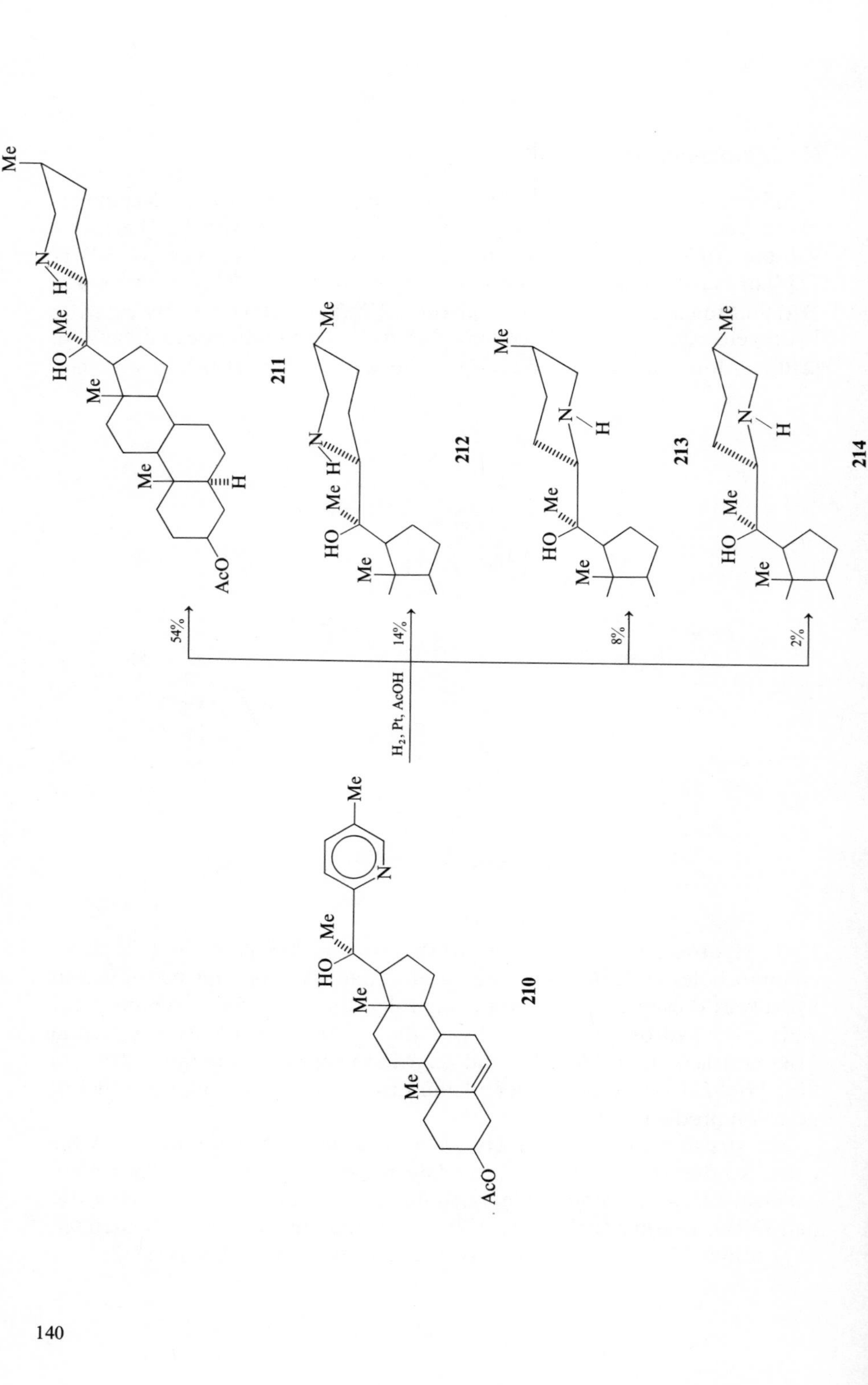

210 211 212 213 214

215

H_2, Pt, AcOH

74%

4%

216

217

218

H_2, Pt, AcOH

33%

219

6%

220

5%
KOH

221

7. Microbiological Modifications

Tomatidine (**9**) (*240, 241*), solasodine (**1**) (*242*), soladulcidine (**161**) (*242*), *N*-methyltomatidine (*243*), *N*-acetyltomatidine (*243*), (22*S*,25*S*)- and (22*R*, 25*S*)-22,26-epimino-5α-cholestane-3β,16β-diol (*244*), demissidine (*243*), and (25*S*)-26-acetylamino-5α,22α*H*-furostan-3β-ol (*244*) can be dehydrogenated to the corresponding 1,4-diene-3-keto derivatives by *Nocardia restrictus*. In general, however, the yields are low except in the transformations of tomatidine, solasodine, and the epiminocholestanes mentioned. (25*S*)-22β*N*-Spirosola-1,4-dien-3-one was shown to be the major metabolite of tomatidine (**9**); minor amounts of (25*S*)-5α,22β*N*-spirosol-1-en-3-one and (25*S*)-22β*N*-spirosol-4-en-3-one were also isolated (*241*). *N*-Acetyltomatidine is mainly dehydrogenated to its saturated 3-keto derivative (*243*). Demissidine is converted to 22α*H*,25β*H*-solanid-4-en-3-one and 22α*H*,25β*H*-solanida-1,4-dien-3-one (*243*).

The primary amino group of (22ξ,25*S*)-26-amino-5α-cholestane-3β,16β, 22-triol was acetylated by *N. restrictus* (*244*). Side-chain-cleaving enzymes, induced by cholesterol in *Arthrobacter simplex*, are reported to be capable of splitting the spiroaminoketal side chain of tomatidine (**9**). Yields of 2.5%–4.0% of androsta-1,4-diene-3,17-dione were obtained in the presence of α,α′-dipyridyl (*245*). Tomatidine (**9**), *N*-acetyltomatidine, soladulcidine (**161**), and (22*R*,25*S*)-22,26-epimino-5α-cholestane-3β,16β-diol were transformed by *N. restrictus* into the corresponding 3α-hydroxy compounds (*246*).

Solanidine (**3**) was hydroxylated by *Helicostylum piriforme* to yield 11α-hydroxysolanidine. It readily formed an amorphous diacetate. Kiliani oxidation afforded two products, the first of which exhibits UV and IR absorption indicating the presence of a Δ^4-3-oxo and a six-membered ring ketone system. The second compound possesses a six-membered ketonic function. The prominent peak in the MS of the fermentation product at *m/e* 150 (see Table VI) excludes a hydroxy group in ring F. A fragment at *m/e* 220 instead of 204 in solanidine (see Table VI) and the difference from the known 12-hydroxysolanidines proved the presence of an 11-hydroxy group. From consideration both of the molecular rotation difference between the compound and solanidine (ΔM_D +2°), and of the ease of acetylation, an α-orientation was assigned to the 11-hydroxy group (*247*).

C. Degradation

1. Degradation of Spirosolanes

A simple conversion of solasodine (**1**) into 3β-acetoxypregna-5,16-dien-20-one (**222**) is relevant to the production of hormones. This has been achieved by reaction of the *N*,*O*-diacetate of **1** with acid followed by oxidation and acid-catalyzed cleavage of the side chain (*3*). Recently, in addition to 16β-acyloxy-20-oxopregnane **223** as the main product of dichromate oxidation of the intermediate **224**, the α-hydroxy ester **225** and

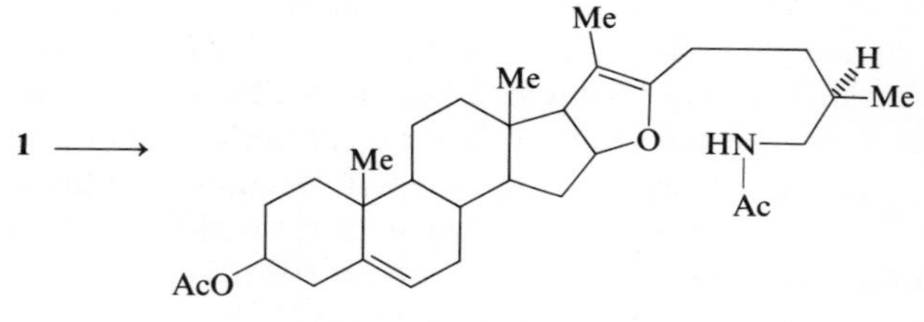

1 ⟶ **224**

$Na_2Cr_2O_7$, AcOH

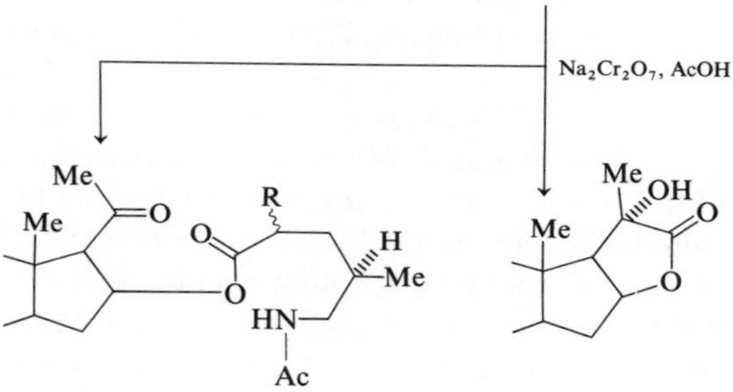

223 R = H
225 R = OH

226

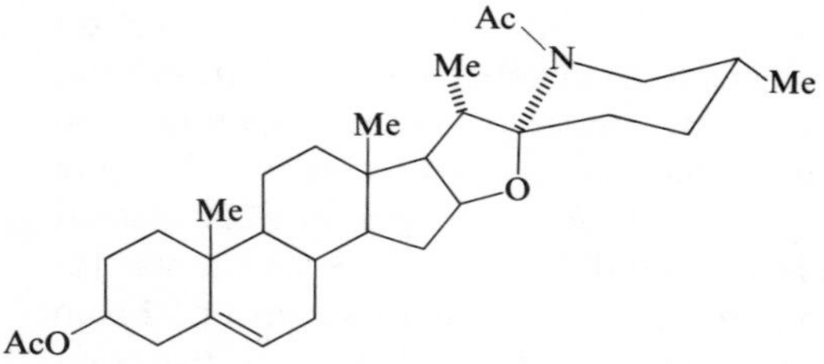

227 $\xrightarrow{Na_2Cr_2O_7}$ **228**

1. MeMgI
2. HCl

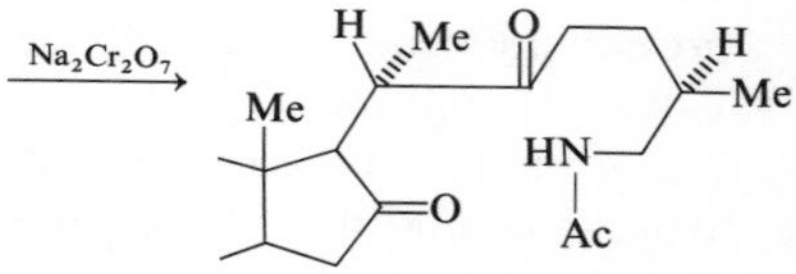

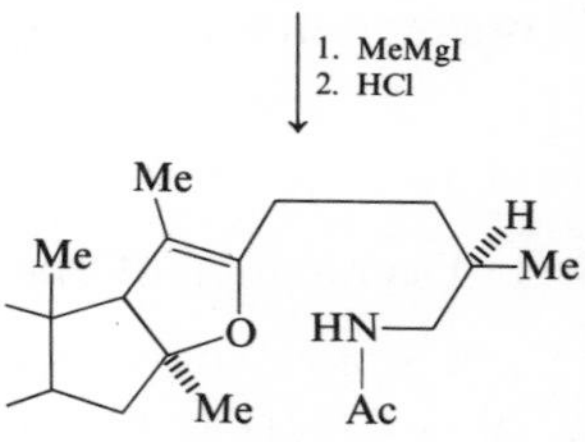

229

lactone **226** have been discovered as by-products [(*248*); concerning the kinetics, cf. (*249*)].

According to a recent paper (*250*; cf. *251*), *N*,*O*-diacetylsolasodine (**227**) was oxidized by sodium dichromate in aqueous acetic acid to the diketone **228**, which could be transformed by Grignard addition and treatment with HCl into the 16-alkyl-substituted furostadiene derivative **229**.

An alternative method involves the reaction of *O*-acetylsolasodine (**55**) with phosgene in a basic medium (*252*). In triethylamine, to which dimethylamine was later added, a major product (**230**) and two by-products (**231** and **232**) were formed. Each of these compounds was readily transformed into

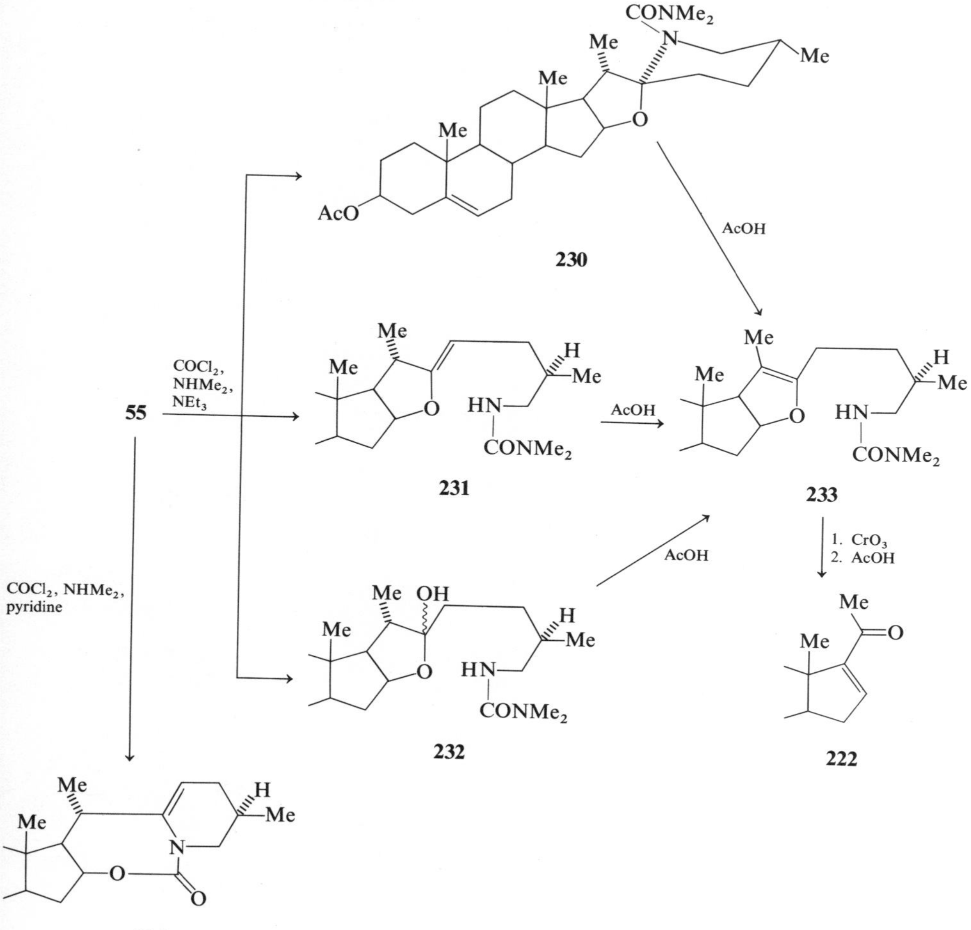

CONMe$_2$ Me Me Me Me AcO **230**

AcOH

$COCl_2$, $NHMe_2$, NEt_3

55

Me Me H Me O HN CONMe$_2$ **231**

AcOH

Me Me H Me O HN CONMe$_2$ **233**

1. CrO_3
2. AcOH

AcOH

$COCl_2$, $NHMe_2$, pyridine

Me OH Me H Me O HN CONMe$_2$ **232**

Me Me O **222**

Me Me H Me N O O **234**

furostadiene **233** by heating with glacial acetic acid. Oxidation of **233** with chromium trioxide in aqueous acetic acid, followed by treatment with boiling acetic acid, gave 3β-acetoxypregna-5,16-dien-20-one (**222**). Reaction of *O*-acetylsolasodine (**55**) with phosgene and dimethylamine in pyridine, on the other hand, gave **234** as the main product.

An improved degradation of *O*-acetyl-*N*-nitrososolasodine (**235**) to 3β-acetoxypregna-5,16-dien-20-one (**222**) was described (*253*). Treating a suspension of **235** in methanol with *p*-toluenesulfonic acid at 65°, and then neutralizing and evaporating, resulted in a residue consisting of 2% acetyldiosgenin and mixtures of hemiketals and ketals **236** and **237** (30%), **238** and **239** (15%), and **240** and **241** (30%). By heating the residue in acetic acid, derivatives **242**–**244** were formed, oxidation of which with sodium dichro-

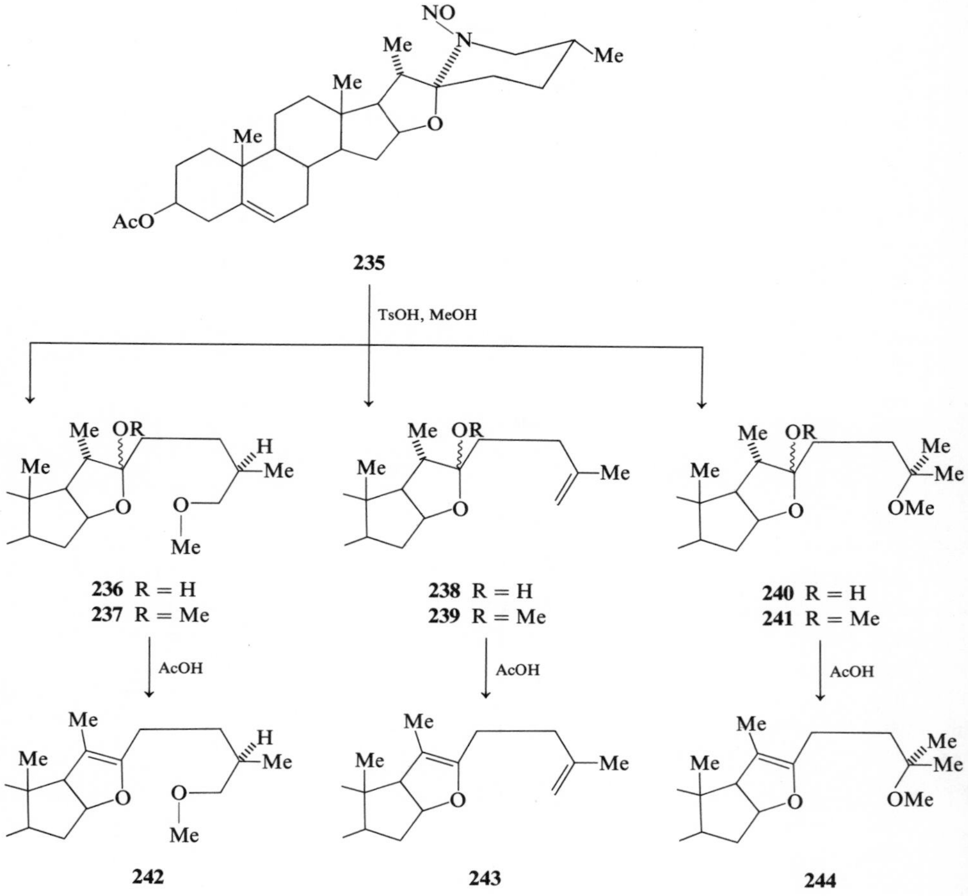

NO
Me
Me
Me
Me
O
AcO

235

TsOH, MeOH

Me OR H Me Me O O Me

236 R = H
237 R = Me

Me OR Me Me O

238 R = H
239 R = Me

Me OR Me Me Me O OMe

240 R = H
241 R = Me

AcOH

Me H Me Me O O Me

242

Me Me Me O

243

Me Me Me Me O OMe

244

mate followed by side-chain cleavage by treatment with boiling acetic acid gave **222**. Thus, solasodine was degraded in an overall yield of 60%.

2. Fragmentation of *N*-Chloro- and *N*-Nitroso-22,26-epiminocholestanes

An entirely novel method for degrading spirosolanes leads to 20-chloropregnanes in a three-step sequence (reduction, N-chlorination, and photolysis). The intermediate *N*-chloro derivatives of tetrahydrosolasodine A (**245**) as well as dihydrotomatidine A and B (**246**, **247**) can easily be obtained by reduction of natural spirosolanes (*3*) (see also Section IV,B,6) and subsequent N-chlorination using *N*-chlorosuccinimide. By UV irradiation of the *N*-chloroamines in trifluoroacetic acid followed by $NaHCO_3$ treatment of the photolysis products, high yields were afforded of a mixture of both stereoisomeric 20-chloro-5α-pregnane-3β,16β-diols (**248**, **249**), which was separated by chromatography. Compounds **248** and **249** were converted into the corresponding diacetates **250** and **251** directly obtainable by photochemical fragmentation of (22*S*,25*R*)-3β,16β-diacetoxy-22,26-chloroepimino-5α-cholestane (**252**) (*254*, *255*).

The structures of the 20-chloropregnanes were assigned by physical methods (cf. *256*, *257*), especially by X-ray analysis of the bis-4-bromobenzoyl derivative of **248** (*258*). The C_6 side-chain moieties (C-22 to C-27) which arise by photofragmentation of **245** and **246** have been identified as *R*- and *S*-5-methyl-1-piperideine, respectively. After catalytic hydrogenation, they yielded *R*- and *S*-β-pipecoline (*259*; cf. *260*). A Δ^4-3-oxo function did not prevent the photofragmentation. Thus, photolysis of (22*S*,25*R*)-22,26-chloroepimino-16β-hydroxycholest-4-en-3-one, readily accessible via partial Oppenauer oxidation of the 22,26-epiminocholest-5-ene-3β,16β-diol (**107**) and subsequent N-chlorination, proceeds to give both 20-stereoisomeric 20-chloro-16β-hydroxypregn-4-en-3-ones, which have been dehydrogenated with 2,3-dichloro-5,6-dicyano-1,4-benzoquinone to the corresponding pregna-1,4-dien-3-ones. Therefore, this reaction sequence opens a degradation pathway of solasodine (**1**) to new 20-chloropregnanes with Δ^4-and $\Delta^{1,4}$-3-keto partial structures (*261*). The *N*-chloro derivatives of the 16-unsubstituted (22*S*,25*R*)- and (22*R*,25*S*)-22,26-epimino-5α-cholestan-3β-ols formed solanidanes (*3*) by photochemically induced Hofmann–Löffler–Freytag cyclization employing reaction conditions similar to those described above for the photochemical fragmentation.

An attempt to apply this reaction to the stereoisomeric (22*S*,25*S*)- and (22*R*,25*R*)-22,26-chloroepimino-5α-cholestan-3β-ol (**253**, **254**) furnished an inseparable mixture of the stereoisomeric 20-chloro-5α-pregnan-3β-ols (**255**, **256**) in both cases. Removal of chlorine by reaction with Raney nickel led to the known 5α-pregnan-3β-ol (*255*, *262*, *263*). Treatment of

(20S)-3β-acetoxy-5α-pregnan-20-ol with $SOCl_2$ or PCl_5 afforded the stereoisomeric 3β-acetoxy-20-chloro-5α-pregnanes (*264*). By contrast, chlorination of the corresponding (20R)-alcohol did not yield 20-chloro-5α-pregnan-3β-ols, but 17a-stereoisomeric 3β-acetoxy-17a-chloro-17α-methyl-D-homo-5α-androstanes (*265*).

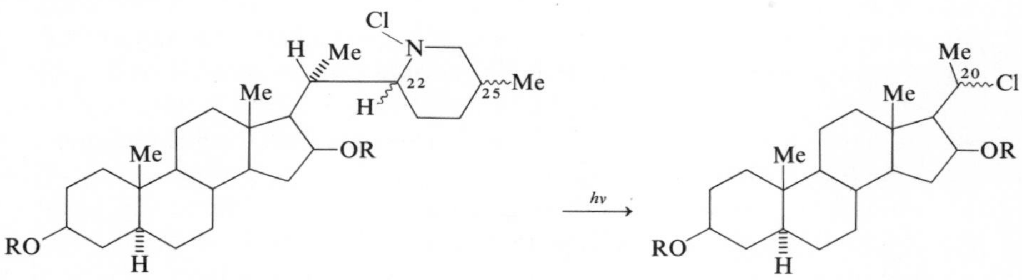

245 R = H, 22*S*,25*R*
246 R = H, 22*S*,25S
247 R = H, 22*R*,25*S*
252 R = Ac, 22*S*,25*R*

248 R = H, 20*S*
249 R = H, 20*R*
250 R = Ac, 20*S*
251 R = Ac, 20*R*

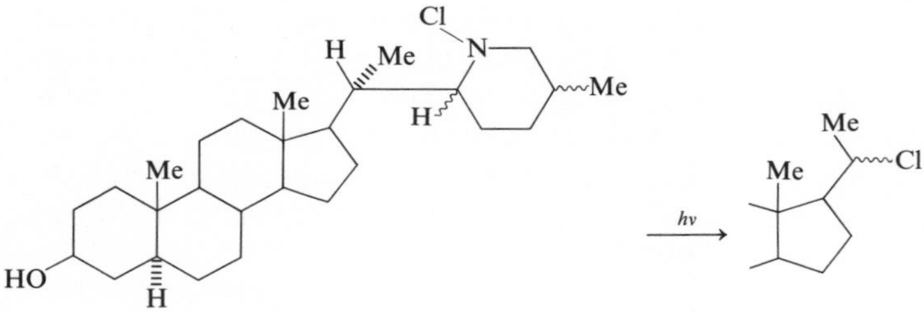

253 22*S*,25*S*
254 22*R*,25*R*

255 20*S*
256 20*R*

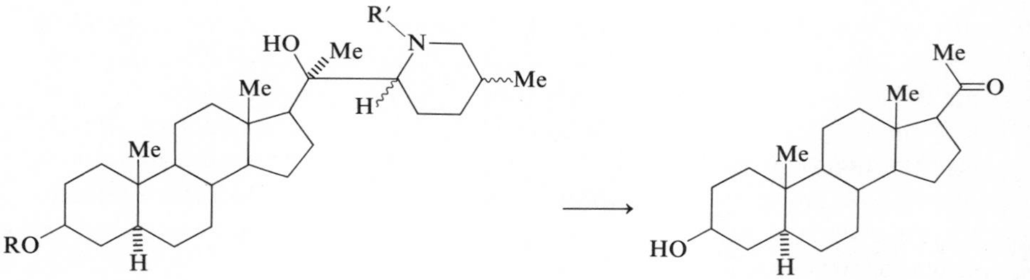

257 R = H, R′ = Cl, 22*S*,25*S*
268 R = Ac, R′ = Cl, 22*S*,25*S*
269 R = Ac, R′ = Cl, 22*S*,25*R*
270 R = Ac, R′ = Cl, 22*R*,25*R*
271 R = Ac, R′ = Cl, 22*R*,25*S*
263 R = H, R′ = NO, 22*S*,25*S*

260

Surprisingly, UV irradiation of 20-hydroxylated 22,26-chloroepiminocholestanes (**257**–**259**) led to 20-oxopregnanes (**260** and **261**; **259** gave the dehydration product **262**) (*238*, *239*). In general, 22,26-nitrosoepiminocholestanes are converted to azomethines during UV irradiation in an acidic medium (*3*); however the corresponding 20-hydroxy compounds **263**–**266** yielded the 20-oxopregnanes **260**, **261**, and **267**, respectively. Alkaline treatment of the 22,26-chloroepimino-20-hydroxycholestanes **257**–**259** and **268**–**274** did not afford the corresponding azomethines, but the

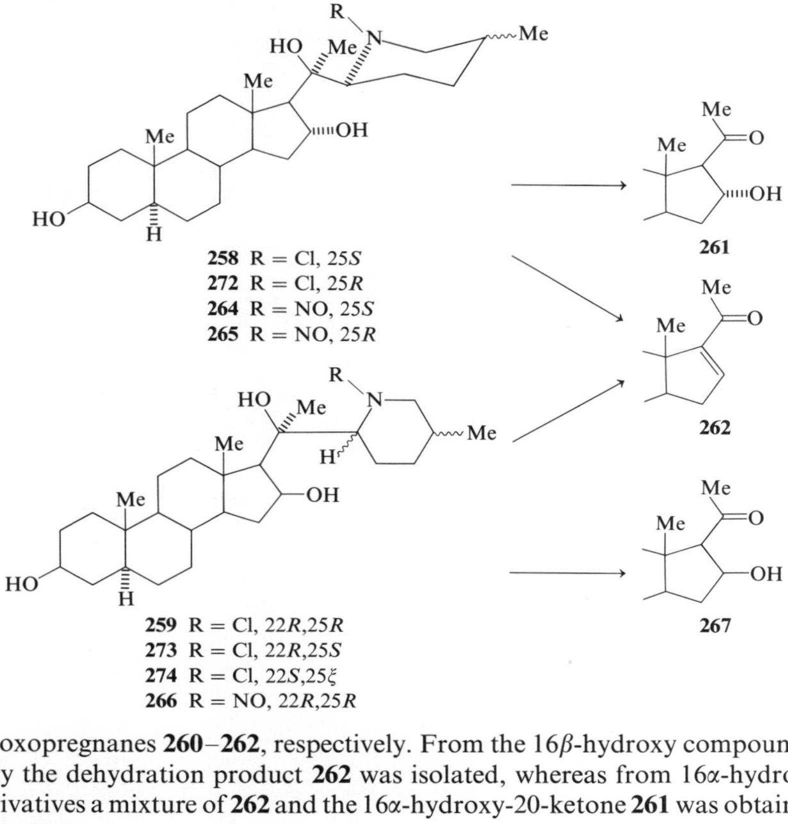

258 R = Cl, 25*S*
272 R = Cl, 25*R*
264 R = NO, 25*S*
265 R = NO, 25*R*

261

262

259 R = Cl, 22*R*,25*R*
273 R = Cl, 22*R*,25*S*
274 R = Cl, 22*S*,25*ξ*
266 R = NO, 22*R*,25*R*

267

20-oxopregnanes **260**–**262**, respectively. From the 16*β*-hydroxy compounds, only the dehydration product **262** was isolated, whereas from 16*α*-hydroxy derivatives a mixture of **262** and the 16*α*-hydroxy-20-ketone **261** was obtained (*238*, *239*).

3. Degradation of Solanidanes

Contrary to an earlier report *O*-acetyldemissidine (**275**) reacted smoothly with cyanogen bromide in refluxing chloroform to yield bromocyanamide **276**. In agreement with the postulated mechanism of the von Braun degradation, reaction of CN with the nitrogen lone pair of electrons at the *β* face of the molecule in the *S*-configuration is expected to be followed by α-attack of the bromide ion. That the displacement of the quaternized nitrogen occurs on C-16 to give a 16α-bromosteroid followed from identification of the LAH reduction products obtained from **276**. The major product was shown to be identical with known **277**, while the minor one was demissidine (**278**). Compound **276** was hydrolyzed in refluxing aqueous ethanol to afford a mixture of **275** and **278**. The 5-unsaturated *O*-acetylsolanidine behaved in an analogous manner (*266*). Reaction of the 16α-bromosteroid **276** with NaN_3 gave the 16β-azidosteroid **279**, which was reduced by LAH to the 16β-amine **280**. The corresponding 5-unsaturated compound was also

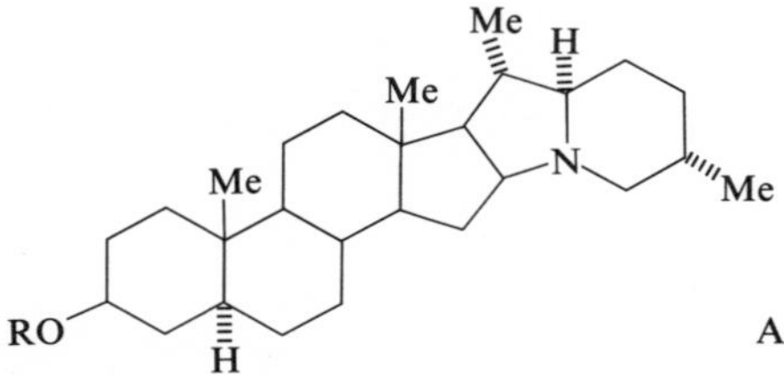

275 R = Ac
278 R = H Demissidine

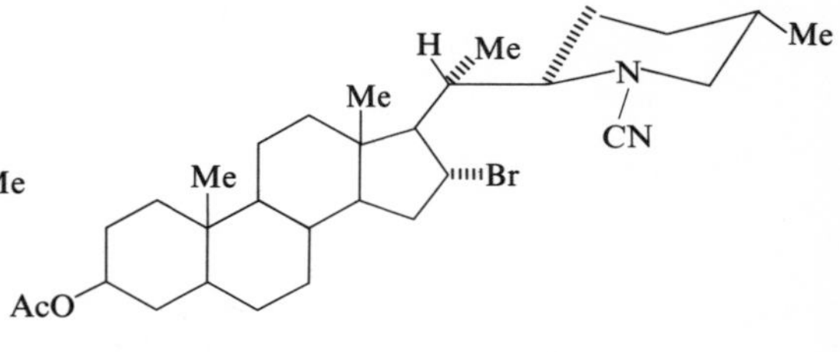

276 5αH
281 Δ^5

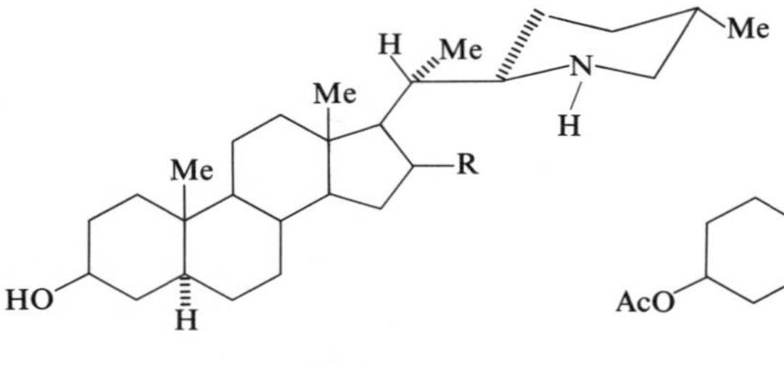

277 R = H
280 R = NH_2

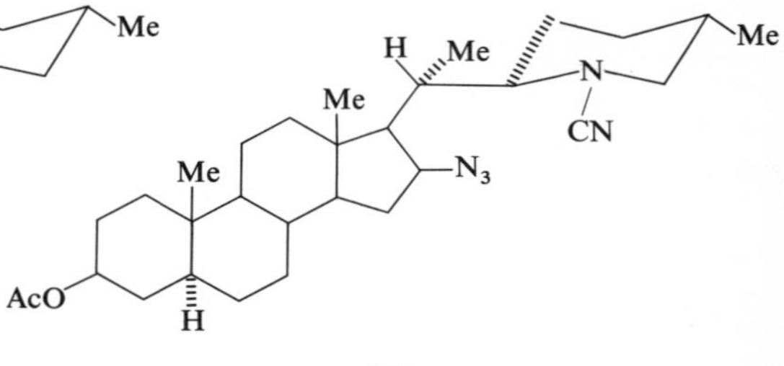

279

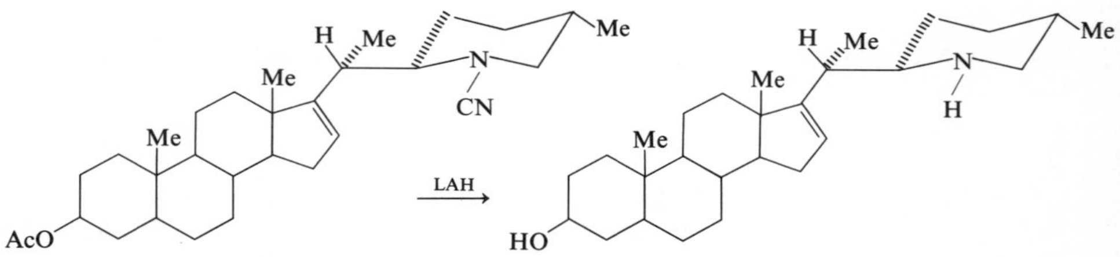

LAH →

282 5αH
283 Δ^5

284 5αH
285 Δ^5

synthesized (*267*). Treatment of **276** or the analogous Δ^5 compound **281** with refluxing dimethylaniline afforded the Δ^{16}-cyanamides **282** and **283**, respectively, which have been converted by reduction with LAH into the 22,26-epiminocholestane derivatives **284** and **285**, respectively (*268*).

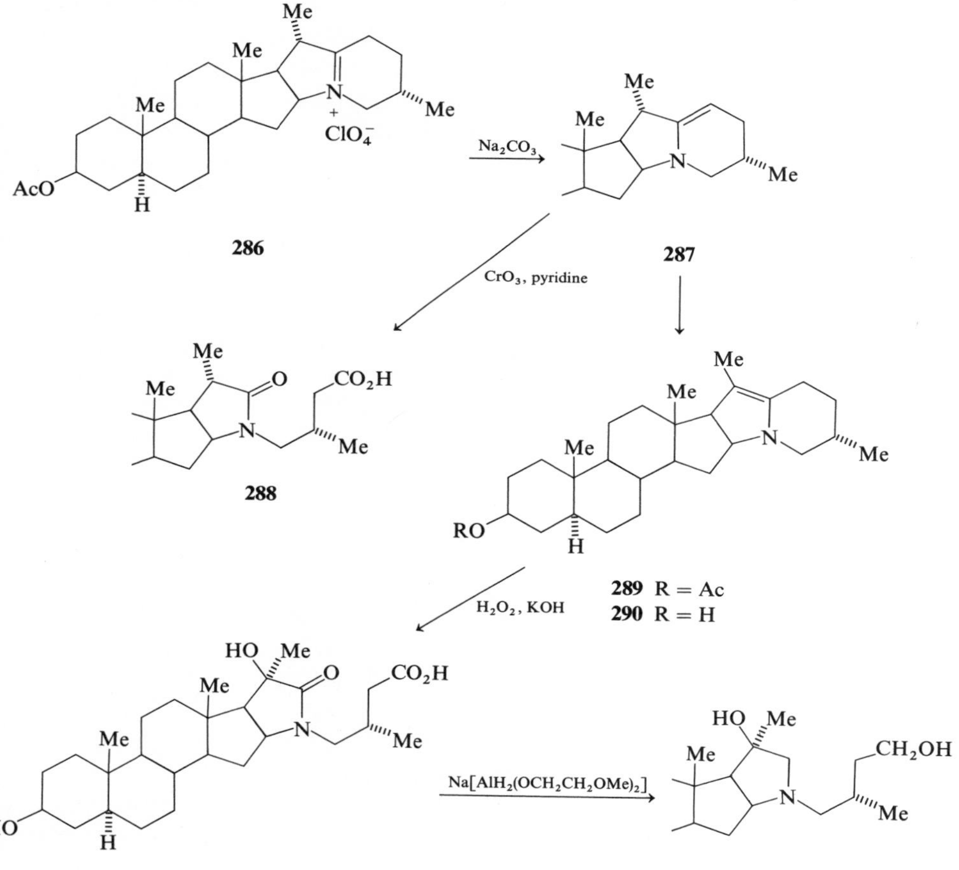

Me Me Me N + ClO_4^- Me AcO H

286

Na_2CO_3 →

Me Me N Me

287

CrO_3, pyridine

Me Me O CO_2H N Me

288

Me Me Me N Me RO H

289 R = Ac
290 R = H

H_2O_2, KOH

HO Me Me O CO_2H Me N Me HO H

291

$Na[AlH_2(OCH_2CH_2OMe)_2]$ →

HO Me Me CH_2OH N Me

292

Me Ph Me O Ph Me N Me AcO H

293

O_3 →

Me Me O CO_2H N Me

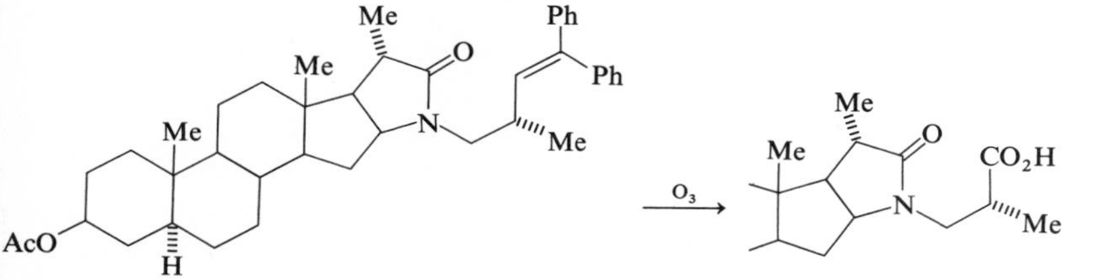

294

3β-Acetoxy-5α,25β*H*-solanid-22(*N*)-ene perchlorate (**286**) (*269*) was converted by sodium carbonate to the 22-unsaturated product **287** (*270*), which in turn was oxidized by chromium trioxide in pyridine to (25*S*)-3β-acetoxy-22-oxo-22,23-seco-5α-solanidan-23-oic acid (**288**) (*271*; cf. *269*). Compound **287** could be rearranged to the 20(22)-unsaturated compound **289** by vacuum sublimation (*272*) or by organic solvents, preferably in the presence of carboxylic acids. In alcohols, **287** was not rearranged, and in halogen-containing solvents by-products were observed (*273*). The above-mentioned reactions were also carried out starting from other imonium salts (e.g., 3β-alcohol with or without a 5-double bond, 5α-3-ketone, 5β-3-ketone, 4-unsaturated 3-ketone). 5α,25β*H*-Solanid-20(22)-en-3β-ol (**290**) was oxidized to the 20-hydroxy-22,23-secosolanidane **291**, which was reduced to triol **292** by means of sodium dihydrobis(2-methoxyethoxy)aluminate. This sequence of reactions could also be carried out in the presence of a 5-double bond (*274*).

The methyl ester of **288**, treated with phenylmagnesium bromide, followed by *O*(3)-reacetylation and dehydration of the tertiary hydroxy group of the resulting diphenylcarbinol, afforded compound **293**, which was degraded by ozonolysis to the nor-acid **294** (*275*).

V. Biochemistry

The biochemistry, especially the biosynthesis, of plant steroids in general and of steroid alkaloids in particular, has been reviewed occasionally in the last decade; for instance, see Heftmann (*276–278*), Schreiber (*35*, *279*), Schütte (*280*, *281*), and Roddick (*36*).

Following the general pathway of steroid biosynthesis in plants (*277*, *281–284*, *374*), starting from acetyl coenzyme A via the principal intermediates mevalonic acid, isopentenyl pyrophosphate, farnesyl pyrophosphate, squalene, cycloartenol, and cholesterol, the latter compound or a biogenetic equivalent of it is the precursor of both the C_{27}-steroid sapogenins and alkaloids. According to recent knowledge, the biosynthesis and metabolism of these sapogenins and alkaloids, occurring together in plants, are closely related. In addition to the spirostanols already summarized (*3*), some further nitrogen-free steroid sapogenins have been isolated from *Solanum* and *Veratrum* species containing steroid alkaloids (see Table VII).

According to earlier studies, acetate, mevalonate, and cholesterol (*3*) as well as cycloartenol (and lanosterol) (*294*) are significantly incorporated into tomatidine, solasodine, solanidine, solanocapsine, and/or spirostanols. Cholest-4-en-3-one and 26-hydroxycholesterol were shown to be the first products of cholesterol metabolism in potato plants (*295*). The (25*R*)- and

TABLE VII

NITROGEN-FREE STEROID SAPOGENINS (SPIROSTANOLS) AND RELATED COMPOUNDS IN *SOLANUM* SPECIES AND OTHER PLANTS CONTAINING C_{27}-STEROID ALKALOIDS[a]

Steroid sapogenin	Species	Reference
Solaspigenin (**295**)	*S. hispidum* Pers.	(*285*)
Neosolaspigenin (**296**)	*S. hispidum* Pers.	(*285*)
Solagenin (**297**)	*S. hispidum* Pers.	(*286*)
Hispigenin (**298**)	*S. hispidum* Pers.	(*287*)
Isocaelagenin (**299**)	"*S. jamaicense*"	(*288*)
Andesgenin (**300**)	*S. hypomalacophyllum* Bitt.	(*289*)
Barogenin (**301**)	*S. tuberosum* L.	(*290*)
Dormantinone (**302**)	*Veratrum grandiflorum* (Max.) Loesen	(*291*)
Dormantinol (**303**)	*V. grandiflorum* (Max.) Loesen	(*292*)

[a] Generally reported since 1967; supplement of the corresponding compilation (*3*).

(25*S*)-stereoisomers of the latter metabolite are converted stereospecifically into soladulcidine (**161**) (*296*) and tomatidine (**9**) (*297*, cf. *298*, *299*), respectively. 20-Hydroxycholesterol is incorporated into both tigogenin and solasodine (**1**) (*300*), whereas 16β-hydroxy- (*301*), 16β,26-dihydroxy- (*302*), and 16β-hydroxy-22-oxo-5α-cholestanol (*303*), as well as 22-oxocholesterol (*304*), are convertible into spirostanes but not into spirosolanes (*305*).

It should be mentioned again that analogous steroids barogenin (**301**), dormantinone (**302**), and dormantinol (**303**), have been isolated from *S. tuberosum* (*290*) and *V. grandiflorum* (*291*, *292*), respectively (see Table VII).

In other investigations, radioactively labeled (25*R*)-26-aminocholesterol (*306*) administered to *S. laciniatum* was incorporated to a high extent into solasodine (**1**), whereas the corresponding 16β-hydroxy derivatives (25*R*)-26-aminocholest-5-ene-3β,16β-diol and its *N*-acetyl derivative (*306*) show only small incorporations (*307*). These results suggested that in the biosynthesis of C_{27}-steroid alkaloids the introduction of nitrogen occurs imme-

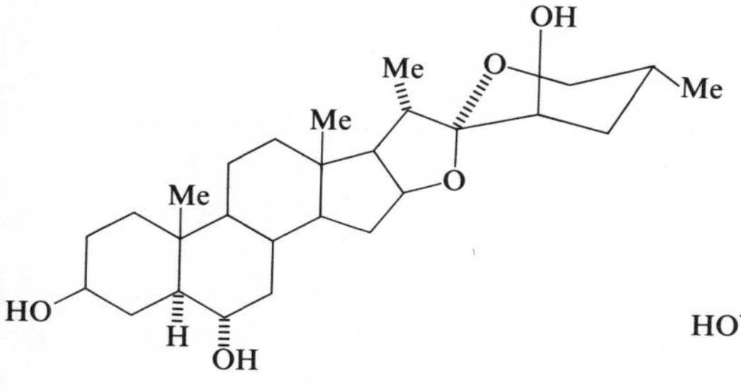

295 Solaspigenin

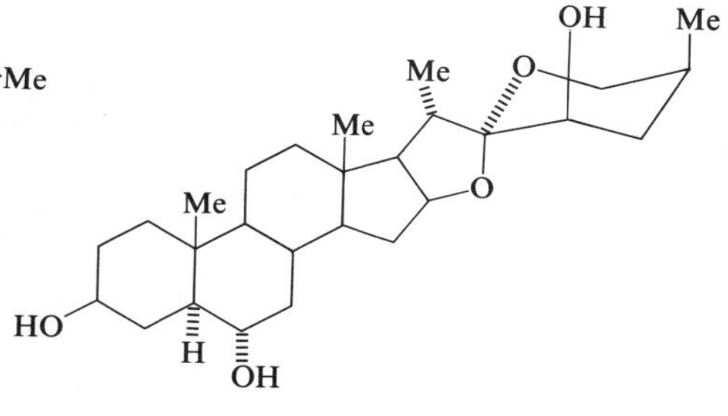

296 Neosolaspigenin

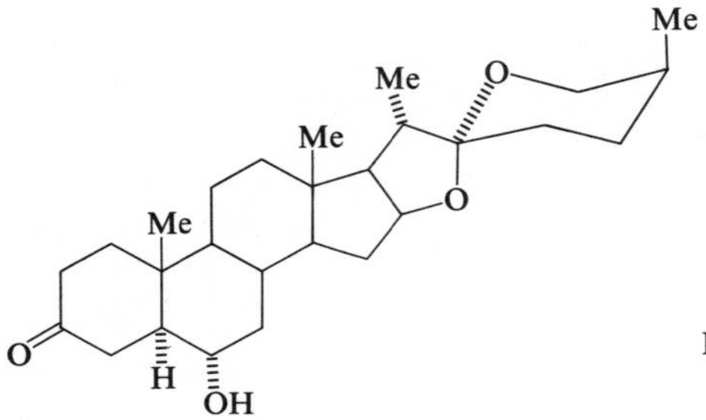

297 Solagenin

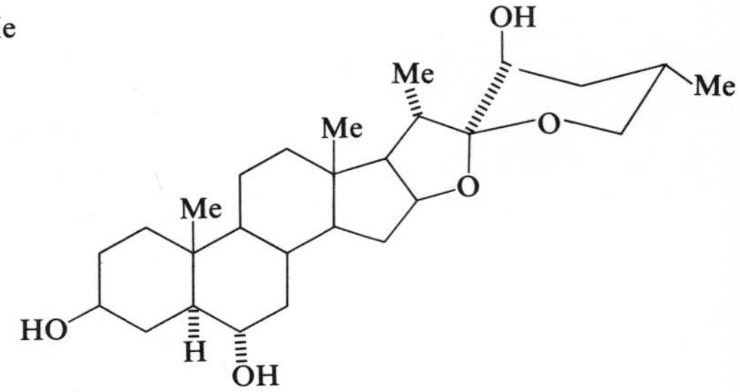

298 Hispigenin

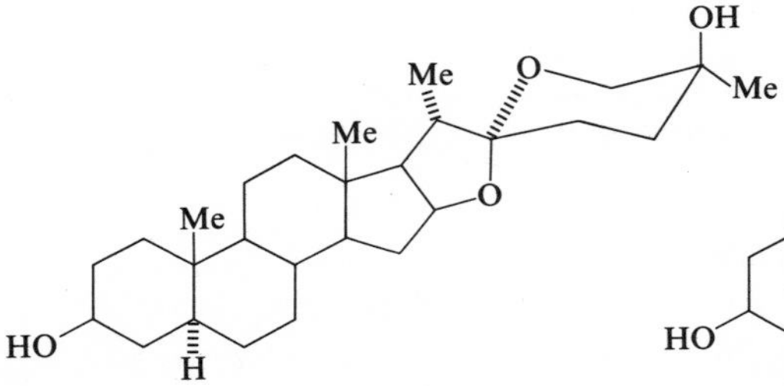

299 Isocaelagenin

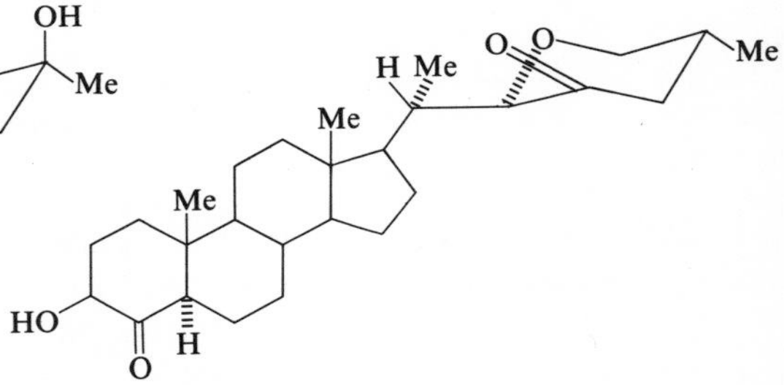

300 Andesgenin

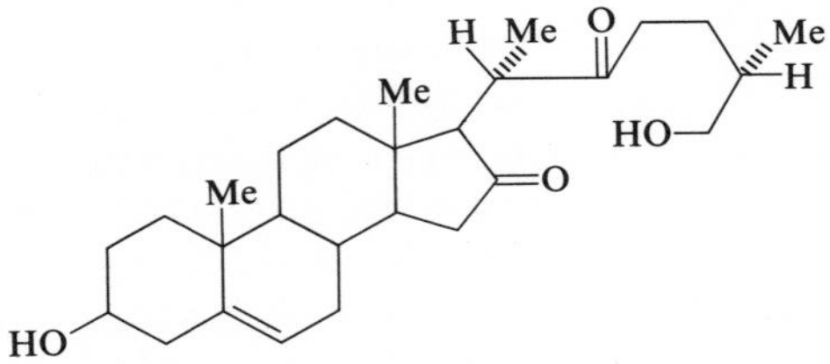

301 Barogenin

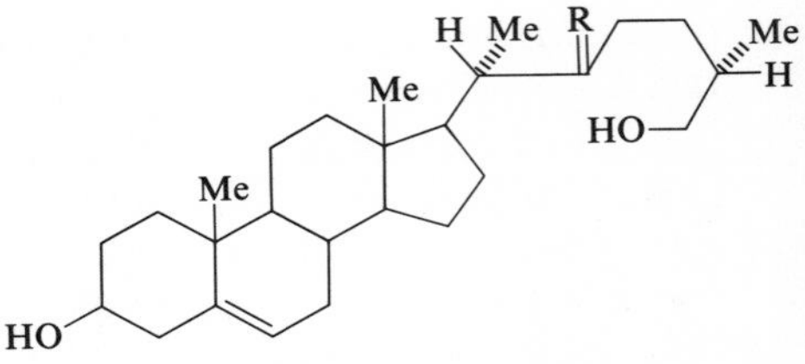

302 R = O Dormantinone

303 R = OH, H Dormantinol

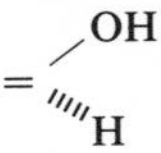

diately after the hydroxylation at C-26 (*307*; cf. *296*). In the biosynthesis of solanidine (**3**) in *V. grandiflorum*, the amino acid arginine was shown to be the nitrogen source (*308*). The weak incorporation of (25*R*)-26-aminocholest-5-ene-3β,16β-diol into solasodine (**1**) also demonstrates that the 16-hydroxy group is introduced only after the formation of ring F, i.e., after formation of the 22,26-epiminocholestane structure (*307*, *309*). This conclusion is confirmed by the isolation of 16-unsubstituted 22,26-epiminocholest-22(*N*)-ene derivatives as endogenous alkamines from plant material, e.g., solacongestidine (**2**) and verazine (**123**) (see Section IV,A,2). The 16β-hydroxy derivatives of **2** and **123** are not stable in this form, but they are stereospecifically cyclized to the spirosolanes soladulcidine (**161**) and tomatidenol (**101**), respectively (*3*). Accordingly, labeled solacongestidine (**2**) and (22*S*)-dihydrosolacongestidine administered to *S. dulcamara* as well as (22*S*,25*R*)-22,26-epiminocholest-5-en-3β-ol and its 16β-hydroxy derivative administered to *S. laciniatum*, were converted to soladulcidine (**161**) and solasodine (**1**), respectively (*305*).

The biosynthesis of alkaloids of the solanidane type via verazine (**123**), its 16α-hydroxy derivative etioline (**21**) and the corresponding (22*R*)-22,*N*-dihydro derivative teinemine (**32**) (see Section IV,A,2) in *V. grandiflorum* has been investigated by Kaneko *et al.* (*308*, *310*; cf. *292*). During the biosynthesis of tomatidine (**9**) in *Lycopersicon pimpinellifolium*, the 16β-hydrogen atom of cholesterol is inverted to the 16α-position; the same hydrogen atom is lost during the biosynthesis of solanidine (**3**) in *S. tuberosum* (*311*). Solacongestidine (**2**), its 16α-hydroxy derivative solafloridine (**41**), and/or 23-oxosolacongestidine (**39**) are considered to be precursors of solanocapsine (**4**) (*35*, *279*).

Without any doubt, the 3β-aminospirostanes [e.g., jurubidine (**5**); see Section IV,A,5) are biosynthesized analogously to the corresponding nitrogen-free spirostanols.

The distribution of solasodine glycosides in leaves, flowers, and fruits of *S. laciniatum* has been studied by Weiler *et al.* (*371*) using radioimmunoassay. The highest levels of alkaloids are found in empty anthers and in the inner pericarp of green fruits.

The subcellular localization of steroidal glycoalkaloids in the vegetative organs of *L. esculentum* and *S. tuberosum* has been studied by Roddick (*312*; cf. *313*, *314*). In both species, the alkaloids were found to accumulate mainly in the soluble fraction, with smaller amounts in the microsomal fraction. Labeled tomatine is biosynthesized in cultured excised tomato roots in the presence of [^{14}C]mevalonic acid lactone (*315*; cf. *316*).

The biosynthesis of spirosolane alkaloids in tissue cultures of *Solanum* species has been reported (*317–320*, *375–377*, cf. *321–325*). Still, other authors were able to find only diosgenin and not any solasodine (**1**) in tissue cultures of *S. laciniatum* (*326*; cf. *316*). The presence and content of spirostanols and/or spirosolanols may be dependent on the ratio of plant growth substances, e.g., 2,4-D, kinetin, and gibberellic acid (*318*). It is reported that the addition of cholesterol increases the solasodine content (*320*).

The isolation of 3β-hydroxy-5α-pregn-16-en-20-one from *L. pimpinellifolium* leads to the suggestion that this pregnane may be an endogenous degradation product of tomatidine (**9**) and that its biological breakdown occurs in a manner analogous to its chemical breakdown (cf. *3*, *35*). This hypothesis has been confirmed by Heftmann and Schwimmer (*327*; cf. *328*), who found that [4-^{14}C]tomatine, when incubated in whole ripe tomato fruits, was rapidly converted into a glycoside, probably the lycotetraoside, of 3β-hydroxy-5α-[4-^{14}C]pregn-16-en-20-one. Independently, the 3(*O*)-β-lycotetraoside of 3β-hydroxy-5α-pregn-16-en-20-one has been obtained by chemical degradation of tomatine (*329*), and the 3(*O*)-β-chacotrioside of 3β-hydroxypregna-5,16-dien-20-one has been obtained by isolation from fruits of *S. vespertilio* Ait. (*330*). The last-mentioned pregnane (aglycone) has also been isolated from *V. grandiflorum* (*331*), 3β-hydroxypregna-5,16-dien-20-one as has progesterone from *S. verbascifolium* (*349*), and—for the first time—3β-hydroxy-5α-pregnan-16-one from *S. hainanense* (*332*). In analogy to the spirosolane alkaloids and their glycosides, the nitrogen-free spirostanol and furostanol glycosides are biologically degradable, too, to Δ^{16}-pregnen-20-one derivatives (*333*, *334*; cf. *335*). As intermediates of this biodegradation, glycosides of 20,22-secofurostane derivatives have been isolated (*335*), confirming the analogy to Marker's chemical degradation procedure. A new steroid lactone, 20*S*-hydroxyvespertilin, the deacetyl derivative of **226**, has been isolated from *S. vespertilio* (*336*), possibly also a biological degradation product of spirostanes and/or spirosolanes.

The enzymatic glycosylation of steroid alkaloids by extracts from *S. laciniatum* (*337*) as well as in the potato tuber (*338*) has been investigated. A fungal β-glucosidase capable of hydrolyzing α-tomatine and demissine has been isolated and purified and its properties studied (*339*).

VI. Tables of Physical Constants

The following tables list all the glycosides (Table VIII), sugar-free alkamines (Tables IX–XVI) and their derivatives possessing spirosolane (Table IX), epiminocholestane (Table X), 26-aminofurostane (Table XI), solanidane (Table XII), solanocapsine (Table XIII), and 3-aminospirostane structures (Table XIV) reported between 1967 and 1979. Table XV includes further nitrogenous steroids with other basic skeletons derived from the above-mentioned compounds or obtained as intermediates in the course of their syntheses. Table XVI compiles alkaloids the structures of which are not completely known. Substances already discussed in Volumes III, VII, or X of this series are only mentioned where alterations require a repeated citation (noted in footnotes to the tables). Wherever a 16β,23-epoxy structure was mentioned in Volume X, Chapter 1, Table XIII, the prefix of the names must be changed into "16α,23-epoxy." Tables IX–XV list nonglycosidic compounds classified according to their basic skeleton, arranged according to their molecular formulas in order of increasing number first of C atoms, then of H and X atoms (X indicates other elements in alphabetical order). Isomeric compounds are listed alphabetically. When they differ merely in the position of substituents, they are arranged in ascending order of the position in question. In the case of stereoisomers, α is ordered before β, and *S* before *R*. Compounds for which neither melting point nor specific rotation was given have not been included in the tables.

TABLE VIII

GLYCOSIDES AND THEIR DERIVATIVES

Compound	Formula; melting point (°C); [α]D (solvent) (reference)
Solasodine glycosides	
γ-Solasonine	$C_{33}H_{53}NO_7$; 265–266; −78° (MeOH) (*160*)
N-Nitroso derivative	$C_{33}H_{52}N_2O_8$; 253–254 (*160*)
Tetra-*O*-acetyl derivative	$C_{41}H_{61}NO_{11}$; 268–270; −50° ($CHCl_3$) (*160*)
Solasodine bioside	$C_{39}H_{63}NO_{11}$; 261 (dec.); −89° (*161*)
Picrate	178 (dec.) (*161*)
Picrolonate	209–210 (dec.) (*161*)
Solamargine[a]	
N-Nitroso derivative	$C_{45}H_{72}N_2O_{16}$; 260 (*97*)
Solasodine trioside	$C_{45}H_{73}NO_{15}$; 278–279; −103° (MeOH) (*95*)
Solakhasianine	251–256 (dec.) (*90*)
Khasianine	$C_{39}H_{63}NO_{11}$; 226–228; −95° (MeOH) (*383*)
Solapersine	$C_{49}H_{79}NO_{20}$; 282–284; −48.1° (MeOH) (*102*)
Solaplumbine	$C_{39}H_{63}NO_{11}$; 180–181; −90° (MeOH) (*126*)
Hexaacetyl derivative	$C_{51}H_{75}NO_{17}$; 152–154 (*126*)
Hexamethyl derivative	$C_{45}H_{75}NO_{11}$; 138–142 (*126*)
Solaplumbinine	$C_{33}H_{53}NO_6$; 184–185, −39.5° (MeOH) (*126*)
Solaradixine[a]	$C_{51}H_{83}NO_{21}$; 275–278; −69.7° (pyridine), −51.9° (MeOH) (*54*)
Tridecamethyl derivative	$C_{64}H_{109}NO_{21}$; +43.1° (MeOH) (*54*)
Solashabanine	$C_{57}H_{93}NO_{26}$; 270–273 (*55*)
Solaradinine	$C_{63}H_{103}NO_{31}$; 227–230; −45.8° (50% EtOH) (*55*)
Solasurine	$C_{39}H_{63}NO_{11}$; 228–230 (*111*)
Picrate	165–167 (*111*)
Solatifoline	292–295, −119° (pyridine?) (*104*)
Soladulcidine glycoside	
α-Soladulcine	$C_{50}H_{83}NO_{21}$; 265–269; −56° (MeOH) (*107*)
Demissidine glycoside	
Commersonine	$C_{51}H_{85}NO_{21}$; 230–232; −17° (pyridine) (*60*)
Other glycosides	
Isojurubidine glycoside SG1	$C_{33}H_{57}NO_8$; 190–198; −60° (pyridine) (*100*)
Isojurubidine glycoside SG2	$C_{33}H_{57}NO_8$; 201–203; −30° (pyridine) (*100*)
Isojurubidine glycoside SG3	$C_{33}H_{57}NO_8$; 156–160; −53° (pyridine) (*100*)
Isojurubidine glycoside SG4	$C_{33}H_{57}NO_8$; 150–158; −32° (pyridine) (*100*)
Isojuripine (isojuripidine glycoside SG5)	$C_{33}H_{57}NO_9$; 162–171; −20° (pyridine) (*100*)
Solacongestine	$C_{39}H_{65}NO_{11}$ (*61*)
α-Solacongestinine	$C_{44}H_{73}NO_{14}$ (*61*)
β-Solacongestinine	$C_{38}H_{63}NO_{10}$ (*61*)
Glucoveracintine	$C_{32}H_{51}NO_6$; $[\alpha]_{546}$ +26° (MeOH) (*140*)
Edpetilinine	$C_{33}H_{55}NO_6$; 262–264; −10.9° (pyridine) (*131*)
Veralosine	$C_{35}H_{55}NO_8$; 213–215; −147.7° (MeOH) (*141*)
Rhinoline	$C_{34}H_{57}NO_7$; 255–257; −53.2° (EtOH) (*350*)
Veralodinine	$C_{35}H_{53}NO_9$; 226–228; −95.4° ($CHCl_3$) (*365*)

[a] See introductory paragraph of this Section.

TABLE IX
SPIROSOLANE DERIVATIVES

Formula	Compound; melting point (°C); $[\alpha]_D$ (solvent) (reference)
$C_{27}H_{39}NO_2$	(25*S*)-22β*N*-Spirosola-1,4-dien-3-one; 242–245; +16° ($CHCl_3$) (*241*)
$C_{27}H_{39}NO_2$	(25*R*)-22α*N*-Spirosola-1,4-dien-3-one; 210–214; −61° ($CHCl_3$) (*242*)
$C_{27}H_{40}ClD_2NO_2$	(25*R*)-*N*-Chloro-15ξ,17α-dideutero-22α*N*-spirosol-5-en-3β-ol; 154–155 (*357*)
$C_{27}H_{40}ClNO_2$	(25*R*)-*N*-Chloro-22α*N*-spirosol-4-en-3-one; 168–170 (dec.); −16.0° ($CHCl_3$) (*78*)
$C_{27}H_{40}N_2O_3$	(25*R*)-*N*-Nitroso-22α*N*-spirosol-4-en-3-one; 159; +29.0° ($CHCl_3$) (*78*)
$C_{27}H_{41}D_2NO_2$	(25*R*)-15ξ,17α-Dideutero-22α*N*-spirosol-5-en-3β-ol; 200–202 (*357*)
$C_{27}H_{41}NO_2$	(25*S*)-5α,22β*N*-Spirosol-1-en-3-one; 179–182; +25° ($CHCl_3$) (*241*)
$C_{27}H_{41}NO_2$	(25*S*)-22β*N*-Spirosol-4-en-3-one;[a] 202–204; +53° ($CHCl_3$) (*241*)
$C_{27}H_{41}NO_2$	(25*R*)-22α*N*-Spirosol-4-en-3-one (solasodenone) (**12**)[a]; 178; +28.0° ($CHCl_3$) (*78*); 183–184 (*196*); 184–185 (*221*); guanylhydrazone dihydrochloride; 246–248 (dec.) (*340*)
$C_{27}H_{42}N_2O_2$	(25*R*)-3-Oximino-22α*N*-spirosol-4-ene (**175**); 226–227 (*227*); 225–226 (*221*)
$C_{27}H_{43}NO$	(25*R*)-5α,22α*N*-Spirosol-2-ene; 192–195 (*341*)
$C_{27}H_{43}NO$	(25*R*)-5ξ,22α*N*-Spirosol-3-ene; 161–162 (*175*)
$C_{27}H_{43}NO$	(25*R*)-22α*N*-Spirosol-4-ene; 155–157 (*175*); 129–131 (*342*)
$C_{27}H_{43}NO$	(25*R*)-22α*N*-Spirosol-5-ene; 159–161 (*343*)
$C_{27}H_{43}NO_2$	(25*R*)-5β,22α*N*-Spirosolan-3-one (**168**); 164–165 (*196*); 160–161 (*221*)
$C_{27}H_{43}NO_3$	(25*R*)-22α*N*-Spirosol-5-ene-3β,12β-diol (solanaviol) (**13**); 229–232.5; −113° ($CHCl_3$) (*53*)
$C_{27}H_{44}N_2O_2$	(25*R*)-3-Oximino-5β,22α*N*-Spirosolane; 156–157 (*221*)
$C_{27}H_{45}NO$	(25*R*)-5α,22α*N*-Spirosolane; 171–171.5 (*175*); 149–151 (*342*)
$C_{27}H_{45}NO$	(25*R*)-5β,22α*N*-Spirosolane; 134–136 (*175*); 141–143 (*342*)
$C_{27}H_{45}NO_2$	(25*S*)-5α,22β*N*-Spirosolan-3α-ol; 218–223; +8° ($CHCl_3$) (*246*)
$C_{27}H_{45}NO_2$	(25*R*)-5α,22α*N*-Spirosolan-3α-ol; 230–234; −53° ($CHCl_3$) (*246*); 199–199.5 (*196*); 199–200 (*221*)
$C_{27}H_{45}NO_2$	(25*R*)-5β,22α*N*-Spirosolan-3α-ol; 206–207 (*196*); 205–206 (*221*)
$C_{27}H_{45}NO_3$	(25*R*)-22α*N*-Spirosolane-3β,12β-diol; 131–135, 216–219.5 (*53*)
$C_{27}H_{46}N_2O$	(25*S*)-3β-Amino-5α,22β*N*-spirosolane (soladunalinidine) (**8**); amorphous; +1.3° ($CHCl_3$) (*68, 356*); picrate, 235–255 (*68*)
$C_{28}H_{40}N_2O_2$	(25*R*)-22α*N*-Spirosola-2,4-dien[2,3-*d*]isoxazol (**193**); 214–215 (*233*)
$C_{28}H_{41}NO_2$	(25*S*)-*N*-Methyl-22β*N*-spirosola-1,4-dien-3-one; 180–182 (*243*)
$C_{28}H_{41}NO_3$	(25*R*)-2-Hydroxymethylene-22α*N*-spirosol-4-en-3-one (**190**); 212–214 (*233*)
$C_{28}H_{41}N_3O$	(25*R*)-22α*N*-Spirosola-2,4-diene[2,3-*d*]pyrazol (**196**); 311–313 (*233*)
$C_{28}H_{42}N_2O_2$	(25*R*)-5α,22α*N*-Spirosol-2-en[2,3-*d*]isoxazol (**194**); 228–229 (*233*)
$C_{28}H_{42}N_2O_2$	(25*R*)-5β,22α*N*-Spirosol-2-en[2,3-*d*]isoxazol (**195**); 234–235 (*233*)
$C_{28}H_{43}NO_3$	(25*R*)-2-Hydroxymethylene-5α,22α*N*-spirosolan-3-one (**191**); 219–220 (*233*)
$C_{28}H_{43}NO_3$	(25*R*)-2-Hydroxymethylene-5β,22α*N*-spirosolan-3-one (**192**); 248–250 (*233*)

(*Continued*)

TABLE IX (*Continued*)

Formula	Compound; melting point (°C); $[\alpha]_D$ (solvent) (reference)
$C_{28}H_{43}N_3O$	(25*R*)-5α,22α*N*-Spirosol-2-ene[2,3-*d*]pyrazol (**197**); 238–240 (*233*)
$C_{28}H_{43}N_3O$	(25*R*)-5β,22α*N*-Spirosol-2-ene[2,3-*d*]pyrazol (**198**); 294–295 (*233*)
$C_{28}H_{43}N_3OS$	(25*R*)-2′-Amino-5α,22α*N*-spirosol-2-ene[3,2-*d*]thiazol (**200**); 305–306 (*235*)
$C_{28}H_{45}NO_2$	(25*R*)-6-Methyl-22α*N*-spirosol-5-en-3β-ol (**171**); 198–200; −93° ($CHCl_3$) (*222*)
$C_{28}H_{45}NO_2$	(25*R*)-*N*-Methyl-22α*N*-spirosol-5-en-3β-ol (**174**) (?); 115/160; −58° ($CHCl_3$) (*225*); 128–133; −121.3° ($CHCl_3$) (*357*); methiodide: 245–250 (*225*)
$C_{28}H_{45}NO_2$	(25*R*)-*N*-Methyl-22β*N*-spirosol-5-en-3β-ol (?); 180–185; −20° ($CHCl_3$) (*225*)
$C_{29}H_{39}F_3N_2O_2$	(25*R*)-3′-Trifluoromethyl-22α*N*-spirosola-2,4-dien[2,3-*d*]pyrazol; 273–275 (*234*)
$C_{29}H_{40}F_3NO_3$	(25*R*)-2ξ-Trifluoroacetyl-22α*N*-spirosol-4-en-3-one; 140–142 (*234*)
$C_{29}H_{40}F_3N_3O$	(25*R*)-3′-Trifluoromethyl-22α*N*-spirosola-2,4-diene[2,3-*d*]pyrazol; 228–230 (*234*)
$C_{29}H_{41}NO_3$	(25*S*)-*N*-Acetyl-22β*N*-spirosola-1,4-dien-3-one; 271–274 (*243*)
$C_{29}H_{43}D_2NO_3$	(25*R*)-*N*-Acetyl-15ξ,17α-dideutero-22α*N*-spirosol-5-en-3β-ol; 212–213 (*357*)
$C_{29}H_{43}NO_3$	(25*R*)-*N*-Acetyl-3α,5α-cyclo-22α*N*-spirosolan-6-one (**170**); 182–185, +17° ($CHCl_3$) (*222*)
$C_{29}H_{43}NO_3$	(25*R*)-*N*-Acetyl-22α*N*-spirosol-4-en-3-one; 124; +40.1° ($CHCl_3$) (78); 219–220 (*221*)
$C_{29}H_{43}NO_4$	(25*R*)-3β-Acetoxy-22α*N*-spirosol-5-en-12-one; 198–204 (*53*)
$C_{29}H_{43}N_3O$	(25*R*)-1′-Methyl-22α*N*-spirosola-2,4-diene[2,3-*d*]pyrazol; 202–203 (*234*)
$C_{29}H_{45}NO_3$	(25*R*)-*N*-Acetyl-3α,5α-cyclo-22α*N*-spirosolan-6β-ol; 144–145 (175); 166–167, +14° ($CHCl_3$) (*222*)
$C_{29}H_{45}NO_3$	(25*R*)-*N*-Acetyl-5β,22α*N*-spirosolan-3-one; 210–211 (*221*)
$C_{29}H_{45}NO_4$	(25*R*)-3β-Acetoxy-22α*N*-spirosol-5-en-12β-ol; 225–229 (*53*)
$C_{29}H_{46}N_2O_3$	(25*R*)-*N*-Acetyl-3-oximino-5β,22α*N*-spirosolane; 245–246 (*221*)
$C_{29}H_{47}NOS_2$	(25*R*)-3,3-Ethylenedimercapto-5β,22α*N*-spirosolane; 241–243 (*175*)
$C_{29}H_{47}NO_3$	(25*S*)-*N*-Acetyl-5α,22β*N*-spirosolan-3α-ol; 198–202; +15° ($CHCl_3$) (*246*)
$C_{30}H_{45}NO_4$	(25*R*)-3β-Acetoxy-*N*-formyl-22α*N*-spirosol-5-ene A; 200–201 (*226*)
$C_{30}H_{45}NO_4$	(25*R*)-3β-Acetoxy-*N*-formyl-22α*N*-spirosol-5-ene B; 194–197 (*226*)
$C_{30}H_{47}NO_3$	(25*R*)-3β-Acetoxy-6-methyl-22α*N*-spirosol-5-ene; 223–224; −103° ($CHCl_3$) (*222*); 223–224; −140° ($CHCl_3$) (*223*)
$C_{30}H_{47}NO_3$	(25*R*)-3β-Acetoxy-*N*-methyl-22α*N*-spirosol-5-ene; 168–180; −40° ($CHCl_3$) (*225*)
$C_{31}H_{44}N_2O_4$	(25*R*)-3′-Ethoxycarbonyl-22α*N*-spirosola-2,4-dien[2,3-*d*]isoxazol; 274–275 (*234*)
$C_{31}H_{45}NO_5$	(25*R*)-2ξ-Ethoxyglyoxyl-22α*N*-spirosol-4-en-3-one; 154–155 (*234*)

(*Continued*)

TABLE IX (*Continued*)

Formula	Compound; melting point (°C); $[\alpha]_D$ (solvent) (reference)
$C_{31}H_{45}N_3O_3$	(25*R*)-3′-Ethoxycarbonyl-22α*N*-spirosola-2,4-diene[2,3-*d*]pyrazol 226–228 (*234*)
$C_{31}H_{49}NO_4$	(25*R*)-3α-Acetoxy-*N*-acetyl-5β,22α*N*-spirosolane; 122–123 (*221*)
$C_{31}H_{50}N_2O_3$	(25*S*)-*N*-Acetyl-3β-acetylamino-5α,22β*N*-spirosolane (**10**); 286–288, +4.6° ($CHCl_3$) (*68, 356*)
$C_{32}H_{49}NO_4$	(25*R*)-3β-Acetoxy-*N*-acetyl-6-methyl-22α*N*-spirosol-5-ene; 151–153; −64° ($CHCl_3$) (*222*)
$C_{33}H_{50}N_2O_2$	(25*R*)-*N*-Acetyl-3-(1-pyrrolidyl)-22α*N*-spirosola-3,5-diene; 245–248 (*175*)
$C_{34}H_{45}N_3O$	(25*R*)-1′-Phenyl-22α*N*-spirosola-2,4-diene[2,3-*d*]pyrazol; 216–218 (*234*)
$C_{34}H_{48}N_2O_4S$	(25*R*)-3-[O-(4-Toluenesulfonyl)oximino]-22α*N*-spirosol-4-ene; 220 (*227*)
$C_{34}H_{50}N_2O_2$	(25*S*)-3β-Salicylidenamino-5α,22β*N*-spirosolane; 227–229; +18.1° ($CHCl_3$) (*68, 356*)
$C_{34}H_{50}N_2O_2$	(25*S*)-3β-(4-Hydroxybenzylidenamino)-5α,22β*N*-spirosolane; 238–240; + 12.2° ($CHCl_3$) (*68, 356*)
$C_{35}H_{48}N_4OS$	(25*R*)-2′-(2-Benzylidenehydrazino)-5α,22α*N*-spirosol-2-ene[3,2-*d*]-thiazol; 298–300 (*235*)
$C_{36}H_{51}NO_5S$	(25*R*)-*N*-Acetyl-3β-(4-toluenesulfonyloxy)-22α*N*-spirosol-5-ene (**169**); 138–142; −40° ($CHCl_3$) (*222*)
$C_{37}H_{50}N_4OS$	(25*R*)-2′-[2-(3-Phenylprop-2-enylidene)hydrazino]-5α,22α*N*-spirosol-2-ene[3,2-*d*]thiazol; 270–271 (*235*)
$C_{38}H_{52}N_4OS$	(25*R*)-2′-[2-(1-Methyl-3-phenylprop-2-enylidene)hydrazino]-5α,22α*N*-spirosol-2-ene[3,2-*d*]thiazol; 189–191 (*235*)
$C_{41}H_{52}N_4OS$	(25*R*)-2′-[2-Diphenylmethylene)hydrazino]-5α,22α*N*-spirosol-2-ene-[3,2-*d*]thiazol; 258–259 (*235*)

[a] See introductory paragraph of this Section.

TABLE X
EPIMINOCHOLESTANE DERIVATIVES[a]

Formula	Compound; melting point (°C); $[\alpha]_D$ (solvent) (reference)
$C_{26}H_{39}NO$	(22ξ)-22,25-Epimino-27-norcholesta-4,25(*N*)-dien-3-one (**65**); 153–157; +107° ($CHCl_3$), +121° (EtOH) (*139*)
$C_{26}H_{41}NO$	(22ξ)-22,25-Epimino-27-norcholesta-5,25(*N*)-dien-3β-ol (veracintine) (**64**); 198; +9.5° ($CHCl_3$) (*139*); 196–201; +7.5° ($CHCl_3$) (*138*)
$C_{26}H_{43}NO$	(22ξ,25ξ)-22,25-Epimino-27-norcholest-5-en-3β-ol; 103–105; −27° ($CHCl_3$) (*138*)
$C_{26}H_{45}NO$	(22ξ,25ξ)-22,25-Epimino-27-nor-5α-cholestan-3β-ol; 247; +19.5° (EtOH) (*139*); 247–250; +10.7° (EtOH) (*138*)
$C_{27}H_{39}NO_2$	(20ξ,25*R*)-23,26-Epiminocholesta-4,23(*N*)-diene-3,22-dione;[b] 142–144.5; +99.0° ($CHCl_3$) (*344*; cf. *172*)
$C_{27}H_{39}NO_2$	(20ξ,25*R*)-23,26-Epiminocholesta-5,23(*N*)-diene-3,22-dione;[b] 169–172; +17.4° ($CHCl_3$) (*344*; cf. *172*)
$C_{27}H_{39}NO_2$	(20*S*)-20-(5-Methyl-2-pyridyl)pregn-5-ene-3β,16β-diol; 247–249; −155.4° ($CHCl_3$) (*220*)
$C_{27}H_{41}NO$	(20*S*)-20-(5-Methyl-2-pyridyl)-5α-pregnan-3β-ol; 261–261 (*62*)
$C_{27}H_{41}NO_2$	(20ξ,25*R*)-22,26-Epimino-3β-hydroxycholesta-5,22(*N*)-dien-23-one (**62**); 215–220 (*172*)
$C_{27}H_{41}NO_2$	(22*S*,25*S*)-22,26-Epimino-16β-hydroxycholesta-1,4-dien-3-one; 220–226; −11° (MeOH) (*244*)
$C_{27}H_{41}NO_2$	(22*R*,25*S*)-22,26-Epimino-16β-hydroxycholesta-1,4-dien-3-one; 270–280, +14° (MeOH) (*244*)
$C_{27}H_{41}NO_2$	(20ξ,25*R*)-23,26-Epimino-3β-hydroxycholesta-5,23(*N*)-dien-22-one (tomatillidine) (**52**);[b] 219–222; −18.1° ($CHCl_3$) (*344*; cf. *172*); semicarbazone, 239–240 (*344*; cf. *172*)
$C_{27}H_{41}NO_2$	(20*S*)-20-(5-Methyl-2-pyridyl)-5α-pregnane-3β,16α-diol; 226–227; +80.4° ($CHCl_3$) (*217*)
$C_{27}H_{41}NO_3$	(25*S*)-22,26-Epimino-12α,16α-dihydroxycholesta-4,22(*N*)-dien-3-one; 235–238 (*147*)
$C_{27}H_{41}NO_3$	(20ξ,25ξ)-23,26-Epimino-3β-hydroxy-5α-cholest-23(*N*)-en-4,22-dione (solamaladine) (**63**); 178–180, $[\alpha]_{578}$ +39.5° (EtOH) (*83*)
$C_{27}H_{42}ClD_2NO_2$	(22*S*,25*R*)-22,26-Chloroepimino-15ξ,17α-dideuterocholest-5-ene-3β,16β-diol; 253–259 (dec.) (*357*)
$C_{27}H_{42}N_2O_2$	3β-Amino-22,26-epimino-5α-cholesta-22(*N*),23,25-triene-16α,23-diol (solaseaforthine) (**14**); amorphous; +22° (MeOH) (*109*)
$C_{27}H_{42}N_2O_2$	(20*R*)-3β-Amino-22,26-epimino-5α-cholesta-22(*N*),23,25-triene-16α,23-diol (isosolaseaforthine) (**15**); amorphous; +26° (MeOH) (*109*)
$C_{27}H_{43}ClDNO_2$	(22*S*,25*R*)-22,26-Chloroepimino-22-deuterocholest-5-ene-3β,16β-diol; 253–259 (*358*)
$C_{27}H_{43}D_2NO_2$	(22*S*,25*R*)-15ξ,17α-Dideutero-22,26-epiminocholest-5-ene-3β,16β-diol; 264–265; −67.3° ($CHCl_3$) (*357*)
$C_{27}H_{43}NO$	(22*S*,25*R*)-22,26-Epiminocholesta-3,5-dien-16β-ol;[b] 193–194 (*237*, *343*)

(*Continued*)

TABLE X (*Continued*)

Formula	Compound; melting point (°C); $[\alpha]_D$ (solvent) (reference)
$C_{27}H_{43}NO$	(25*R*)-22,26-Epiminocholesta-5,22(*N*)-dien-3β-ol;[b] 132–136; +6° (*345*); 138–140; +5.3° (MeOH) (*358*)
$C_{27}H_{43}NO$	(20ξ,25*R*)-22,26-Epiminocholesta-5,22(*N*)-dien-3β-ol (**61**); 169–171 (*172*)
$C_{27}H_{43}NO$	(25*R*)-22,26-Epimino-5α-cholest-22(*N*)-en-3-one; 139–142; +51° ($CHCl_3$) (*62*)
$C_{27}H_{43}NO_2$	(25*S*)-22,26-Epiminocholesta-5,22(*N*)-dien-3β,16α-diol (etioline) (**21**); 153–156; −4.2° ($CHCl_3$) (*80*)
$C_{27}H_{43}NO_2$	(22*S*;25*S*)-22,26-Epimino-3β-hydroxycholest-5-en-26-one (**121**); 241–243 (*215*)
$C_{27}H_{43}NO_2$	(25*R*)-22,26-Epimino-3β-hydroxy-5α-cholest-22(*N*)-en-23-one (23-oxosolacongestidine) (**39**); 213–223/300 (dec.); +33.0° ($CHCl_3$) (*62*)
$C_{27}H_{43}NO_2$	(22*S*,25*R*)-22,26-Epimino-16β-hydroxycholest-4-en-3-one;[b] 203–204; −49.3° (*261*)
$C_{27}H_{43}NO_2$	(20ξ,25*R*)-23,26-Epimino-3β-hydroxy-5α-cholest-23(*N*)-en-22-one (5,6-dihydrotomatillidine) (**53**);[b] 179–181; +21.4° ($CHCl_3$) (*344*; cf. *172*)
$C_{27}H_{43}NO_2$	(20ξ,25*R*)-23,26-Epimino-3β-hydroxy-5α-cholest-23(*N*)-en-22-one ("24-oxosolacongestidine") (**40**); 158–162; +40.9° ($CHCl_3$) (*62*; cf. *172*)
$C_{27}H_{43}NO_3$	(25*S*)-22,26-Epiminocholesta-5,22(*N*)-diene-3β,12α,16α-triol (hakurirodine) (**24**); 194–197; −163° (*147*)
$C_{27}H_{43}NO_3$	(22*S*,25*S*)-22,26-Epimino-3β,16β-dihydroxycholest-5-en-26-one; 262–264; +67.6° (209); 262–264; +60° (*210*)
$C_{27}H_{43}NO_3$	(22*R*,25*S*)-22,26-Epimino-3β,16β-dihydroxycholest-5-en-26-one (**99**); 250–253; + 79.6° (*209*)
$C_{27}H_{44}DNO_2$	(22*S*,25*R*)-22-Deutero-22,26-epiminocholest-5-ene-3β,16β-diol; 265–266 (*358*)
$C_{27}H_{44}ClNO$	(22*S*,25*R*)-22,26-Chloroepiminocholest-5-en-3β-ol; 220–245 (dec.); −32° ($CHCl_3$) (*345*)
$C_{27}H_{44}DNO$	(22*R*,25*S*)-16β-Deutero-22,26-epiminocholest-5-en-3β-ol; 209–212 (*266*)
$C_{27}H_{45}NO$	(22*S*,25*R*)-22,26-Epiminocholest-4-en-16β-ol; 159–161 (*342*)
$C_{27}H_{45}NO$	(22*S*,25*S*)-22,26-Epiminocholest-5-en-3β-ol (veramiline) (**29**);[b] 198–200; −49° (EtOH) (*137*)
$C_{27}H_{45}NO$	(22*S*,25*R*)-22,26-Epiminocholest-5-en-3β-ol; 235–238; +19° ($CHCl_3$) (*345*)
$C_{27}H_{45}NO$	(22*R*,25*S*)-22,26-Epiminocholest-5-en-3β-ol; 218–219; −45° ($CHCl_3$) (*266*)
$C_{27}H_{45}NO$	(25*R*)-22,26-Epimino-5α-cholest-22(*N*)-en-3β-ol (solacongestidine) (**2**);[b] 168–172; +34.8° ($CHCl_3$) (*358*)
$C_{27}H_{45}NO_2$	(22*S*,25*S*)-22,26-Epiminocholest-5-ene-3β,16α-diol (isoteinemine) (**33**); 217–220 (*148*)
$C_{27}H_{45}NO_2$	(22*R*,25*S*)-22,26-Epiminocholest-5-ene-3β,16α-diol (teinemine) (**32**); 204–209; −35.8° ($CHCl_3$) (*148*)

(*Continued*)

TABLE X (*Continued*)

Formula	Compound; melting point (°C); $[\alpha]_D$ (solvent) (reference)
$C_{27}H_{45}NO_2$	(25*S*)-22,26-Epimino-5α-cholest-22(*N*)-ene-3β,16α-diol (25-isosolafloridine) (**22**); 164.5–166.5; +44.8° ($CHCl_3$) (*58, 355*); 158–159; +56.4° ($CHCl_3$) (*214*); hydrochloride, 325–330 (*58, 355*)
$C_{27}H_{45}NO_2$	(25*R*)-22,26-Epimino-5α-cholest-22(*N*)-ene-3β,16α-diol (solafloridine) (**41**);[b] 172–175; +122.7° ($CHCl_3$) (*62*); 167–168; +127° ($CHCl_3$) (*214*); hydrochloride; 280–288 (*62*)
$C_{27}H_{45}NO_2$	(22*S*,25*R*)-22,26-Epimino-16β-hydroxy-5β-cholestan-3-one; 169–170.5 (*196*); 169–170 (*221*)
$C_{27}H_{45}NO_3$	(22*R*,25*S*)-22,26-Epiminocholest-5-ene-3β,12α,16α-triol (baikeine) (**36**); 153–153.5; −97.9° ($CHCl_3$) (*149*); hydrochloride; 285 (*149*); picrate; 177.5–178 (*149*)
$C_{27}H_{45}NO_3$	(22*S*,23*S*,25*S*)-22,26-Epiminocholest-5-ene-3β,16β,23-triol (**130**); 259–262 (dec.); −36.4° (pyridine) (*216*)
$C_{27}H_{45}NO_3$	(22*R*,23*S*,25*R*)-22,26-Epiminocholest-5-ene-3β,16β,23-triol (**137**); 257–260; −45.8° ($CHCl_3$) (*216*)
$C_{27}H_{45}NO_4$	(22*R*,23*S*,25*R*)-22,26-Epimino-3β,16α,23-trihydroxy-5α-cholestan-4-one (deacetylsolaphyllidine) (**46**); 270–273; +50° (MeOH) (69); 272–276; +48.3° (MeOH) (*81*)
$C_{27}H_{46}ClNO$	(22*R*,25*R*)-22,26-Chloroepimino-5α-cholestan-3β-ol (**254**); 276–280 (dec.); +58.2° ($CHCl_3$) (*263*)
$C_{27}H_{46}ClNO_2$	(22*S*,25*S*)-22,26-Chloroepimino-5α-cholestane-3β,16α-diol; 280 (dec.); −71.6° ($CHCl_3$) (*214*)
$C_{27}H_{46}ClNO_2$	(20*R*,22*S*,25*S*)-22,26-Chloroepimino-5α-cholestane-3β,20-diol (**257**); 137–144 (dec.); +22.7° (*238*)
$C_{27}H_{46}ClNO_3$	(20*R*,22*S*,25*S*)-22,26-Chloroepimino-5α-cholestane-3β,16α,20-triol (**258**); 174–183 (dec.); +22.4° ($CHCl_3$) (*239*)
$C_{27}H_{46}ClNO_3$	(20*R*,22*S*,25*R*)-22,26-Chloroepimino-5α-cholestane-3β,16α,20-triol (**272**); 102–104 (dec.); +83.7° ($CHCl_3$) (*239*)
$C_{27}H_{46}ClNO_3$	(20*R*,22*S*,25ξ)-22,26-Chloroepimino-5α-cholestane-3β,16β,20-triol (**274**); 113–119 (dec.); +31.8° ($CHCl_3$) (*238*)
$C_{27}H_{46}ClNO_3$	(20*R*,22*R*,25*S*)-22,26-Chloroepimino-5α-cholestane-3β,16β,20-triol (**273**); 100–107 (dec.); −10.1° ($CHCl_3$) (*238*)
$C_{27}H_{46}ClNO_3$	(20*R*;22*R*;25*R*)-22,26-Chloroepimino-5α-cholestane-3β,16β,20-triol (**259**); 133–139 (dec.); −16.4° ($CHCl_3$) (*238*)
$C_{27}H_{46}N_2O$	(22*R*,25*S*)-16β-Amino-22,26-epiminocholest-5-en-3β-ol; 233–245 (dec.); −50.6° ($CHCl_3$) (*267*); bishydrochloride; −43.1° (MeOH) (*267*)
$C_{27}H_{46}N_2O$	(25*S*)-3β-Amino-22,26-epimino-5α-cholest-22(*N*)-en-16α-ol (solacallinidine) (**23**); 175–178; +51.3° ($CHCl_3$) (*58, 355*)
$C_{27}H_{46}N_2O_3$	(22*S*,25*S*)-22,26-Nitrosoepimino-5α-cholestane-3β,16α-diol; 211–213 (dec.); −2.0° ($CHCl_3$) (*214*)
$C_{27}H_{46}N_2O_3$	(22*S*,25*R*)-22,26-Nitrosoepimino-5α-cholestane-3β,16α-diol; 267–269 (dec.); +9.8° ($CHCl_3$) (*214*)
$C_{27}H_{46}N_2O_3$	(22*R*,25*R*)-22,26-Nitrosoepimino-5α-cholestane-3β,16β-diol; 260–265; −4.6° (dioxane) (*180*)
$C_{27}H_{46}N_2O_3$	(20*R*,22*S*,25*S*)-22,26-Nitrosoepimino-5α-cholestane-3β,20-diol (**263**); 275–277; +5.8° ($CHCl_3$) (*238*)

(*Continued*)

TABLE X (*Continued*)

Formula	Compound; melting point (°C); $[\alpha]_D$ (solvent) (reference)
$C_{27}H_{46}N_2O_4$	(20*R*,22*S*,25*S*)-22,26-Nitrosoepimino-5α-cholestane-3β,16α,20-triol (**264**); 283–284 (dec.); −19.0° ($CHCl_3$) (*239*)
$C_{27}H_{46}N_2O_4$	(20*R*,22*S*;25*R*)-22,26-Nitrosoepimino-5α-cholestane-3β,16α,20-triol (**265**); 243–245 (dec.); −25.6° ($CHCl_3$) (*239*)
$C_{27}H_{46}N_2O_4$	(20*R*,22*R*,25*R*)-22,26-Nitrosoepimino-5α-cholestane-3β,16β,20-triol (**266**); 278–279; +17.8° ($CHCl_3$) (*238*)
$C_{27}H_{47}NO$	(22*S*,25*S*)-22,26-Epimino-5α-cholestan-3β-ol;[b] 179–182; +8° (EtOH) (*137*)
$C_{27}H_{47}NO$	(22*S*,25*R*)-22,26-Epimino-5α-cholestan-3β-ol (**112**);[b] 239–243; +11.2° ($CHCl_3$) (*212*)
$C_{27}H_{47}NO$	(22*S*,25*R*)-22,26-Epimino-5α-cholestan-16β-ol; 140–142 (*342*)
$C_{27}H_{47}NO$	(22*S*,25*R*)-22,26-Epimino-5β-cholestan-16β-ol; 139–142 (*342*)
$C_{27}H_{47}NO_2$	(22*R*,25*S*)-22,26-Epimino-5α-cholestane-3α,16β-diol; 210–212; −3° (MeOH) (*246*)
$C_{27}H_{47}NO_2$	(22*S*,25*S*)-22,26-Epimino-5α-cholestane-3β,16α-diol (**116**); 206–207; +22.4° ($CHCl_3$) (*214*); 201–202; +14.8° ($CHCl_3$) (*355*)
$C_{27}H_{47}NO_2$	(22*S*,25*R*)-22,26-Epimino-5α-cholestane-3β,16α-diol (**42**);[b] 280–284; +24.6° ($CHCl_3$) (*211*); 285–287 (dec.); +33.6° ($CHCl_3$) (*214*)
$C_{27}H_{47}NO_2$	(22*R*,25*S*)-22,26-Epimino-5α-cholestane-3β,16α-diol (**30**); 202–204 (*148*); 155–157 (*149*); 233–235; −8.7° ($CHCl_3$) (*355*)
$C_{27}H_{47}NO_2$	(22*S*,25*R*)-22,26-Epimino-5β-cholestane-3α,16β-diol; 214–215 (*196*, *221*)
$C_{27}H_{47}NO_2$	(20*R*,22*S*,25*S*)-22,26-Epimino-5α-cholestane-3β,20-diol; 219–220.5; +2.2° ($CHCl_3$) (*238*); hydroiodide; 270–271 (dec.) (*186*)
$C_{27}H_{47}NO_2$	(20*R*,22*S*,25*R*)-22,26-Epimino-5α-cholestane-3β,20-diol; 235–238; +10.8° ($CHCl_3$) (*238*)
$C_{27}H_{47}NO_2$	(20*R*,22*R*,25*S*)-22,26-Epimino-5α-cholestane-3β,20-diol; 243–246; +10.4° ($CHCl_3$) (*238*)
$C_{27}H_{47}NO_2$	(20*R*,22*R*,25*R*)-22,26-Epimino-5α-cholestane-3β,20-diol; 214–216; +20.6° ($CHCl_3$) (*238*)
$C_{27}H_{47}NO_2$	(20ξ,22ξ,23ξ,25*R*)-23,26-Epimino-5α-cholestane-3β,22-diol;[b] 229–231, 239–241; −2.5° (MeOH) (*344*; cf. *172*)
$C_{27}H_{47}NO_3$	(20*R*,22*S*,25*S*)-22,26-Epimino-5α-cholestane-3β,16α,20-triol; 265–269; +5.7° ($CHCl_3$) (*239*); hydroiodide; 208–214 (dec.) (*185*)
$C_{27}H_{47}NO_3$	(20*R*,22*S*,25*R*)-22,26-Epimino-5α-cholestane-3β,16α,20-triol; 296–300 (dec.); −13.1° ($CHCl_3$) (*239*)
$C_{27}H_{47}NO_3$	(20*R*,22*S*,25ξ)-22,26-Epimino-5α-cholestane-3β,16β,20-triol (**221**); 281–285; −6.5° ($CHCl_3$) (*238*)
$C_{27}H_{47}NO_3$	(20*R*,22*R*,25*S*)-22,26-Epimino-5α-cholestane-3β,16β,20-triol; 288–292; +20.7° ($CHCl_3$) (*238*)
$C_{27}H_{47}NO_3$	(20*R*,22*R*,25*R*)-22,26-Epimino-5α-cholestane-3β,16β,20-triol; 245–248; +11.2° ($CHCl_3$) (*238*)
$C_{27}H_{47}NO_3$	(22*S*,23*S*,25*S*)-22,26-Epimino-5α-cholestane-3β,16β,23-triol; 257–260; −0.4° (pyridine) (*216*)
$C_{27}H_{47}NO_3$	(22*S*,23*R*,25*S*)-22,26-Epimino-5α-cholestane-3β,16β,23-triol; 260–262; +1.3° (pyridine) (*216*; cf. *188*); hydrobromide; 308–310 (*188*)
$C_{27}H_{47}NO_3$	(22*R*,23*S*,25*S*)-22,26-Epimino-5α-cholestane-3β,16β,23-triol; 248 (dec.); ±0.0° (pyridine) (*216*; cf. *188*)

(*Continued*)

TABLE X (*Continued*)

Formula	Compound; melting point (°C); $[\alpha]_D$ (solvent) (reference)
$C_{27}H_{47}NO_3$	(22*R*,23*S*,25*R*)-22,26-Epimino-5α-cholestane-3β,16β,23-triol (**136**); 281–283; +27.4° (pyridine) (*216*)
$C_{27}H_{47}NO_3$	(22*R*,23*R*,25*S*)-22,26-Epimino-5α-cholestane-3β,16β,23-triol; 241 (dec.); −14.0° (pyridine) (*216*)
$C_{27}H_{47}NO_3$	(22*R*,23*R*,25*R*)-22,26-Epimino-5α-cholestane-3β,16β,23-triol; 263–264; +34.8° (pyridine) (*216*)
$C_{27}H_{47}NO_4$	(22*R*,23*S*,25*R*)-22,26-Epimino-5α-cholestane-3β,4ξ,16α,23-tetraol; 310–312; +218° (MeOH) (*69*)
$C_{27}H_{48}N_2O$	(22*R*,25*S*)-16β-Amino-22,26-epimino-5α-cholestan-3β-ol (**280**); 222–227; ±0° ($CHCl_3$) (*267*); bishydrochloride; 270 (dec.); +1.8° (MeOH) (*267*)
$C_{27}H_{48}N_2O_2$	(22*R*,23*S*,25*R*)-3β-Amino-22,26-epimino-5α-cholestane-16α,23-diol (**67**); 287–289 (*76*)
$C_{28}H_{41}NO_3$	(25*R*)-22,26-Epimino-*N*,16β-carboxycholesta-5,22-dien-3β-ol; −305° ($CHCl_3$) (*252*)
$C_{28}H_{43}NO_3$	(22*S*,25*R*)-22,26-Formylepimino-3β-hydroxycholest-5-en-16-one; 234–235; −85° ($CHCl_3$) (*345*)
$C_{28}H_{43}N_5O$	(22*R*,25*S*)-16β-Azido-22,26-cyanoepiminocholest-5-en-3β-ol; 175–178; +55.5° ($CHCl_3$) (*267*)
$C_{28}H_{45}NO_2$	(22*S*,25*R*)-22,26-Formylepiminocholest-5-en-3β-ol; 229–232; +8° ($CHCl_3$) (*345*)
$C_{28}H_{45}NO_3$	(22*S*, 25*R*)-22,26-Formylepiminocholest-5-ene-3β,16β-diol; 146–147; −21° ($CHCl_3$) (*345*)
$C_{28}H_{45}N_5O$	(22*R*,25*S*)-16β-Azido-22,26-cyanoepimino-5α-cholestan-3β-ol; 140–143/187–190; +82.8° (MeOH) (*267*)
$C_{28}H_{45}N_5O$	(22*R*,25*S*)-22,26-Epimino-3β-hydroxy-5α-cholestanyl-16β → 1′, *N* → 5′-tetrazole; 272–275; −61.2° (MeOH) (*267*)
$C_{28}H_{47}NO_2$	(22*S*,25*S*)-22,26-Methylepiminocholest-5-ene-3β,16β-diol (hapepunine) (**34**); 196.5–198.5; −72.6° (*128*)
$C_{28}H_{47}NO_2$	(22*S*,25*R*)-22,26-Methylepiminocholest-5-ene-3β, 16β-diol; 260–264.5 (*128*)
$C_{28}H_{47}NO_2$	(22*R*,25*S*)-22,26-Methylepiminocholest-5-ene-3β,16β-diol; −36.8° (MeOH) (*367*)
$C_{28}H_{47}NO_2$	(22*R*,25*R*)-22,26-Methylepiminocholest-5-ene-3β,16β-diol; 260–264 (*367*)
$C_{28}H_{47}NO_2$	(25*S*)-22,26-Methylepimino-5α-cholest-22-ene-3β,16α-diol; 155–160; +17.7° ($CHCl_3$) (*355*); hydroiodide; 300–305 (dec.) (*355*)
$C_{28}H_{47}NO_3$	(22*S*,25*S*)-22,26-Methylepiminocholest-5-ene-3β,16β,18-triol (anrakorinine); 248–251; −50.4° ($CHCl_3$) (*367*)
$C_{29}H_{42}D_3NO_3$	(22*S*,25*R*)-22,26-Acetylepimino-15α,15β,17α-trideutero-3β-hydroxycholest-5-en-16-one; 218–221 (*357*)
$C_{29}H_{43}NO_2$	(20*S*)-3β-Acetoxy-20-(5-methyl-2-pyridyl)-5α-pregnane (**38**); 164–166 (*62*)
$C_{29}H_{43}NO_3$	(20ξ,25*R*)-3β-Acetoxy-23,26-epiminocholesta-5,23(*N*)-dien-22-one,[b] 141–144; +9.6° ($CHCl_3$) (*344*; cf. *172*)
$C_{29}H_{43}NO_4$	(20ξ,25ξ)-3β-Acetoxy-23,26-epimino-5α-cholest-23(*N*)-en-4,22-dione; 202–208, $[\alpha]_{578}$ −69° (EtOH) (*83*)

(*Continued*)

TABLE X (*Continued*)

Formula	Compound; melting point (°C); $[\alpha]_D$ (solvent) (reference)
$C_{29}H_{45}D_2NO_3$	(22*S*,25*R*)-22,26-Acetylepimino-15ξ,17α-dideuterocholest-5-ene-3β,16β-diol; 244–246; −45.8° ($CHCl_3$) (*357*)
$C_{29}H_{45}NO_2$	(22*S*,25*R*)-22,26-Acetylepiminocholesta-3,5-dien-16β-ol; 109–110 (*343*)
$C_{29}H_{45}NO_3$	(20ξ,25*R*)-3β-Acetoxy-23,26-epimino-5α-cholest-23(*N*)-en-22-one; 200–203 (*62*; cf. *172*)
$C_{29}H_{45}NO_3$	(22*S*,25*R*)-22,26-Acetylepimino-3β-hydroxycholest-5-en-16-one (**109**);[b] 207–209; −146.0° ($CHCl_3$) (*212*); 212–214; −149.2° ($CHCl_3$) (*357*)
$C_{29}H_{45}NO_3$	(22*R*,25*S*)-22,26-Ethylepimino-16α-hydroxycholest-4-ene-3,12-dione; 275 (dec.) (*149*)
$C_{29}H_{46}ClNO_2$	(22*S*,25*R*)-3β-Acetoxy-22,26-chloroepiminocholest-5-ene; 275–279 (dec.) (*172*)
$C_{29}H_{47}NO_2$	(22*S*,25*R*)-3β-Acetoxy-22,26-epiminocholest-5-ene (**60**); 256–257 (*172*)
$C_{29}H_{47}NO_2$	(25*R*)-3β-Acetoxy-22,26-epimino-5α-cholest-22(*N*)-ene (**37**); 185–195 (*62*)
$C_{29}H_{47}NO_2$	(22ξ,25ξ)-22,26-Acetylepimino-5ξ-cholestan-3-one; 129–134 (*108*)
$C_{29}H_{47}NO_2$	(22*S*,25*R*)-22,26-Acetylepiminocholest-5-en-3β-ol; 247–249; −4° ($CHCl_3$) (*345*)
$C_{29}H_{47}NO_3$	(22*S*,25*R*)-3β-Acetoxy-22,26-epiminocholest-5-en-16β-ol (**56**); 211.5–213; −67.1° ($CHCl_3$) (*211*)
$C_{29}H_{47}NO_3$	(22*S*,25*R*)-22,26-Acetylepiminocholest-5-ene-3β,16β-diol (**108**);[b] 216–217; −45.9° ($CHCl_3$) (*212*); 250–250.5 (*343*); 244–246; −45.1° ($CHCl_3$) (*357, 358*)
$C_{29}H_{47}NO_3$	(22*R*,25*S*)-22,26-Acetylepimino-3β-hydroxy-5α-cholestan-16-one;[b] 177–179 (*149*)
$C_{29}H_{47}NO_4$	(22*R*,25*S*)-22,26-Acetylepiminocholest-5-ene-3β,12α,16α-triol; 141–143; −63.3° ($CHCl_3$) (*149*)
$C_{29}H_{47}NO_5$	(22*R*,23*S*,25*R*)-16α-Acetoxy-22,26-epimino-3β,23-dihydroxy-5α-cholestan-4-one (solaphyllidine) (**45**); 165–170; −24.0° (MeOH) (*69*); 165–170; −25.4° (MeOH) (*81*)
$C_{29}H_{48}ClNO_3$	(20*R*,22*S*,25*S*)-3β-Acetoxy-22,26-chloroepimino-5α-cholestan-20-ol (**268**); 102–105 (dec.); +26.8° ($CHCl_3$) (*238*)
$C_{29}H_{48}ClNO_3$	(20*R*,22*S*,25*R*)-3β-Acetoxy-22,26-chloroepimino-5α-cholestan-20-ol (**269**); 130–136 (dec.); +31.7° ($CHCl_3$) (*238*)
$C_{29}H_{48}ClNO_3$	(20*R*,22*R*,25*S*)-3β-Acetoxy-22,26-chloroepimino-5α-cholestan-20-ol (**271**); 275–278 (dec.); −22.1° ($CHCl_3$) (*238*)
$C_{29}H_{48}ClNO_3$	(20*R*,22*R*,25*R*)-3β-Acetoxy-22,26-chloroepimino-5α-cholestan-20-ol (**270**); 250–255 (dec.); −40.2° (*238*)
$C_{29}H_{48}N_2O_2$	(25*S*)-16α-Acetoxy-3β-amino-22,26-epimino-5α-cholest-22(*N*)-ene; −115.6° ($CHCl_3$) (*355*)
$C_{29}H_{49}NO_2$	(22*S*,25*R*)-22,26-Acetylepimino-5α-cholestan-3β-ol (**111**); 266–269; +9.1° ($CHCl_3$) (*212*)
$C_{29}H_{49}NO_3$	(22*S*,25*R*)-3β-Acetoxy-22,26-epimino-5α-cholestan-16β-ol (**102**); 184–185, −15.0° ($CHCl_3$) (*211*)
$C_{29}H_{49}NO_3$	(20*R*,22*S*,25*S*)-3β-Acetoxy-22,26-epimino-5α-cholestan-20-ol (**211**); 215–217; +5.1° ($CHCl_3$) (*238*)

(*Continued*)

TABLE X (*Continued*)

Formula	Compound; melting point (°C); $[\alpha]_D$ (solvent) (reference)
$C_{29}H_{49}NO_3$	(20*R*,22*S*,25*R*)-3β-Acetoxy-22,26-epimino-5α-cholestan-20-ol (**212**); 201–203; −2.3° ($CHCl_3$) (*238*)
$C_{29}H_{49}NO_3$	(20*R*,22*R*,25*S*)-3β-Acetoxy-22,26-epimino-5α-cholestan-20-ol (**214**); 230–233; −8.9° ($CHCl_3$) (*238*)
$C_{29}H_{49}NO_3$	(20*R*,22*R*,25*R*)-3β-Acetoxy-22,26-epimino-5α-cholestan-20-ol (**213**); 195–197; −4.7° ($CHCl_3$) (*238*)
$C_{29}H_{49}NO_3$	(22*S*,25*S*)-22,26-Acetylepimino-5α-cholestane-3β,16α-diol; 273–276 (dec.); +24.4° ($CHCl_3$) (*214*)
$C_{29}H_{49}NO_3$	(22*R*,25*S*)-22,26-Acetylepimino-5α-cholestane-3β,16α-diol; 234.5–235 (*149*)
$C_{29}H_{49}NO_3$	(20*R*,22*S*,25*S*)-22,26-Acetylepimino-5α-cholestane-3β,20-diol; 256–259; −2.4° ($CHCl_3$) (*238*)
$C_{29}H_{49}NO_3$	(22*R*,25*S*)-22,26-Ethylepiminocholest-5-ene-3β,12α,16α-triol; 267–271 (*149*); methiodide; 297 (dec.) (*149*)
$C_{29}H_{49}NO_4$	(20*R*,22*S*,25*S*)-3β-Acetoxy-22,26-epimino-5α-cholestane-16α,20-diol; 288–293; −6.3° ($CHCl_3$) (*239*)
$C_{29}H_{49}NO_4$	(20*R*,22*S*,25*R*)-3β-Acetoxy-22,26-epimino-5α-cholestane-16α,20-diol (**217**); 285–291; −18.4° ($CHCl_3$) (*239*)
$C_{29}H_{49}NO_4$	(22*R*,25*S*)-22,26-Acetylepimino-5α-cholestane-3β,12α,16α-triol; 220–222; +10.3° ($CHCl_3$) (*149*)
$C_{29}H_{49}NO_4$	(20*R*,22*S*,25*S*)-22,26-Acetylepimino-5α-cholestane-3β,16α,20-triol; 284–287; +10.3° ($CHCl_3$) (*239*)
$C_{29}H_{49}NO_4$	(20*R*,22*S*,25*R*)-22,26-Acetylepimino-5α-cholestane-3β,16α,20-triol; 206–210 (dec.); −18.2° ($CHCl_3$) (*239*)
$C_{29}H_{49}NO_4$	(20*R*,22*R*,25*R*)-22,26-Acetylepimino-5α-cholestane-3β,16β,20-triol; 264–266; +12.9° ($CHCl_3$) (*238*)
$C_{29}H_{49}NO_5$	(22*R*,23*S*,25*R*)-16α-Acetoxy-22,26-epimino-5α-cholestane-3β,4ξ,23-triol; 246–248; −46.5° (MeOH) (*81*)
$C_{29}H_{51}NO_2$	(22ξ,25ξ)-22,26-Epimino-3,3-dimethoxy-5ξ-cholestane (solaquidine) (**48**); 278–281 (*108*)
$C_{29}H_{51}NO_2$	(22*R*,25*S*)-22,26-Ethylepimino-5α-cholestane-3β,16α-diol; 215–218 (*149*)
$C_{30}H_{43}NO_4$	(25*R*)-3β-Acetoxy-22,26-epimino-*N*,16β-carboxycholesta-5,22-diene (**234**); 264–267; −285° ($CHCl_3$) (*252*)
$C_{30}H_{43}NO_5$	3β,16α-Diacetoxy-23,26-epimino-27-nor-5α-cholesta-23,25-dien-22-one (**205**); 224–226; −48° ($CHCl_3$) (*236*)
$C_{30}H_{43}NO_5$	(20*R*)-3β,16α-Diacetoxy-23,26-epimino-27-nor-5α-cholesta-23,25-dien-22-one; 216–218; −80° ($CHCl_3$) (*236*)
$C_{30}H_{45}BrN_2O_2$	(22*R*,25*S*)-3β-Acetoxy-16α-bromo-22,26-cyanoepiminocholest-5-ene; 185/245–260 (*266*)
$C_{30}H_{45}NO_3$	(22ξ)-3β-Acetoxy-22,25-acetylepimino-27-norcholesta-5,24-diene; amorphous; −98° (EtOH) (*138*)
$C_{30}H_{45}NO_4$	(25*R*)-3β-Acetoxy-22,26-epimino-*N*,16β-carboxy-5α-cholest-22-ene; 288.5–291; −225° ($CHCl_3$) (*252*)
$C_{30}H_{45}NO_5$	(22*S*,25*R*)-22,26-Formylepimino-3β,16β-diformyloxycholest-5-ene; 192–194, −8° ($CHCl_3$) (*345*)
$C_{30}H_{45}N_5O_2$	(22*R*,25*S*)-3β-Acetoxy-16β-azido-22,26-cyanoepiminocholest-5-ene; 202–210; +38.1° ($CHCl_3$) (*267*)

(*Continued*)

TABLE X (*Continued*)

Formula	Compound; melting point (°C); $[\alpha]_D$ (solvent) (reference)
$C_{30}H_{47}BrN_2O_2$	(22*R*,25*S*)-3β-Acetoxy-16α-bromo-22,26-cyanoepimino-5α-cholestane (**276**); 155/250-280 (dec.) (*266*)
$C_{30}H_{47}NO_2S_2$	(22*S*,25*R*)-16,16-Ethylenedimercapto-22,26-formylepiminocholest-5-en-3β-ol; 221–224; −58° ($CHCl_3$) (*345*)
$C_{30}H_{47}NO_4$	(22*S*,25*R*)-3β-Acetoxy-22,26-epimino-*N*,16β-carboxy-5α-cholestane; 312–315; −138.9° ($CHCl_3$) (211); 323.5–324.5 (*252*)
$C_{30}H_{47}NO_4$	(22*R*,25*R*)-3β-Acetoxy-22,26-epimino-*N*,16β-carboxy-5α-cholestane; 276–278 (*252*)
$C_{30}H_{47}N_5O_2$	(22*R*,25*S*)-3β-Acetoxy-16β-azido-22,26-cyanoepimino-5α-cholestane (**279**); 220–225; +72.8° ($CHCl_3$) (*267*)
$C_{30}H_{52}N_2O_2$	(22*R*,23*S*,25*R*)-3β-Dimethylamino-22,26-methylepiminocholest-5-ene-16α,23-diol (**80**); 236–238 (*110*)
$C_{30}H_{54}N_2O_2$	(22*R*,23*S*,25*R*)-3β-Dimethylamino-22,26-methylepimino-5α-cholestane-16α,23-diol (**81**); 225–226 (*110*)
$C_{31}H_{43}NO_4$	(20*S*)-3β,16β-Diacetoxy-20-(5-methyl-2-pyridyl)-pregn-5-ene; 220–222; −7.2° ($CHCl_3$) (*220*)
$C_{31}H_{45}NO_4$	(20ξ,25*R*)-3β-Acetoxy-23,26-acetylepiminocholesta-5,23-dien-22-one;[b] 155.5–157; −88.5° ($CHCl_3$) (*344*; cf. *172*)
$C_{31}H_{45}NO_4$	(20*S*)-3β,16α-Diacetoxy-20-(5-methyl-2-pyridyl)-5α-pregnane (**146**),[b] 231–232 (dec.); −96.5° ($CHCl_3$) (*217*)
$C_{31}H_{45}NO_5$	(20ξ,25ξ)-3β-Acetoxy-23,26-acetylepimino-5α-cholest-23-en-4,22-dione; 227–235 (*83*)
$C_{31}H_{45}NO_5$	(25*R*)-3β-Acetoxy-22,26-formylepimino-16β-formyloxycholesta-5,22-diene; 174–176 (*226*)
$C_{31}H_{45}NO_5$	(25*S*)-3β,16B-Diacetoxy-22,26-epiminocholesta-5,22(*N*)-dien-23-one (**128**); 167–170; −51.7° ($CHCl_3$) (*216*)
$C_{31}H_{45}NO_5$	(25*R*)-3β,16β-Diacetoxy-22,26-epiminocholesta-5,22(*N*)-dien-23-one;[b] 186–187; +18.9° ($CHCl_3$) (*192*); 181–185 (*172*)
$C_{31}H_{45}NO_5$	(25*R*)-3β,16β-Diacetoxy-23,26-epiminocholesta-5,23(*N*)-dien-22-one; 184 (*172*)
$C_{31}H_{47}NO_3$	(25*R*)-3β-Acetoxy-22,26-acetylepiminocholesta-5,22-diene;[b] 167–169; +4° ($CHCl_3$) (*345*)
$C_{31}H_{47}NO_3$	(22*S*,25*R*)-16β-Acetoxy-22,26-acetylepiminocholesta-3,5-diene; 127–127.2 (*343*)
$C_{31}H_{47}NO_4$	(22*S*,25*R*)-3β-Acetoxy-22,26-acetylepiminocholest-5-en-16-one; 151–152.5; −152.7° ($CHCl_3$) (*357*)
$C_{31}H_{47}NO_4$	(20ξ,25*R*)-3β-Acetoxy-23,26-acetylepimino-5α-cholest-23-en-22-one; 184–187 (*62*; cf. *172*)
$C_{31}H_{47}NO_4$	(22*S*,25*R*)-16β-Acetoxy-22,26-acetylepiminocholest-4-en-3-one;[b] amorphous; +80.0° (*261*)
$C_{31}H_{47}NO_4$	(25*S*)-3β,16β-Diacetoxy-22,26-epiminocholesta-5,22(*N*)-diene (**127**); 169–171; −27.7° ($CHCl_3$) (*216*)
$C_{31}H_{47}NO_5$	(25*R*)-3β,16α-Diacetoxy-22,26-epimino-5α-cholest-22(*N*)-en-23-one (**144**); 196–198.5; −27.5° (*219*)
$C_{31}H_{47}NO_5$	(25*R*)-3β,16α-Diacetoxy-23,26-epimino-5α-cholest-23(*N*)-en-22-one (**147**); 159–162; −43.3° ($CHCl_3$) (*217*; cf. *172*)

(*Continued*)

TABLE X (*Continued*)

Formula	Compound; melting point (°C); $[\alpha]_D$ (solvent) (reference)
$C_{31}H_{48}N_2O_4$	(22*S*,25*R*)-16β-Acetoxy-22,26-acetylepimino-3-oximinocholest-4-ene; 126–129 (*227*)
$C_{31}H_{49}NO_2S_2$	(22*S*,25*R*)-22,26-Acetylepimino-16, 16-ethylenedimercaptocholest-5-en-3β-ol (**110**); 165–170/223–226; −80.2° ($CHCl_3$) (*212*); 234–237; −65° ($CHCl_3$) (*345*)
$C_{31}H_{49}NO_3$	(22*S*,25*S*)-3β-Acetoxy-22,26-acetylepiminocholest-5-ene; 156–157; −21° (EtOH) (*137*)
$C_{31}H_{49}NO_3$	(22*S*,25*R*)-3β-Acetoxy-22,26-acetylepiminocholest-5-ene; 179–180 (*358*)
$C_{31}H_{49}NO_3$	(22*S*,25*R*)-16β-Acetoxy-22,26-acetylepiminocholest-5-ene; 109–110 (*343*)
$C_{31}H_{49}NO_4$	(22*S*,25*R*)-16β-Acetoxy-22,26-acetylepimino-5β-cholestan-3-one; 184–186 (196); 182–183 (*221*)
$C_{31}H_{49}NO_4$	(22*S*,25*R*)-3β-Acetoxy-22,26-acetylepiminocholest-5-en-16β-ol; 170.5–172; −52.4° ($CHCl_3$) (*357*)
$C_{31}H_{49}NO_4$	(25*S*)-3β,16α-Diacetoxy-22,26-epimino-5α-cholest-22(*N*)-ene; 158–161; −127° (*214*); 166–168; −129.0° ($CHCl_3$) (*355*)
$C_{31}H_{49}NO_4$	(25*R*)-3β,16α-Diacetoxy-22,26-epimino-5α-cholest-22(*N*)-ene (**143**); 182–183; −53.3° ($CHCl_3$) (*214*); 173–175; −56.6° (*219*)
$C_{31}H_{49}NO_4$	(22*S*,25*R*)-3β,16β-Diacetoxy-22,26-epiminocholest-5-ene; 172–173 (*358*)
$C_{31}H_{49}NO_5$	(22*R*,25*S*)-3β-Acetoxy-22,26-acetylepiminocholest-5-ene-12α,16α-diol (*N*-acetylbaikeidine); 223.5–225; −103° ($CHCl_3$) (*149*)
$C_{31}H_{49}NO_5$	(22*R*,25*S*)-3β-Acetoxy-22,26-acetylepimino-16α-hydroxy-5α-cholestan-12-one; 258–261 (*149*)
$C_{31}H_{49}NO_5$	(22*S*,23*S*,25*S*-3β,16β-Diacetoxy-22,26-epiminocholest-5-en-23-ol (**129**); 222–223, +10.4° ($CHCl_3$) (*216*)
$C_{31}H_{49}NO_5$	(22*R*,23*S*,25*R*)-3β,16β-Diacetoxy-22,26-epiminocholest-5-en-23-ol; 176–179; −12.8° ($CHCl_3$) (*216*)
$C_{31}H_{50}ClNO_4$	(22*S*,25*S*)-3β,16α-Diacetoxy-22,26-chloroepimino-5α-cholestane; 160–164 (dec.); −117° ($CHCl_3$) (*214*)
$C_{31}H_{50}ClNO_4$	(22*S*,25*R*)-3β,16α-Diacetoxy-22,26-chloroepimino-5α-cholestane; 280 (dec.); −105° ($CHCl_3$) (*214*)
$C_{31}H_{50}ClNO_5$	(20*R*,22*S*,25*S*)-3β,16α-Diacetoxy-22,26-chloroepimino-5α-cholestan-20-ol; 128–135 (dec.); −20.8° ($CHCl_3$) (*239*)
$C_{31}H_{50}ClNO_5$	(22*R*,23*S*,25*R*)-3β,16α-Diacetoxy-22,26-chloroepimino-5α-cholestan-23-ol; 270–280 (dec.); −43.8° (pyridine) (*217*)
$C_{31}H_{50}ClNO_5$	(22*S*,23*S*,25*S*)-3β,16β-Diacetoxy-22,26-chloroepimino-5α-cholestan-23-ol; +57.6° (dioxane) (*216*)
$C_{31}H_{50}ClNO_5$	(22*S*,23*R*,25*S*)-3β,16β-Diacetoxy-22,26-chloroepimino-5α-cholestan-23-ol; +36.0° (dioxane) (*216*; cf. *188*)
$C_{31}H_{50}ClNO_5$	(22*R*,23*S*,25*S*)-3β,16β-Diacetoxy-22,26-chloroepimino-5α-cholestan-23-ol; −2.0° (dioxane) (*216*; cf. *188*)
$C_{31}H_{50}ClNO_5$	(22*R*,23*S*,25*R*)-3β,16β-Diacetoxy-22,26-chloroepimino-5α-cholestan-23-ol; 280 (dec.); +25.5° (pyridine) (*192*)
$C_{31}H_{50}ClNO_5$	(22*R*,23*R*,25*S*)-3β,16β-Diacetoxy-22,26-chloroepimino-5α-cholestan-23-ol; −19.4° (dioxane) (*216*)

(*Continued*)

TABLE X (*Continued*)

Formula	Compound; melting point (°C); $[\alpha]_D$ (solvent) (reference)
$C_{31}H_{50}ClNO_5$	(22*R*,23*R*,25*R*)-3β,16β-Diacetoxy-22,26-chloroepimino-5α-cholestan-23-ol; 260–270 (dec.); −21.7° (pyridine) (*192*)
$C_{31}H_{50}N_2O_3$	(22*R*,25*S*)-16β-Acetylamino-22,26-acetylepiminocholest-5-en-3β-ol; 222–226/260–263; −38.8° ($CHCl_3$) (*267*)
$C_{31}H_{50}N_2O_4$	(22*S*,25*R*)-16β-Acetoxy-22,26-acetylepimino-3-oximino-5β-cholestane; 222–223 (*221*)
$C_{31}H_{50}N_2O_5$	(22*S*;25*S*)-3β,16α-Diacetoxy-22,26-nitrosoepimino-5α-cholestane; 209–211; −48.4° ($CHCl_3$) (*214*)
$C_{31}H_{50}N_2O_5$	(22*S*,25*R*)-3β,16α-Diacetoxy-22,26-nitrosoepimino-5α-cholestane (**142**); 215–217; −15.1° ($CHCl_3$) (*214*)
$C_{31}H_{50}N_2O_6$	(22*R*,23*S*,25*R*)-3β,16α-Diacetoxy-22,26-nitrosoepimino-5α-cholestan-23-ol; 222–224; +20.6° (dioxane) (*217*)
$C_{31}H_{50}N_2O_6$	(22*R*,23*S*,25*R*)-3β,16β-Diacetoxy-22,26-nitrosoepimino-5α-cholestan-23-ol; 211–213; +53.0° (dioxane) (*192*)
$C_{31}H_{50}N_2O_6$	(22*R*,23*R*,25*R*)-3β,16β-Diacetoxy-22,26-nitrosoepimino-5α-cholestan-23-ol; amorphous; +15.4° (dioxane) (*192*)
$C_{31}H_{51}NO_4$	(20*R*,22*S*,25*S*)-3β-Acetoxy-22,26-acetylepimino-5α-cholestan-20-ol; 208–210; +5.7° ($CHCl_3$) (*238*)
$C_{31}H_{51}NO_4$	(22*S*,25*R*)-16β-Acetoxy-22,26-acetylepimino-5β-cholestan-3α-ol; 167–168 (*221*)
$C_{31}H_{51}NO_4$	(22*S*,25*S*)-3β,16α-Diacetoxy-22,26-epimino-5α-cholestane (**115**); 198–200; −76.6° ($CHCl_3$) (*214*)
$C_{31}H_{51}NO_4$	(22*S*,25*R*)-3β,16α-Diacetoxy-22,26-epimino-5α-cholestane (**114**); 179–181; −60.8° ($CHCl_3$) (*214*)
$C_{31}H_{51}NO_5$	(22*R*,25*S*)-3β-Acetoxy-22,26-acetylepimino-5α-cholestane-12α,16α-diol; 131–132; −39.4° ($CHCl_3$) (*149*)
$C_{31}H_{51}NO_5$	(20*R*,22*R*,25*R*)-3β-Acetoxy-22,26-acetylepimino-5α-cholestane-16β,20-diol; 147–150; +9.4° ($CHCl_3$) (*238*)
$C_{31}H_{51}NO_5$	(20*R*,22*S*,25*S*)-3β,16α-Diacetoxy-22,26-epimino-5α-cholestan-20-ol (**216**); 196–199; −55.8° ($CHCl_3$) (*239*)
$C_{31}H_{51}NO_5$	(20*R*,22*R*,25*S*)-3β,16β-Diacetoxy-22,26-epimino-5α-cholestan-20-ol (**220**); 252–254; −4.2° ($CHCl_3$) (*238*)
$C_{31}H_{51}NO_5$	(20*R*,22*R*,25*R*)-3β,16β-Diacetoxy-22,26-epimino-5α-cholestan-20-ol (**219**); 233–236; +35.5° ($CHCl_3$) (*238*)
$C_{31}H_{51}NO_5$	(22*R*,23*S*,25*R*)-3β,16α-Diacetoxy-22,26-epimino-5α-cholestan-23-ol (**145**); 179–180; −53.7° (pyridine) (*217*)
$C_{31}H_{51}NO_5$	(22*S*,23*S*,25*S*)-3β,16β-Diacetoxy-22,26-epimino-5α-cholestan-23-ol (**132**); 197–200; +45.3° ($CHCl_3$) (*216*)
$C_{31}H_{51}NO_5$	(22*S*,23*R*,25*S*)-3β,16β-Diacetoxy-22,26-epimino-5α-cholestan-23-ol (**133**); 242–244; +24.8° ($CHCl_3$) (*216*; cf. *188*)
$C_{31}H_{51}NO_5$	(22*R*,23*S*,25*S*)-3β,16β-Diacetoxy-22,26-epimino-5α-cholestan-23-ol (**135**); 186–189; +20.9° ($CHCl_3$) (*216*; cf. *188*)
$C_{31}H_{51}NO_5$	(22*R*,23*S*,25*R*)-3β,16β-Diacetoxy-22,26-epimino-5α-cholestan-23-ol (**163**); 163–165; +37.8° (pyridine) (*192*); 173–175; +36.4° (pyridine) (*220*)
$C_{31}H_{51}NO_5$	(22*R*,23*R*,25*S*)-3β,16β-Diacetoxy-22,26-epimino-5α-cholestan-23-ol (**134**); 218–221; +2.7° ($CHCl_3$) (*216*)

(*Continued*)

TABLE X (*Continued*)

Formula	Compound; melting point (°C); $[\alpha]_D$ (solvent) (reference)
$C_{31}H_{51}NO_5$	(22*R*,23*R*,25*R*)-3β,16β-Diacetoxy-22,26-epimino-5α-cholestan-23-ol (**164**); 223–225; +31.6° (pyridine) (*192*)
$C_{31}H_{52}N_2O_3$	(22*R*,25*S*)-16β-Acetylamino-22,26-acetylepimino-5α-cholestan-3β-ol;[b] 223–224; +8.2° ($CHCl_3$) (*267*)
$C_{32}H_{51}NO_4$	(22*S*,25*S*)-3β,16β-Diacetoxy-22,26-methylepiminocholest-5-ene; 207–212 (*128*)
$C_{33}H_{49}NO_5$	(25*S*)-3β,16α-Diacetoxy-22,26-acetylepiminocholesta-5,22-diene; 194–196 (*146*)
$C_{33}H_{49}D_2NO_5$	(22*S*,25*R*)-3β,16β-Diacetoxy-22,26-acetylepimino-15ξ,17α-dideuterocholest-5-ene; 171–173 (*358*)
$C_{33}H_{50}DNO_5$	(22*S*,25*R*)-3β,16β-Diacetoxy-22,26-acetylepimino-22-deuterocholest-5-ene; 171–172 (*358*)
$C_{33}H_{50}N_2O_4$	(20*R*)-16α,23-Diacetoxy-3β-dimethylamino-22,26-epimino-5α-cholesta-22(*N*)23,25-triene; 202–204 (*109*)
$C_{33}H_{50}N_2O_5$	(22*R*,25*R*)-16α-Acetoxy-3β-acetylamino-22,26-acetylepiminocholest-5-en-23-one (**82**); 226–227 (*110*)
$C_{33}H_{51}NO_5$	(22*R*,25*S*)-3β,16α-Diacetoxy-22,26-acetylepiminocholest-5-ene; 147–148.5 (*148*)
$C_{33}H_{51}NO_5$	(25*S*)-3β,16α-Diacetoxy-22,26-acetylepimino-5α-cholest-22-ene; 228–229; +87.5° ($CHCl_3$) (*58, 355*)
$C_{33}H_{51}NO_5$	(25*R*)-3β,16β-Diacetoxy-22,26-acetylepimino-5α-cholest-22-ene; 141–143; +134.0° ($CHCl_3$) (*192*)
$C_{33}H_{51}NO_6$	(25*R*)-3β,16α-Diacetoxy-22,26-acetylepimino-5α-cholestan-23-one (**158**); 194.5–196.5; −37.7° (*219*)
$C_{33}H_{52}N_2O_4$	(22*R*,25*S*)-3β-Acetoxy-16β-acetylamino-22,26-acetylepiminocholest-5-ene; 233–236; −40.9° ($CHCl_3$) (*267*)
$C_{33}H_{52}N_2O_4$	(25*S*)-16α-Acetoxy-3β-acetylamino-22,26-acetylepimino-5α-cholest-22-ene; 152–153; +63.5° ($CHCl_3$) (*58, 355*)
$C_{33}H_{53}NO_5$	(22*S*,25*R*)-3α,16β-Diacetoxy-22,26-acetylepimino-5β-cholestane; 101–102 (*221*)
$C_{33}H_{53}NO_5$	(22*S*,25*S*)-3β,16α-Diacetoxy-22,26-acetylepimino-5α-cholestane; 158–161; −28.9° ($CHCl_3$) (*214*)
$C_{33}H_{53}NO_5$	(22*R*,25*S*)-3β,16α-Diacetoxy-22,26-acetylepimino-5α-cholestane; 227–228; −57.4° ($CHCl_3$) (*355*)
$C_{33}H_{53}NO_5$	(20ξ,22ξ,23ξ,25*R*)-3β,22-Diacetoxy-23,26-acetylepimino-5α-cholestane (**49**–**51**); 129–133 (*172*); 122–125 (*172*); 157–159 (*172*)
$C_{33}H_{53}NO_6$	(20*R*,22*S*,25*S*)-3β,16α-Diacetoxy-22,26-acetylepimino-5α-cholestan-20-ol; 199–202; −15.5° ($CHCl_3$) (*239*)
$C_{33}H_{53}NO_6$	(20*R*,22*S*,25*R*)-3β,16α-Diacetoxy-22,26-acetylepimino-5α-cholestan-20-ol; 104–108; −20.2° (*239*)
$C_{33}H_{53}NO_6$	(20*R*,22*R*,25*R*)-3β,16β-Diacetoxy-22,26-acetylepimino-5α-cholestan-20-ol; 228–229; +22.9° ($CHCl_3$) (*238*)
$C_{33}H_{54}N_2O_4$	(22*R*,25*S*)-3β-Acetoxy-16β-acetylamino-22,26-acetylepimino-5α-cholestane;[b] 213–215; −21.2° (EtOH), ±0° ($CHCl_3$) (*267*)
$C_{33}H_{54}N_2O_4$	(22*R*,25*S*)-16α-Acetoxy-3β-acetylamino-22,26-acetylepimino-5α-cholestane; 200–201; −52.5° ($CHCl_3$) (*355*)

(*Continued*)

TABLE X (*Continued*)

Formula	Compound; melting point (°C); $[\alpha]_D$ (solvent) (reference)
$C_{33}H_{54}N_2O_5$	(22*R*,25*S*)-16β-Ethoxycarbonylamino-22,26-(*N*-ethoxycarbonyl-epimino)cholest-5-en-3β-ol; 197–202; −11.3° ($CHCl_3$) (*267*)
$C_{33}H_{56}N_2O_5$	(22*R*,25*S*)-16β-Ethoxycarbonylamino-22,26-(*N*-ethoxycarbonyl-epimino)-5α-cholestan-3β-ol; 170–174; +23.7° ($CHCl_3$) (*267*)
$C_{34}H_{50}N_2O_2$	(25*S*)-22,26-Epimino-3β-salicylidenamino-5α-cholest-22(*N*)-en-16α-ol; 215–217; +59.3° ($CHCl_3$) (*355*)
$C_{35}H_{47}NO_5$	(22*R*,25*R*)-*N*-Benzyloxycarbonyl-22,26-epimino-5α-cholestane-3,16,23-trione (**154**); 136–138; −85.2° (pyridine) (*217*)
$C_{35}H_{51}NO_4$	(22*S*,25*R*)-*N*-Benzyloxycarbonyl-22,26-epimino-3β-hydroxy-5α-cholestan-16-one; 193–195; −86.3° ($CHCl_3$) (*211*)
$C_{35}H_{51}NO_7$	(25*S*)-3β,12α,16α-Triacetoxy-22,26-acetylepiminocholesta-5,22-diene; 137–138 (*147*)
$C_{35}H_{53}NO_7$	(22*R*,25*S*)-3β,12α,16α-Triacetoxy-22,26-acetylepiminocholest-5-ene; 230–234 (*149*)
$C_{35}H_{53}NO_8$	(22*R*,23*S*,25*R*)-3β,16α,23-Triacetoxy-22,26-acetylepimino-5α-cholestan-4-one; 158–163; −24.0° (MeOH) (*69*); 204–206; −25.4° (MeOH) (*81*)
$C_{35}H_{55}NO_6$	(22*R*,25*S*)-3β,12α,16α-Triacetoxy-22,26-ethylepiminocholest-5-ene; 201–202.5 (*149*)
$C_{35}H_{55}NO_8$	(22ξ,23ξ,25*R*)-3β,16β,23-Triacetoxy-22,26-acetylepimino-5α-cholestan-22-ol; 131–136; +11.8° (pyridine) (*192*)
$C_{35}H_{56}N_2O_6$	(22*R*,23*S*,25*R*)-16α,23-Diacetoxy-3β-acetylamino-22,26-acetylepimino-5α-cholestane; 151–153 (*76*)
$C_{37}H_{51}NO_5$	(22*S*,25*R*)-3β-Acetoxy-*N*-benzyloxycarbonyl-22,26-epiminocholest-5-en-16-one (**58**); 162–164 (*172*)
$C_{37}H_{53}NO_5$	(22*S*,25*R*)-3β-Acetoxy-*N*-benzyloxycarbonyl-22,26-epimino-5α-cholestan-16-one (**104**); 156–158; −83.3° ($CHCl_3$) (*211*)
$C_{37}H_{53}NO_5$	(22*S*,25*R*)-3β-Acetoxy-*N*-benzyloxycarbonyl-22,26-epiminocholest-5-en-16β-ol (**57**); 124–127 (*172*)
$C_{37}H_{55}NO_5$	(22*S*,25*R*)-3β-Acetoxy-*N*-benzyloxycarbonyl-22,26-epimino-5α-cholestan-16β-ol (**103**); 127–129; −12.9° ($CHCl_3$) (*211*)
$C_{37}H_{57}NO_9$	(22*R*,23*S*,25*R*)-3β,4ξ,16α,23-Tetraacetoxy-22,26-acetylepimino-5α-cholestane; 210–211 (*81*)
$C_{38}H_{54}N_2O_6S$	(22*S*,25*R*)-16β-Acetoxy-22,26-acetylepimino-3-[*O*-(4-toluenesulfonyl)oximino]cholest-4-ene; 125–128 (*227*)
$C_{39}H_{55}NO_4S_2$	(22*S*,25*R*)-3β-Acetoxy-*N*-benzyloxycarbonyl-22,26-epimino-16,16-ethylenedimercaptocholest-5-ene (**59**); 115–118 (*172*)
$C_{39}H_{55}NO_7$	(22*R*,25*R*)-3β,16α-Diacetoxy-*N*-benzyloxycarbonyl-22,26-epimino-5α-cholestan-23-one (**149**); amorphous; −11.8° (pyridine) (*217*)
$C_{39}H_{55}NO_7$	(22*R*,25*R*)-3β,16β-Diacetoxy-*N*-benzyloxycarbonyl-22,26-epimino-5α-cholestan-23-one (**165**); amorphous; +11.5° (pyridine) (*192*)
$C_{39}H_{57}NO_4S_2$	(22*S*,25*R*)-3β-Acetoxy-*N*-benzyloxycarbonyl-22,26-epimino-16,16-ethylenedimercapto-5α-cholestane (**105**); 144–146 (*211*)
$C_{39}H_{57}NO_7$	(22*R*,23*S*,25*R*)-3β,16α-Diacetoxy-*N*-benzyloxycarbonyl-22,26-epimino-5α-cholestan-23-ol (**148**); 177–178; −25.5° (pyridine) (*217*)
$C_{39}H_{57}NO_7$	(22*R*,23*S*,25*R*)-3β,16β-Diacetoxy-*N*-benzyloxycarbonyl-22,26-epimino-5α-cholestan-23-ol; 177–178; +9.6° (pyridine) (*192*)

(*Continued*)

TABLE X (*Continued*)

Formula	Compound; melting point (°C); $[\alpha]_D$ (solvent) (reference)
$C_{39}H_{57}NO_7$	(22*R*,23*R*,25*R*)-3β,16β-Diacetoxy-*N*-benzyloxycarbonyl-22,26-epimino-5α-cholestan-23-ol; amorphous; +9.8° (pyridine) (*192*)
$C_{41}H_{59}NO_6S_2$	(22*R*,25*R*)-3β,16β-Diacetoxy-*N*-benzyloxycarbonyl-22,26-epimino-23,23-ethylenedimercapto-5α-cholestane; 200–205; +26.5° (pyridine) (*192*)
$C_{42}H_{69}NO_5$	(25*S*)-22,26-Pivaloylepimino-3β,16α-dipivaloyloxy-5α-cholest-22-ene; 202–205; +103.2° ($CHCl_3$) (*355*)

[a] Including the respective 27-norcholestane derivatives and the 20-(5-methyl-2-pyridyl)pregnanes equal to 22,26-epiminocholesta-22(*N*),23,25-trienes.
[b] See introductory paragraph of this Section.

TABLE XI
26-AMINOFUROSTANE DERIVATIVES

Formula	Compound; melting point (°C); $[\alpha]_D$ (solvent) (reference)
$C_{29}H_{43}NO_3$	(25*R*)-26-Acetylamino-3α, 5α-cyclofurost-20(22)-en-6-one; 87–88; +19° ($CHCl_3$) (*222*)
$C_{29}H_{45}NO_3$	(25*R*)-26-Acetylaminofurosta-5,20(22)-dien-3β-ol; 184–186; −21° ($CHCl_3$) (*222*)
$C_{29}H_{46}N_2O_5$	(25*R*)-26-Acetylamino-3β-nitryloxy-22ξ*H*-furost-5-ene; 140–141; −27° ($CHCl_3$) (*346*)
$C_{29}H_{47}NO_3$	(25*R*)-26-Acetylamino-22ξ*H*-furost-5-en-3β-ol; 189–191; −62° ($CHCl_3$) (*346*)
$C_{29}H_{49}NO_3$	(25*S*)-26-Acetylamino-5α,22α*H*(?)-furostan-3β-ol;[a] 174–176; −5° ($CHCl_3$) (*244*)
$C_{30}H_{47}NO_3$	(25*R*)-26-Acetylamino-16α-methylfurosta-5,20(22)-dien-3β-ol; 176–178 (*250*)
$C_{30}H_{47}NO_3$	(25*R*)-26-(*N*-Acetyl-methylamino)-furosta-5,20(22)-dien-3β-ol; 159–163 (*225*)
$C_{31}H_{49}NO_3$	(25*R*)-26-Acetylamino-16α-ethylfurosta-5,20(22)-dien-3β-ol; 167–170 (*250*)
$C_{32}H_{49}NO_4$	(25*R*)-3β-Acetoxy-26-acetylamino-16α-methylfurosta-5,20(22)-diene (**229**); 132–134 (*250*)
$C_{32}H_{50}N_2O_4$	(25*R*)-3β-Acetoxy-26-(*N'*,*N'*-dimethylureido)furosta-5,20(22)-diene (**233**); 154–159; −30.2° ($CHCl_3$) (*252*)
$C_{32}H_{50}N_2O_4$	(25*R*)-3β-Acetoxy-26-(*N'*,*N'*-dimethylureido)furosta-5,22-diene (**231**); 132–140 (*252*)
$C_{32}H_{53}NO_4$	(25*R*)-26-Acetylamino-22ξ-ethoxy-16α-methylfurost-5-en-3β-ol; 156–158 (*250*)
$C_{33}H_{51}NO_4$	(25*R*)-3β-Acetoxy-26-acetylamino-16α-ethylfurosta-5,20(22)-diene; 145–146.5 (*250*)
$C_{37}H_{51}NO_4$	(25*R*)-3β-Acetoxy-26-acetylamino-16α-phenylfurosta-5,20(22)-diene; 155–158 (*250*)

[a] See introductory paragraph of this Section.

TABLE XII
SOLANIDANE DERIVATIVES

Formula	Compound; melting point (°C); $[\alpha]_D$ (solvent) (reference)
$C_{27}H_{39}NO$	25*βH*-Solanida-4,20(22)-dien-3-one; 189–193; +190.4° (C_6H_6) (272)
$C_{27}H_{39}NO \cdot HClO_4$	25*βH*-Solanida-4,22(*N*)-dien-3-one perchlorate; 283–287; +11.9° (acetone/water) (*270*)
$C_{27}H_{39}NO_2$	22*αH*,25*βH*-Solanid-4-ene-3,11-dione; 245–249 (*247*)
$C_{27}H_{41}NO \cdot HClO_4$	22*αH*,25*βH*-Solanida-5,16(*N*)-dien-3*β*-ol perchlorate; 302–312; −143° (MeOH) (*347*)
$C_{27}H_{41}NO$	25*βH*-Solanida-5,20(22)-dien-3*β*-ol; 198–204; +110° (C_6H_6) (*272*)
$C_{27}H_{41}NO \cdot HClO_4$	25*βH*-Solanida-5,22(*N*)-dien-3*β*-ol perchlorate; 286–291; −89.8° (MeOH) (*270*)
$C_{27}H_{41}NO$	22*αH*,25*βH*-Solanid-4-en-3-one, quanylhydrazone dihydrochloride; 320–322 (dec.) (*340*)
$C_{27}H_{41}NO$	5*α*,25*βH*-Solanid-20(22)-en-3-one; 194–203; +165.8° (C_6H_6) (*272*)
$C_{27}H_{41}NO$	5*β*,25*βH*-Solanid-20(22)-en-3-one; 180–183; +192.2° (C_6H_6) (*272*)
$C_{27}H_{41}NO \cdot HClO_4$	5*α*,25*βH*-Solanid-22(*N*)-en-3-one perchlorate; 250–252; −54.2° (MeOH) (*270*)
$C_{27}H_{41}NO \cdot HClO_4$	5*β*,25*βH*-Solanid-22(*N*)-en-3-one perchlorate; 222–225; −46.0° (MeOH) (*270*)
$C_{27}H_{41}NO_2$	3*β*-Hydroxy-22*αH*,25*βH*-solanid-5-en-11-one; 204–207 (*247*)
$C_{27}H_{41}NO_2$	3*β*-Hydroxy-22*αH*,25*βH*-solanid-5-en-26-one (**124**);[a] 222–224; −26.5° (*207*)
$C_{27}H_{41}NO_2$	3*β*-Hydroxy-22*βH*,25*βH*-solanid-5-en-26-one;[a] 243–245; −31.2° (*207*)
$C_{27}H_{41}NO_2$	5*α*,22*αH*,25*βH*-Solanidane-3,12-dione (**27**); 228–231; +110° ($CHCl_3$) (*147*)
$C_{27}H_{43}NO$	5*α*,25*βH*-Solanid-20(22)-en-3*β*-ol (**290**); 197–199; +172° (ether) (*272*)
$C_{27}H_{43}NO$	(22*R*,25*S*)-*N*(16→18)-abeo-Solanid-5-en-3*β*-ol (procevine) (**69**); 235–237; −12.2° ($CHCl_3$) (*150*)
$C_{27}H_{43}NO_2$	22*αH*,25*βH*-Solanid-5-ene-3*β*,11*α*-diol; 240–244; −25.0° ($CHCl_3$) (*247*)
$C_{27}H_{43}NO_2$	22*αH*,25*βH*-Solanid-5-ene-3*β*,12*β*-diol; 231–234; −32.3° ($CHCl_3$) (*366*)
$C_{27}H_{43}NO_2$	22*αH*,25*βH*-Solanid-5-ene-3*β*,16*α*-diol; 156–166; −39.4° ($CHCl_3$) (*347*)
$C_{27}H_{43}NO_2$	22*βH*,25*αH*-Solanid-5-ene-3*β*,23*β*-diol (**141**); 134; −2.6° ($CHCl_3$) (*216*)
$C_{27}H_{43}NO_2$	22*αH*,25*βH*-Solanid-5-ene-3*β*,27-diol; 261–265; −19.4° (MeOH) (*368*)
$C_{27}H_{45}ClN_2$	3*α*-Chloroamino-5*α*,22*αH*,25*βH*-solanidane; +31.1° (dioxane) (*180*)
$C_{27}H_{45}ClN_2$	3*α*-Chloroamino-5*β*,22*αH*,25*βH*-solanidane; +40.6° (dioxane) (*180*)
$C_{27}H_{45}ClN_2$	3*β*-Chloroamino-5*α*,22*αH*,25*βH*-solanidane; +24.1° (dioxane) (*180*)
$C_{27}H_{45}ClN_2$	3*β*-Chloroamino-5*β*,22*αH*,25*βH*-solanidane; +29.4° (dioxane) (*180*)
$C_{27}H_{45}NO_2$	5*α*,22*αH*,25*βH*-Solanidane-3*β*,12*α*-diol; 217–219 (*147*)
$C_{27}H_{45}NO_2$	5*α*,22*αH*,25*βH*-Solanidane-3*β*,16*α*-diol; 163–167; +1.1° ($CHCl_3$) (*347*)
$C_{27}H_{45}NO_2$	5*α*,22*αH*,25*βH*-Solanidane-3*β*,23*α*-diol (**139**); 214–218; +3.3° ($CHCl_3$) (*216*; cf. *188*)

(*Continued*)

TABLE X (*Continued*)

Formula	Compound; melting point (°C); $[\alpha]_D$ (solvent) (reference)
$C_{27}H_{45}NO_2$	5α,22α*H*,25β*H*-Solanidane-3β,23β-diol (**138**);[a] 221–224; +31.3° ($CHCl_3$) (*216*)
$C_{27}H_{45}NO_2$	5α,22β*H*,25α*H*-Solanidane-3β,23β-diol (**140**); 127–131; +34.6° ($CHCl_3$) (*216*)
$C_{29}H_{43}NO_2 \cdot HClO_4$	3β-Acetoxy-22α*H*,25β*H*-solanida-5,16(*N*)-diene perchlorate; 276–280; −150.5° (MeOH) (*347*)
$C_{29}H_{43}NO_2$	3β-Acetoxy-25β*H*-solanida-5,20(22)-diene; 184–186; +42.4° (C_6H_6) (*272*)
$C_{29}H_{43}NO_2 \cdot HClO_4$	3β-Acetoxy-25β*H*-solanida-5,22(*N*)-diene perchlorate; 270–278; −110.9° (MeOH) (*270*)
$C_{29}H_{43}NO_3$	3β-Acetoxy-22α*H*,25β*H*-solanid-5-en-26-one (**125**); 230–232; −34.3° (*207*)
$C_{29}H_{43}NO_3$	3β-Acetoxy-22β*H*,25β*H*-solanid-5-en-26-one; 239–240; −28.2° (*207*)
$C_{29}H_{45}NO_2$	(22*R*,25*S*)-*N*(16→18)-abeo-3β-Acetoxysolanid-5-ene; 166–168 (*150*)
$C_{29}H_{45}NO_2 \cdot HClO_4$	3β-Acetoxy-5α,22α*H*,25β*H*-solanid-16(*N*)-ene perchlorate; 260–266; −94.2° (MeOH) (*347*)
$C_{29}H_{45}NO_2$	3β-Acetoxy-5α,25β*H*-solanid-20(22)-ene (**289**); 170–175; +120° (C_6H_6) (*272*); 168–171; +131° (*n*-butylacetate) (*273*)
$C_{29}H_{45}NO_2 \cdot HClO_4$	3β-Acetoxy-5α,25β*H*-solanid-22(*N*)-ene perchlorate (**286**); 254–257; −73° (MeOH) (*270*)
$C_{29}H_{45}NO_3$	3β-Acetoxy-22α*H*,25β*H*-solanid-5-en-16α-ol; 151–155; −60.4° (C_6H_6) (*347*)
$C_{29}H_{47}NO_3$	3β-Acetoxy-5α,22α*H*,25β*H*-solanidan-16α-ol; 166–169; −11.2° (C_6H_6) (*347*)
$C_{29}H_{48}N_2$	3β-Dimethylamino-5α,25β*H*-solanid-20(22)-ene; 178–185; +127.4° (C_6H_6) (*272*)
$C_{29}H_{48}N_2 \cdot 2HClO_4$	3β-Dimethylamino-5α,25β*H*-solanid-22(*N*)-ene bisperchlorate; 295–300; −59.9° (acetone/water) (*270*)
$C_{29}H_{48}N_2S_2$	*N*-Methylmercaptothiocarbonyl-3α-amino-5α,22α*H*,25β*H*-solanidane; 125–130; +34.4° (MeOH) (*182*)
$C_{29}H_{48}N_2S_2$	*N*-Methylmercaptothiocarbonyl-3α-amino-5β,22α*H*,25β*H*-solanidane; +62.5° (MeOH) (*182*)
$C_{29}H_{48}N_2S_2$	*N*-Methylmercaptothiocarbonyl-3β-amino-5α,22α*H*,25β*H*-solanidane; 186–189; +4.6° ($CHCl_3$) (*182*)
$C_{29}H_{48}N_2S_2$	*N*-Methylmercaptothiocarbonyl-3β-amino-5β,22α*H*,25β*H*-solanidane; 166–170; +33.4° (dioxane) (*182*)
$C_{29}H_{50}N_2O$	3β-Dimethylamino-5α,22β*H*,25α*H*-solanidan-23β-ol; 111–112; +41.3° ($CHCl_3$) (*76*)
$C_{31}H_{47}NO_4$	3β,23β-Diacetoxy-22β*H*,25α*H*-solanid-5-ene; 174–177; −9.8° ($CHCl_3$) *216*)
$C_{31}H_{49}NO_4$	3β,23β-Diacetoxy-5α,22α*H*,25β*H*-solanidane;[a] 229–232; +0.9° ($CHCl_3$) (*216*)
$C_{31}H_{50}N_2O_3$	23β-Acetoxy-3β-acetylamino-5α,22β*H*,25α*H*-solanidane; 219–221; +27.2° ($CHCl_3$) (*76*)

[a] See introductory paragraph of this Section.

TABLE XIII
SOLANOCAPSINE AND ITS DERIVATIVES

Formula	Compound; melting point (°C); $[\alpha]_D$ (solvent) (reference)
$C_{27}H_{43}NO_3$	22,26-Epimino-16α,23-epoxy-23β-hydroxy-5α,22α*H*, 25β*H*-cholestan-3-one (**155**); 208–210; +26.6° (pyridine) (*217*)
$C_{27}H_{43}NO_3$	22,26-Epimino-16β,23-epoxy-23α-hydroxy-5α,22α*H*, 25β*H*-cholestan-3-one; 212–215 (dec.); +27.9° (pyridine) (*220*)
$C_{27}H_{44}N_2O_2$	3β-Amino-22,26-epimino-16α,23-epoxy-22α*H*,25β*H*-cholest-5-en-23β-ol (solanoforthine) (**77**); 208–210; −26.6° ($CHCl_3$) (*110*)
$C_{27}H_{44}N_2O_3$	22,26-Epimino-16α,23-epoxy-3-oximino-5α,22α*H*,25β*H*-cholestan-23β-ol (**156**); 225–228 (dec.); +24.8° (pyridine) (*217*)
$C_{27}H_{44}N_2O_4$	16α,23-Epoxy-22,26-nitrosoepimino-5α,22α*H*,25β*H*-cholestane-3β,23β-diol (**151**);[a] 217–219; +169° (pyridine) (*217*)
$C_{27}H_{45}NO_3$	22,26-Epimino-16α,23-epoxy-5α,22α*H*,25β*H*-cholestane-3β,23β-diol (**152**); 206–207; +27.1° (pyridine) (*217*)
$C_{27}H_{46}N_2O_2$	3α-Amino-22,26-epimino-16α,23-epoxy-5α,22α*H*, 25β*H*-cholestan-23β-ol (**157**); 180–185/208–214 (dec.); +27.2° (pyridine) (*217*)
$C_{27}H_{46}N_2O_2$	3β-Amino-22,26-epimino-16β,23-epoxy-5α,22α*H*,25β*H*-cholestan-23α-ol (**160**); 125–129; +12.0° (pyridine) (*192*; cf. *220*)
$C_{27}H_{46}N_2O_2$	3β-Amino-22,26-epimino-16α,23-epoxy-5α,22α*H*, 25β*H*-cholestan-23β-ol (solanocapsine) (**4**);[a] 199–201; +10.7° ($CHCl_3$) (*110*)
$C_{28}H_{46}N_2O_2$	3β-Amino-22,26-epimino-16α,23-epoxy-23β-methoxy-5α,25β*H*-cholest-22(*N*)-ene (solacasine) (**84**); 215–220; +29° (MeOH) (*105*)
$C_{28}H_{48}N_2O_2$	3β-Amino-22,26-epimino-16α,23-epoxy-23β-methoxy-5α,22α*H*,25β*H*-cholestane (dihydrosolacasine) (**83**); picrate; 178–182 (*105*)
$C_{29}H_{46}N_2O_5$	3β-Acetoxy-16α,23-epoxy-22,26-nitrosoepimino-5α,22α*H*,25β*H*-cholestan-23β-ol; 212–214; +144° (pyridine) (*217*)
$C_{29}H_{47}NO_4$	22,26-Acetylepimino-16α,23-epoxy-5α,22α*H*,25β*H*-cholestane-3β, 23β-diol;[a] 290–292 (dec.); −40.8° (pyridine) (*217*)
$C_{30}H_{48}N_2O_2$	*N*(3)-Isopropylidenesolanoforthine; 209–211; −37.7° ($CHCl_3$) (*110*)
$C_{30}H_{50}N_2O_2$	3α-Dimethylamino-16α, 23-epoxy-22,26-methylepimino-22α*H*, 25β*H*-cholest-5-en-23β-ol (**78**); 201–203; +5.3° ($CHCl_3$) (*110*)
$C_{30}H_{52}N_2O_2$	3β-Dimethylamino-16α,23-epoxy-22,26-methylepimino-5α,22α*H*,25β*H*-cholestan-23β-ol (**79**);[a] 221–223; +42.6° ($CHCl_3$) (*110*)
$C_{31}H_{47}NO_4$	3β-Acetoxy-22,26-acetylepimino-16α,23-epoxy-5α,25β*H*-cholest-22-ene;[a] 262–266; +10.4° (*219*); 245; +12.4° (pyridine) (*192*)
$C_{31}H_{47}NO_4$	3β-Acetoxy-22,26-acetylepimino-16β,23-epoxy-5α,25β*H*-cholest-22-ene; 185–187; −184° (pyridine) (*192*)
$C_{31}H_{48}N_2O_4$	3β-Acetylamino-22,26-acetylepimino-16α,23-epoxy-22α*H*,25β*H*-cholest-5-en-23β-ol; 220–221 (*110*)
$C_{31}H_{49}NO_5$	3β-Acetoxy-22,26-acetylepimino-16α,23-epoxy-5α,22α*H*,25β*H*-cholestan-23β-ol (**159**);[a] 202–205 (*219*); 205–207; −53.2° (pyridine) (*217*)
$C_{31}H_{49}NO_5$	3β-Acetoxy-22,26-acetylepimino-16β,23-epoxy-5α,22α*H*,25β*H*-cholestan-23α-ol; 182–185, 205–210; −42.8° (pyridine) (*192*; cf. *220*)

(*Continued*)

TABLE XIII (*Continued*)

Formula	Compound; melting point (°C); $[\alpha]_D$ (solvent) (reference)
$C_{32}H_{50}N_2O_4$	Solacasine diacetate; 186–189 (*105*)
$C_{34}H_{49}BrN_2O_2$	*N*-(2-Bromobenzylidene)-solanocapsine; 237–240; +17.8° (pyridine) (*187*)
$C_{34}H_{49}NO_3$	22,26-Benzylepimino-16α,23-epoxy-23β-hydroxy-5α,22α*H*,25β*H*-cholestan-3-one; 232–233; +63.2° (pyridine) (*217*)
$C_{34}H_{50}N_2O_3$	22,26-Epimino-16α,23-epoxy-3α-salicylidenamino-5α,22α*H*,25β*H*-cholestan-23β-ol; amorphous; −6.2° (pyridine) (*217*)
$C_{34}H_{50}N_2O_3$	22,26-Epimino-16α,23-epoxy-3β-salicylidenamino-5α,22α*H*,25β*H*-cholestan-23β-ol;[a] 240–242; +35.0° ($CHCl_3$) (*358*)
$C_{34}H_{50}N_2O_3$	22,26-Epimino-16β,23-epoxy-3β-salicylidenamino-5α,22α*H*,25β*H*-cholestan-23α-ol; 190–193; +30.7° (pyridine) (*192*; cf. *220*)
$C_{35}H_{49}NO_4$	*N*-Benzyloxycarbonyl-22,26-epimino-16α,23-epoxy-5α,25β*H*-cholest-22-en-3β-ol; 177–179; +5.1° (pyridine) (*217*)
$C_{35}H_{49}NO_5$	*N*-Benzyloxycarbonyl-22,26-epimino-16α,23-epoxy-23β-hydroxy-5α, 22α*H*,25β*H*-cholestan-3-one (**153**); 211–213; +45.4° (pyridine) (*217*)
$C_{35}H_{49}NO_5$	*N*-Benzyloxycarbonyl-22,26-epimino-16β,23-epoxy-23α-hydroxy-5α, 22α*H*,25β*H*-cholestan-3-one; 198–201; −16.4° (pyridine) (*220*)
$C_{35}H_{50}N_2O_5$	*N*-Benzyloxycarbonyl-22,26-epimino-16β,23-epoxy-3-oximino-5α,22α*H*, 25β*H*-cholestan-23α-ol (**167**); amorphous; −32.6° (pyridine) (*192*; cf. *220*)
$C_{35}H_{51}NO_5$	*N*-Benzyloxycarbonyl-22,26-epimino-16α,23-epoxy-5α,22α*H*,25β*H*-cholestane-3β,23β-diol (**150**); 219–221; +78.0° (pyridine) (*217*)
$C_{35}H_{51}NO_5$	*N*-Benzyloxycarbonyl-22,26-epimino-16β,23-epoxy-5α,22α*H*,25β*H*-cholestane-3β,23α-diol; 223–227; −36.4° (pyridine) (*220*)
$C_{35}H_{52}N_2O_3$	*N*-[1-(2-Hydroxyphenyl)ethylidene]solanocapsine; 244–246; +45.6° (dioxane) (*181*)
$C_{36}H_{51}NO_5$	*N*-Benzyloxycarbonyl-22,26-epimino-16β,23-epoxy-23α-methoxy-5α, 22α*H*,25β*H*-cholestan-3-one; 192–195; −26.4° (pyridine) (*192*; cf. *220*)
$C_{36}H_{53}NO_5$	*N*-Benzyloxycarbonyl-22,26-epimino-16α,23-epoxy-23β-methoxy-5α, 22α*H*,25β*H*-cholestan-3β-ol; 118–121; +74.4° (pyridine) (*217*)
$C_{36}H_{53}NO_5$	*N*-Benzyloxycarbonyl-22,26-epimino-16β,23-epoxy-23α-methoxy-5α, 22α*H*,25β*H*-cholestan-3β-ol (**166**); 154–157; −32.5° (pyridine) (*192*; cf. *220*)
$C_{37}H_{51}NO_5$	3β-Acetoxy-*N*-benzyloxycarbonyl-22,26-epimino-16α,23-epoxy-5α,25β*H*-cholest-22-ene; 217–219; +8.0° (pyridine) (*217*)
$C_{37}H_{53}NO_6$	3β-Acetoxy-*N*-benzyloxycarbonyl-22,26-epimino-16α,23-epoxy-5α,22α*H*, 25β*H*-cholestan-23β-ol; 119–122; +24.3° (pyridine) (*217*)
$C_{37}H_{53}NO_6$	3β-Acetoxy-*N*-benzyloxycarbonyl-22,26-epimino-16β,23-epoxy-5α, 22α*H*, 25β*H*-cholestan-23α-ol; 168–170; −38.4° (pyridine) (*220*)
$C_{43}H_{58}N_2O_6$	*N*,*N′*-Dibenzyloxycarbonylsolanocapsine; 169–171; +25.6° (pyridine) (*217*)

[a] See introductory paragraph of this Section.

TABLE XIV

3-AMINOSPIROSTANES AND THEIR DERIVATIVES

Formula	Compound; melting point (°C); $[\alpha]_D$ (solvent) (reference)
$C_{27}H_{39}N_3O_3$	(25*R*)-3β-Azido-22α*O*-spirost-5-en-7-one; 220–222; −150° ($CHCl_3$) *(99)*
$C_{27}H_{41}NO_4$	(25*R*)-3-Oximino-5α,22α*O*-spirostan-6-one (**94**); 223–224; −96° *(99)*
$C_{27}H_{41}N_3O_2$	(25*R*)-3α-Azido-22α*O*-spirost-5-ene; 181–182; −86° ($CHCl_3$) *(99)*
$C_{27}H_{41}N_3O_2$	(25*R*)-3β-Azido-22α*O*-spirost-5-ene (**91**); 156–157; −104° ($CHCl_3$) *(99)*
$C_{27}H_{41}N_3O_3$	(25*R*)-3β-Azido-5α,6α-epoxy-22α*O*-spirostane; 142–143; −112° ($CHCl_3$) *(99)*
$C_{27}H_{41}N_3O_3$	(25*R*)-3β-Azido-5α,22α*O*-spirostan-6-one (**93**); 180–181; −92° ($CHCl_3$) *(99)*
$C_{27}H_{41}N_3O_3$	(25*R*)-3β-Azido-22α*O*-spirost-5-en-7α-ol; 174.5–175; −137° ($CHCl_3$) *(99)*
$C_{27}H_{41}N_3O_3$	(25*R*)-3β-Azido-22α*O*-spirost-5-en-7β-ol; 186–187; −63° ($CHCl_3$) *(99)*
$C_{27}H_{43}NO_3$	(25*R*)-3β-Amino-22α*O*-spirost-5-en-7α-ol; 143–145; −156° ($CHCl_3$) *(99)*
$C_{27}H_{43}NO_3$	(25*R*)-3β-Amino-22α*O*-spirost-5-en-7β-ol; 171–173; −76° ($CHCl_3$) *(99)*
$C_{27}H_{43}N_3O_3$	(25*R*)-3β-Azido-5α,22α*O*-spirostan-6α-ol (**92**); 186–187; −49° ($CHCl_3$) *(99)*
$C_{27}H_{43}N_3O_3$	(25*R*)-3β-Azido-5α,22α*O*-spirostan-6β-ol; 190–192; −79° ($CHCl_3$) *(99)*
$C_{27}H_{45}NO_2$	(25*R*)-3β-Amino-5α,22α*O*-spirostane (isojurubidine) (**86**); 185–187; −63° ($CHCl_3$) *(100)*
$C_{27}H_{45}NO_3$	(25*R*)-3β-Amino-22α*O*-spirostan-5α-ol; 205–206; −77° ($CHCl_3$) *(99)*
$C_{27}H_{45}NO_3$	(25*R*)-3β-Amino-5α,22α*O*-spirostan-6α-ol (isojuripidine) (**88**); 204–205; −47° ($CHCl_3$) *(100)*
$C_{27}H_{45}NO_3$	(25*R*)-3β-Amino-5α,22α*O*-spirostan-6β-ol; 208–210; −86° ($CHCl_3$) *(99)*
$C_{27}H_{45}NO_3$	(25*R*)-3β-Amino-5α,22α*O*-spirostan-7α-ol; 228–232; −78° *(99)*
$C_{27}H_{45}NO_3$	(25*R*)-3β-Amino-5α,22α*O*-spirostan-7β-ol; 181–184; −40° ($CHCl_3$) *(99)*
$C_{27}H_{45}NO_3$	(25*R*)-3β-Amino-5α,22α*O*-spirostan-9α-ol (isopaniculidine) (**87**); 202–204; −70° ($CHCl_3$) *(100)*
$C_{29}H_{47}NO_3$	*N*-Acetylisojurubidine;[a] 262–265; −70° ($CHCl_3$) *(100)*
$C_{29}H_{47}NO_4$	*N*-Acetylisopaniculidine; 282–284; −75° ($CHCl_3$) *(100)*
$C_{30}H_{49}NO_2$	*N*-Isopropylidenisojurubidine; 182–184; −65.9° ($CHCl_3$) *(100)*
$C_{30}H_{49}NO_3$	*N*-Isopropylidenisojuripidine; 200–201; − 56°($CHCl_3$) *(100)*; 204–205; −58° ($CHCl_3$) *(99)*
$C_{30}H_{49}NO_3$	*N*-Isopropylidenisopaniculidine; 196–198; −69° ($CHCl_3$) *(100)*
$C_{31}H_{49}NO_4$	*N,N*-Diacetylisojurubidine; 184–187; −48° ($CHCl_3$) *(100)*
$C_{31}H_{49}NO_5$	(25*R*)-6α-Acetoxy-3β-acetylamino-5α,22α*O*-spirostane; 153–154; −46° ($CHCl_3$) *(100)*
$C_{34}H_{49}NO_2$	*N*-Benzylidenisojurubidine; 198–200; −50° ($CHCl_3$) *(100)*
$C_{34}H_{49}NO_3$	*N*-Benzylidenisojuripidine; 245–247; −26° ($CHCl_3$) *(100)*
$C_{34}H_{49}NO_3$	*N*-Benzylidenisopaniculidine; 224–226; −47° ($CHCl_3$) *(100)*
$C_{34}H_{49}NO_3$	*N*-Salicylidenisojurubidine; 220–221; −36° ($CHCl_3$) *(100)*
$C_{34}H_{49}NO_4$	*N*-Salicylidenisojuripidine; 243–245, −18° ($CHCl_3$) *(100)*
$C_{34}H_{49}NO_4$	*N*-Salicylidenisopaniculidine; 276–278; −26° *(100)*

[a] See introductory paragraph of this Section.

TABLE XV
ADDITIONAL NITROGENOUS STEROIDS DERIVED BY DEGRADATION OR SYNTHESIS OF *SOLANUM* ALKAMINES

Formula	Compound; melting point (°C); $[\alpha]_D$ (solvent) (reference)
$C_{26}H_{41}NO_4$	(25*R*)-5-Oxo-4-nor-3,5-seco-22α*N*-spirosolan-3-oic acid (**187**); 211–212 (*231*); 210–221 (*232*)
$C_{26}H_{43}NO_4$	(22*S*,25*R*)-22,26-Epimino-16β-hydroxy-5-oxo-4-nor-3,5-secocholestan-3-oic acid; 198–199 (*231*)
$C_{26}H_{47}NO_3$	(22*S*,25*R*)-22,26-Epimino-4-nor-3,5-secocholestane-3,5ξ,16β-triol; 225–226 (*231*)
$C_{27}H_{39}NO_4$	(25*S*)-3,22-Dioxo-22,23-secosolanid-4-en-23-oic acid; 226–235; −44.5° (MeOH) (*271*)
$C_{27}H_{41}NO_4$	(25*S*)-3,22-Dioxo-22,23-seco-5α-solanidan-23-oic acid; 266–273; +7.1° (MeOH) (*271*)
$C_{27}H_{41}NO_4$	(25*S*)-3,22-Dioxo-22,23-seco-5β-solanidan-23-oic acid; 226–235; +3.4° (MeOH) (*271*)
$C_{27}H_{41}NO_4$	(25*S*)-3β-Hydroxy-22-oxo-22,23-secosolanid-5-en-23-oic acid; 205–213; −51.2° (MeOH) (*271*)
$C_{27}H_{41}NO_5$	(25*S*)-3β,20β-Dihydroxy-22-oxo-22,23-secosolanid-5-en-23-oic acid; 242; −64.3° (MeOH) (*274*)
$C_{27}H_{42}N_2O_2$	(25*R*)-3-Aza-A-homo-22α*N*-spirosol-4a-en-4-one (**178**); 266–268 (*227*)
$C_{27}H_{43}NO_3$	(25*R*)-4-Oxa-A-homo-5α,22α*N*-spirosolan-3-one (**182**); 198–200 (*229*)
$C_{27}H_{43}NO_3$	(25*R*)-4-Oxa-A-homo-5β,22α*N*-spirosolan-3-one (**183**); 188–190 (*229*)
$C_{27}H_{43}NO_4$	(25*S*)-3β-Hydroxy-22-oxo-22,23-seco-5α-solanidan-23-oic acid;[a] 241–247; −11.4° (MeOH) (*271*)
$C_{27}H_{43}NO_4$	Methyl (25*R*)-5-oxo-4-nor-3,5-seco-22α*N*-spirosolan-3-oate; 237–239 (*231*); 219–221 (*232*); oxime; 204–206 (*232*); 2,4-dinitrophenylhydrazone; 217–219 (*232*)
$C_{27}H_{43}NO_5$	(25*S*)-3β,20β-Dihydroxy-22-oxo-22,23-seco-5α-solanidan-23-oic acid (**291**); 263; −23.8° (MeOH) (*274*)
$C_{27}H_{43}NO_5$	(25*R*)-3β-Hydroxy-5-oxo-5,6-seco-22α*N*-spirosolan-6-oic acid (**186**); 192–193 (*231*)
$C_{27}H_{44}N_2O_2$	(25*R*)-3-Aza-A-homo-5α,22α*N*-spirosolan-4-one (**180**); 236–237.5 (*228*)
$C_{27}H_{44}N_2O_2$	(25*R*)-4-Aza-A-homo-5α,22α*N*-spirosolan-3-one (**181**); 229–231 (*228*)
$C_{27}H_{44}N_2O_2$	(25*R*)-3-Aza-A-homo-5β,22α*N*-spirosolan-4-one (**179**); 230–232 (*227*)
$C_{27}H_{44}N_2O_2$	(22*S*,25*R*)-22,26-Epimino-16β-hydroxy-3-aza-A-homocholest-4a-en-4-one; 141–144 (*227*)
$C_{27}H_{45}NO_3$	(25*S*)-22,23-Secosolanid-5-ene-3β,20β,23-triol; 142–144; −39.2° (MeOH) (*274*)
$C_{27}H_{45}NO_4$	Methyl (22*S*,25*R*)-22,26-epimino-16β-hydroxy-5-oxo-4-nor-3,5-secocholestan-3-oate; 189–191 (*231*)
$C_{27}H_{45}NO_5$	(22*S*,25*R*)-22,26-Epimino-3β,16β-dihydroxy-5-oxo-5,6-secocholestan-7-oic acid; 186–188 (*231*)
$C_{27}H_{46}N_2O$	(25*R*)-3-Aza-A-homo-5α,22α*N*-spirosolane; 218–219.5 (*228*)
$C_{27}H_{46}N_2O$	(25*R*)-4-Aza-A-homo-5α,22α*N*-spirosolane; 199–201 (*228*)

(*Continued*)

TABLE XV (*Continued*)

Formula	Compound; melting point (°C); $[\alpha]_D$ (solvent) (reference)
$C_{27}H_{46}N_2O_2$	(22*S*,25*R*)-22,26-Epimino-16β-hydroxy-3-aza-A-homo-5β-cholestan-4-one; 198–200 (*227*)
$C_{27}H_{47}NO_3$	(25*S*)-22,23-Seco-5α-solanidane-3β,20β,23-triol (**292**); 164–167; −9.9° (MeOH) (*274*)
$C_{27}H_{49}NO_3$	(22*S*,25*R*)-22,26-Epimino-3,4-seco-5α-cholestane-3,4,16β-triol; 119–122 (*229*)
$C_{27}H_{49}NO_3$	(22*S*,25*R*)-22,26-Epimino-3,4-seco-5β-cholestane-3,4,16β-triol; 157–159 (*229*)
$C_{27}H_{49}NO_4$	(22*S*,25*R*)-22,26-Epimino-5,6-secocholestane-3β,5ξ,6,16β-tetraol; 211–212 (*231*)
$C_{28}H_{41}NO_4$	(25*R*)-*N*-Acetyl-4-oxa-22α*N*-spirosol-5-en-3-one (**189**); 88–89 (*230*)
$C_{28}H_{43}NO_6$	Methyl (22*S*,25*S*)-3β-hydroxy-22-nitro-16-oxocholest-5-en-26-oate; 160–161; −132° (*207*)
$C_{28}H_{43}NO_6$	Methyl (22*R*,25*S*)-3β-hydroxy-22-nitro-16-oxocholest-5-en-26-oate (**97**); 163–164; −125° (*207*)
$C_{28}H_{43}NO_6$	Methyl (22ξ,25*R*)-3β-hydroxy-22-nitro-16-oxocholest-5-en-26-oate; 162–163, 166–167 (*209*)
$C_{28}H_{45}NO_5$	Methyl (25*R*)-3β-hydroxy-5-oxo-5,6-seco-22α*N*-spirosolan-6-oate; 199–201 (*231*)
$C_{28}H_{45}NO_6$	Methyl (22*S*,25*S*)-3β,16β-dihydroxy-22-nitrocholest-5-en-26-oate; 126–127; −30° (*209*)
$C_{28}H_{45}NO_6$	Methyl (22*R*,25*S*)-3β,16β-dihydroxy-22-nitrocholest-5-en-26-oate (**98**); 131–132; −39° (*209*)
$C_{28}H_{47}NO_5$	Methyl (22*S*,25*R*)-22,26-epimino-3β,16β-dihydroxy-5-oxo-5,6-secocholestan-7-oate; 196–197 (*231*)
$C_{29}H_{43}NO_5$	(25*S*)-3β-Acetoxy-22-oxo-22,23-secosolanid-5-en-23-oic acid; 197–211; −55.8° (MeOH) (*271*)
$C_{29}H_{45}NO_4$	(25*R*)-26-Acetylamino-3β-hydroxycholest-5-en-16,22-dione; 212–214 (*250*)
$C_{29}H_{45}NO_5$	(25*S*)-3β-Acetoxy-22-oxo-22,23-seco-5α-solanidan-23-oic acid (**288**);[a] 208–211; −18.2° (MeOH) (*271*)
$C_{29}H_{49}NO_3$	(22*S*,25*R*)-26-Acetylaminocholest-5-ene-3β,22-diol; 168–171; −21° ($CHCl_3$) (*345*)
$C_{29}H_{49}NO_3$	(22*R*,25*R*)-26-Acetylaminocholest-5-ene-3β,22-diol; 181–184; +2° ($CHCl_3$) (*345*)
$C_{30}H_{45}NO_5$	(22*S*,25*R*)-16β-Acetoxy-22,26-acetylepimino-4-oxacholest-5-en-3-one; 77–79 (*230*)
$C_{30}H_{45}NO_7$	Methyl (22*S*,25*S*)-3β-acetoxy-22-nitro-16-oxocholest-5-en-26-oate (**118**); 156–158 (*215*)
$C_{30}H_{45}NO_7$	Methyl (22*R*,25*S*)-3β-acetoxy-22-nitro-16-oxocholest-5-en-26-oate; 146–148 (*215*)
$C_{31}H_{45}NO_6$	(25*R*)-3β-Acetoxy-*N*-acetyl-6-oxa-B-homo-22α*N*-spirosol-4-en-7-one (**188**); 81–83 (*230*)

(*Continued*)

TABLE XV (*Continued*)

Formula	Compound; melting point (°C); $[\alpha]_D$ (solvent) (reference)
$C_{31}H_{47}NO_5$	(25*R*)-3β-Acetoxy-26-acetylaminocholest-5-en-16,22-dione (**228**); 178–180 (*250*)
$C_{31}H_{47}NO_6$	3β-Acetoxy-22,26-acetylepimino-16α,23-epoxy-22,23-seco-5α,25β*H*-cholestane-22,23-dione;[a] −1.0° ($CHCl_3$) (*192*)
$C_{31}H_{47}NO_6$	3β-Acetoxy-22,26-acetylepimino-16β,23-epoxy-22,23-seco-5α,25β*H*-cholestane-22,23-dione; 238–241; +53.3° ($CHCl_3$), +63.5° (pyridine) (*192*)
$C_{31}H_{48}N_2O_4$	(22*S*,25*R*)-16β-Acetoxy-22,26-acetylepimino-3-aza-A-homocholest-4a-en-4-one; 150–151 (*227*)
$C_{31}H_{49}NO_5$	(25*R*)-16β-Acetoxy-26-acetylamino-3β-hydroxycholest-5-en-22-one; 184–184.5; +19° ($CHCl_3$) (*346*)
$C_{31}H_{49}N_3O_9$	(22*S*,25*R*)-16β-Acetoxy-26-acetylamino-3β,22-dinitryloxycholest-5-ene; amorphous; +10° ($CHCl_3$) (*346*)
$C_{31}H_{49}N_3O_9$	(22*R*,25*R*)-16β-Acetoxy-26-acetylamino-3β,22-dinitryloxycholest-5-ene; amorphous; +16° ($CHCl_3$) (*346*)
$C_{31}H_{50}N_2O_4$	(22*S*,25*R*)-16β-Acetoxy-22,26-acetylepimino-3-aza-A-homo-5β-cholestan-4-one; 155–156 (*227*)
$C_{31}H_{51}NO_4$	(25*R*)-3β-Acetoxy-26-acetylamino-5α-cholestan-22-one;[a] 138–140 (*62*)
$C_{31}H_{51}NO_4$	(22*S*,25*R*)-3β-Acetoxy-26-acetylaminocholest-5-en-22-ol; 157–159; −24° ($CHCl_3$) (*345*)
$C_{31}H_{51}NO_4$	(22*R*,25*R*)-3β-Acetoxy-26-acetylaminocholest-5-en-22-ol; 160–163; −6° ($CHCl_3$) (*345*)
$C_{31}H_{51}NO_5$	(22*S*,25*R*)-16β-Acetoxy-26-acetylaminocholest-5-ene-3β,22-diol; 173.5–175.5; −5° ($CHCl_3$) (*346*)
$C_{31}H_{51}NO_5$	(22*R*,25*R*)-16β-Acetoxy-26-acetylaminocholest-5-ene-3β,22-diol; 196.5–197 (*346*)
$C_{32}H_{49}NO_6S_2$	Methyl (22*S*,25*S*)-3β-acetoxy-16,16-ethylenedimercapto-22-nitrocholest-5-en-26-oate (**119**); 226–227 (*215*)
$C_{32}H_{49}NO_8$	Methyl (22*S*,25*S*)-3β,16β-diacetoxy-22-nitrocholest-5-en-26-oate; 260–262 (*209*)
$C_{32}H_{49}NO_8$	Methyl (22*R*,25*S*)-3β,16β-diacetoxy-22-nitrocholest-5-en-26-oate; 266–268 (*209*)
$C_{33}H_{49}NO_7$	(22*S*,25*R*)-3β,16β-Diacetoxy-22,26-acetylepimino-6-oxa-B-homocholest-4-en-7-one; 63–64 (*230*)
$C_{33}H_{53}NO_6$	(25*S*)-3β,16α-Diacetoxy-26-acetylamino-5α-cholestan-22-one; 170–171; −64.4° ($CHCl_3$) (*355*)
$C_{33}H_{60}N_2O_3$	3β-(*N*-Acetyl-ethylamino)-24-(*N*-ethyl-2*R*-4-hydroxy-2-methylbutylamino)-22,23-bisnor-5α-cholan-16α-ol (**74**); amorphous; −26.3° ($CHCl_3$) (*192*); hydrochloride; 285–290; −1.6° (EtOH) (*192*)
$C_{34}H_{55}NO_7$	(22*S*,25*R*)-3,5ξ,16β-Triacetoxy-22,26-acetylepimino-4-nor-3,5-secocholestane; 176–177 (*231*)
$C_{37}H_{59}NO_9$	(22*S*,25*R*)-3β,5ξ,6,16β-Tetraacetoxy-22,26-acetylepimino-5,6-secocholestane; 166–167 (*231*)

(*Continued*)

TABLE XV (*Continued*)

Formula	Compound; melting point (°C); $[\alpha]_D$ (solvent) (reference)
$C_{37}H_{63}NO_4$	(22*R*,25*R*)-26-Amino-3β,22-ditetrahydropyranyloxycholest-5-ene; 83–85 (*346*)
$C_{37}H_{63}NO_4$	(22*R*,25*R*)-26-Amino-3β,22-ditetrahydropyranyloxycholest-5-ene; 97–102 (*346*)
$C_{37}H_{64}N_2O_5$	16α-Acetoxy-3β-(*N*-acetyl-ethylamino)-24-(*N*-ethyl-2*R*-4-acetoxy-2-methylbutylamino)-22,23-bisnor-5α-cholane (**73**); 111–112; −37.6° ($CHCl_3$) (*192*); hydrochloride; 225–230 (dec.); −30.7° (EtOH) (*192*)
$C_{39}H_{63}NO_6$	(22*S*,25*R*)-26-Acetylamino-3β,22-ditetrahydropyranyloxycholest-5-en-16-one; 110–114; −148° ($CHCl_3$) (*346*)
$C_{39}H_{63}NO_6$	(22*R*,25*R*)-26-Acetylamino-3β,22-ditetrahydropyranyloxycholest-5-en-16-one; 157–159; −140° ($CHCl_3$) (*346*)
$C_{39}H_{65}NO_5$	(22*S*,25*R*)-26-Acetylamino-3β,22-ditetrahydropyranyloxycholest-5-ene; amorphous; −23° ($CHCl_3$) (*346*)
$C_{39}H_{65}NO_5$	(22*R*,25*R*)-26-Acetylamino-3β,22-ditetrahydropyranyloxycholest-5-ene; 159–161; −12° ($CHCl_3$) (*346*)
$C_{39}H_{65}NO_6$	(22*S*,25*R*)-26-Acetylamino-3β,22-ditetrahydropyranyloxycholest-5-en-16β-ol; amorphous; −46° ($CHCl_3$) (*346*)
$C_{39}H_{65}NO_6$	(22*R*,25*R*)-26-Acetylamino-3β,22-ditetrahydropyranyloxycholest-5-en-16β-ol; amorphous; −32° ($CHCl_3$) (*346*)
$C_{41}H_{67}NO_7$	(22*S*,25*R*)-16β-Acetoxy-26-acetylamino-3β,22-ditetrahydropyranyloxy-cholest-5-ene; amorphous; +8° ($CHCl_3$) (*346*)
$C_{41}H_{67}NO_7$	(22*R*,25*R*)-16β-Acetoxy-26-acetylamino-3β,22-ditetrahydropyranyloxy-cholest-5-ene; amorphous; +21° ($CHCl_3$) (*346*)

[a] See introductory paragraph of this Section.

TABLE XVI
ALKALOIDS WHOSE STRUCTURES ARE NOT COMPLETELY KNOWN

Formula	Compound; melting point (°C); $[\alpha]_D$ (solvent) (reference)
$C_{27}H_{41}NO_3$	Corsevinine; 224–225; −16.0° (*129*)
$C_{27}H_{43}NO_2$	Veralosidine; 153–155; −92.2° (EtOH) (*141*)
$C_{27}H_{43}NO_2$	Petiline; 205–206; −51.1° (EtOH) (*132*); hydrochloride; 288–289 (*132*)
$C_{28}H_{47}NO_2$	Edpetilidinine; 269–271; +42.5° (EtOH) (*130, 348*)
$C_{28}H_{47}NO_2$	Rhinolidine; 199–201; −52.9° (EtOH) (*350*)
$C_{29}H_{43}NO_4$	Veralodisine; 172–174; −92.8° ($CHCl_3$) (*144*)
$C_{29}H_{45}NO_3$	Veralosinine; 161–163; −186.2° ($CHCl_3$) (*141, 142*)
$C_{29}H_{45}NO_4$	Veralosidinine; 220–221; −173.2° ($CHCl_3$) (*143*)

REFERENCES

1. V. Prelog and O. Jeger, *Alkaloids (N.Y.)* **3**, 247 (1953).
2. V. Prelog and O. Jeger, *Alkaloids (N.Y.)* **7**, 343 (1960).
3. K. Schreiber, *Alkaloids (N.Y.)* **10**, 1 (1968).
4. Y. Sato, H. K. Miller, and E. Mosettig, *J. Am. Chem. Soc.* **73**, 5009 (1951).
5. Y. Sato, A. Katz, and E. Mosettig, *J. Am. Chem. Soc.* **73**, 880 (1951).
6. Y. Sato, A. Katz, and E. Mosettig, *J. Am. Chem. Soc.* **74**, 538 (1952).
7. R. Kuhn, I. Löw, and H. Trischmann, *Chem. Ber.* **85**, 416 (1952).
8. K. Schreiber (ed.), *Tagungsber. Dt. Akad. Landwirtschaftswiss. Berlin* **27** (1961).
9. K. Schreiber, G. Adam, O. Aurich, C. Horstmann, H. Ripperger, and H. Rönsch, *Proc. Int. Congr. Horm. Steroids, 2nd, Milan, 1966* p. 344. Excerpta Med. Found., Amsterdam, 1967.
10. J. Redin and O. Proano, *Polytecnica* **2**(1), 249 (1970); *CA* **75**, 88825 (1971).
11. R. Hartmann, *Phytochemistry* **8**, 1319 (1969).
12. C. Djerassi, *Proc. R. Soc. London, Ser. B* **195**, 175 (1976).
13. N. Applezweig, *Chem. Week* July 10, p. 31 (1974).
14. H. Witzel, *Chem. Ind. (Düsseldorf)* **27**, 394 (1975).
15. D. J. Collins, F. W. Eastwood, J. M. Swan, and C. Fryer, *Search* **7**, 378 (1976); *CA* **86**, 34176 (1977).
16. R. N. Chakravarti, *J. Inst. Chem., Calcutta* **48**, 151 (1976); *CA* **85**, 189161 (1976).
17. J. H. Renwick, *Br. J. Prev. Soc. Med.* **26**, 67 (1972).
18. J. Kuć, *Recent Adv. Phytochem.* **9**, 139 (1975); *CA* **86**, 183953 (1977).
19. D. Brown and R. F. Keeler, *J. Agric. Food Chem.* **26**, 564 (1978).
20. R. F. Keeler, S. Young, and D. Brown, *Res. Commun. Chem. Pathol. Pharmacol.* **13**, 723 (1976).
21. D. Brown and R. F. Keeler, *J. Agric. Food Chem.* **26**, 561 (1978).
22. D. Brown and R. F. Keeler, *J. Agric. Food Chem.* **26**, 566 (1978).
23. J. Tomko, A. Vassová, G. Adam, and K. Schreiber, *Tetrahedron* **24**, 4865 (1968).
24. E. Höhne, G. Adam, K. Schreiber, and J. Tomko, *Tetrahedron* **24**, 4875 (1968).
25. J. Tomko, A. Vassová, G. Adam, and K. Schreiber, *Tetrahedron* **24**, 6839 (1968).
26. E. Höhne, I. Seidel, G. Adam, K. Schreiber, and J. Tomko, *Tetrahedron* **28**, 4019 (1972).
27. J. Tomko, Z. Votický, A. Vassová, G. Adam, and K. Schreiber, *Collect. Czech. Chem. Commun.* **33**, 4054 (1968).
28. S. M. Kupchan and A. W. By, *Alkaloids (N.Y.)* **10**, 193 (1968).
29. J. Tomko and Z. Votický, *Alkaloids (N.Y.)* **14**, 1 (1973).
30. IUPAC/IUB 1967 Revised Tentative Rules for Steroid Nomenclature, IUPAC *Inf. Bull.* No. 33, p. 23 (1968); *Biochim. Biophys. Acta* **164**, 453 (1968).
31. K. Schreiber, *Pure Appl. Chem.* **21**, 131 (1970); *CA* **73**, 99086 (1970).
32. Y. Sato, *in* "Chemistry of the Alkaloids" (S. W. Pelletier, ed.), p. 591. Van Nostrand-Reinhold, New York, 1970.
33. R. B. Herbert, *Alkaloids (London)* **3**, 279 (1973); **4**, 383 (1974); **5**, 256 (1975).
34. D. M. Harrison, *Alkaloids (London)* **6**, 285 (1976); **7**, 290 (1977); **8**, 246 (1978); **9**, 238 (1979).
35. K. Schreiber, *Biochem. Soc. Trans.* **2**, 1 (1974); *CA* **81**, 63845 (1974).
36. J. G. Roddick, *Phytochemistry* **13**, 9 (1974).
37. R. K. Puri and J. K. Bhatnagar, *Pharmacos* **19**, 7 (1974).
38. K. Schreiber, *in* "The Biology and Taxonomy of the Solanaceae" (J. G. Hawkes, R. N. Lester, and A. D. Skelding, eds.), Linnean Society Symposium Series, No. 7, p. 193. Academic Press, New York, 1979.

39. V. Bradley, D. J. Collins, F. W. Eastwood, M. C. Irvine, J. M. Swan, and D. E. Symon, *in* "The Biology and Taxonomy of the Solanaceae" (J. G. Hawkes, R. N. Lester, and A. D. Skelding, eds.), Linnean Society Symposium Series, No. 7, p. 203. Academic Press, New York, 1979.
40. G. Willuhn and A. Kun-anake, *Planta Med.* **18**, 354 (1970).
41. S. M. Kupchan, A. P. Davies, S. J. Barboutis, H. K. Schnoes, and A. L. Burlingame, *J. Am. Chem. Soc.* **89**, 5718 (1967).
42. S. M. Kupchan, A. P. Davies, S. J. Barboutis, H. K. Schnoes, and A. L. Burlingame, *J. Org. Chem.* **34**, 3888 (1969).
43. W. C. Evans, A. Ghani, and V. A. Woolley, *J. Chem. Soc.* 2017 (1972).
44. W. C. Evans and A. Somanabandhu, *Phytochemistry* **16**, 1859 (1977).
45. W. C. Evans, *in* "The Biology and Taxonomy of the Solanaceae" (J. G. Hawkes, R. N. Lester, and A. D. Skelding, eds.), Linnean Society Symposium Series, No. 7, p. 241. Academic Press, New York, 1979; see also W. C. Evans and A. Somanabandhu, *Phytochemistry* **19**, 2351 (1980).
46. S. F. Osman, S. F. Herb, T. J. Fitzpatrick, and P. Schmiediche, *J. Agric. Food Chem.* **26**, 1246 (1978).
47. P. G. Kadkade and C. Rolz, *Lloydia* **40**, 217 (1977).
48. P. G. Kadkade and T. R. Madrid, *Naturwissenschaften* **64**, 147 (1977).
49. P. G. Kadkade and C. Rolz, *Phytochemistry* **16**, 1128 (1977).
50. S. K. Banerjee, V. George, and R. Kapoor, *Planta Med.* **25**, 216 (1974).
51. M. Saleh, *Planta Med.* **23**, 377 (1973).
52. V. Bradley, D. J. Collins, P. G. Crabbe, F. W. Eastwood, M. C. Irvine, J. M. Swan, and D. E. Symon, *Aust. J. Bot.* **26**, 723 (1978).
53. K. Kaneko, K. Niitsu, N. Yoshida, and H. Mitsuhashi, *Phytochemistry* **19**, 299 (1980).
54. P. Bite, M. M. Shabana, L. Jókay, and L. Pongrácz-Sterk, *Acta Chim. Acad. Sci. Hung.* **63**, 343 (1970); *CA* **73**, 15160 (1970).
55. P. Bite and M. M. Shabana, *Acta Chim. Acad. Sci. Hung.* **73**, 361 (1972); *CA* **77**, 111578 (1972).
56. A. H. Saber, S. I. Balbaa, and A. Y. Zaky, *Planta Med.* **16**, 191 (1968).
57. W. Döpke, V. Jimenez, and U. Hess, *Pharmazie* **31**, 488 (1976).
58. G. J. Bird, D. J. Collins, F. W. Eastwood, B. M. K. C. Gatehouse, A. J. Jozsa, and J. M. Swan, *Tetrahedron Lett.* 3653 (1976).
59. D. C. Lewis and D. R. Liljegren, *Phytochemistry* **9**, 2193 (1970).
60. S. F. Osman, S. F. Herb, T. J. Fitzpatrick, and S. L. Sinden, *Phytochemistry* **15**, 1065 (1976).
61. R. Katz, N. Aimi, and Y. Sato, *CA* **83**, 193635 (1975).
62. Y. Sato, Y. Sato, H. Kaneko, E. Bianchi, and H. Kataoka, *J. Org. Chem.* **34**, 1577 (1969).
63. A. F. Rizk and E. N. Abou-Zied, *Planta Med.* **18**, 347 (1970).
64. C. Coune and A. Denoël, *Planta Med.* **28**, 168 (1975).
65. C. Coune, *Planta Med.* **31**, 259 (1977).
66. H. Rönsch, K. Schreiber, and H. Stubbe, *Naturwissenschaften* **55**, 182 (1968).
67. K.-E. Rozumek, *Naturwissenschaften* **56**, 334 (1969).
68. G. J. Bird, D. J. Collins, F. W. Eastwood, and J. M. Swan, *Tetrahedron Lett.* 159 (1978).
69. A. Usubillaga, A. Paredes, P. Martinod, and J. Hidalgo, *Planta Med.* **23**, 286 (1973).
70. A. Usubillaga, A. Paredes, P. Martinod, and J. Hidalgo, *Politecnica* **3**, 107 (1973); *CA* **79**, 113186 (1973).
71. J. Kavka, E. Guerreiro, J. C. Gianello, O. S. Giordano, and A. T. D'Arcangelo, *CA* **80**, 24793 (1974).
72. H. H. Appel and L. Branes B., *CA* **69**, 80137 (1968).
73. D. K. Seth, *J. Inst. Chem., Calcutta* **43**, 116 (1971); *CA* **76**, 1802 (1972).

74. I. Mena, C. Timor, I. Corrales, and V. Fuste, *CA* **83**, 4992 (1975).
75. J. Mola, U. Hess, and W. Döpke, *Pharmazie* **28**, 337 (1973).
76. S. C. Pakrashi, A. K. Chakravarty, and E. Ali, *Tetrahedron Lett.* 645 (1977).
77. W. Döpke, U. Hess, and G. Padron, *Pharmazie* **31**, 133 (1976).
78. G. Adam, H. T. Huong, M. Lischewski, and N. H. Khoi, *Phytochemistry* **17**, 1070 (1978).
79. G. Adam, H. T. Huong, M. Lischewski, and N. H. Khoi, *Pharmazie* **34**, 362 (1979).
80. H. Ripperger, *Pharmazie* **32**, 537 (1977).
81. A. Usubillaga, C. Seelkopf, I. L. Karle, J. W. Daly, and B. Witkop, *J. Am. Chem. Soc.* **92**, 700 (1970).
82. A. Usubillaga, personal communication (1979).
83. A. Usubillaga, *Rev. Latinoam. Quim.* **4**, 32 (1973); *CA* **79**, 123632 (1973).
84. D. V. Zaitschek and R. Segal, *Lloydia* **35**, 192 (1972).
85. M. A. Ali, S. Khan, and Z. Kapadia, *CA* **68**, 47010 (1968).
86. P.-M. Hsu and H.-J. Tien, *CA* **84**, 102338 (1976).
87. S. C. Jain and G. L. Sharma, *Planta Med.* **32**, 233 (1977).
88. S. C. Jain and G. L. Sharma, *Planta Med.* **31**, 212 (1977).
89. O. S. Giordano, J. Kavka, J. C. Gianello, and A. T. D'Arcangelo, *CA* **79**, 75849 (1973).
90. P. C. Maiti and S. Mookherjea, *Indian J. Chem.* **6**, 547 (1968); *CA* **70**, 29197 (1969).
91. S. M. Aslanov, *CA* **74**, 95422 (1971).
92. P. Bite and M. M. Shabana, *CA* **77**, 31536 (1972).
93. M. Motidome, M. E. Leekning, and O. R. Gottlieb, *CA* **75**, 95366 (1971).
94. M. B. E. Fayez and A. A. Saleh, *CA* **68**, 102520 (1968).
95. C. Seelkopf, *Arch. Pharm. Ber. Dtsch. Pharm.* **301**, 111 (1968).
96. H. Ripperger, unpublished data (1979).
97. P. Bite and M. M. Shabana, *Acta Chim. Acad. Sci. Hung.* **83**, 91 (1974); *CA* **82**, 82964 (1975).
98. P. Bite and M. M. Shabana, *CA* **87**, 35893 (1977).
99. C. Gandolfi, G. Doria, and R. Longo, *Tetrahedron Lett.* 1677 (1970).
100. S. Cambiaghi, E. Dradi, and R. Longo, *Ann Chim.* (*Rome*) **61**, 99 (1971); *CA* **75**, 36397 (1971).
101. E. N. Novruzov and S. M. Aslanov, *Khim. Prir. Soedin.* 109 (1974); *CA* **81**, 60822 (1974).
102. E. N. Novruzov, S. M. Aslanov, N. M. Ismailov, and A. A. Imanova, *Khim. Prir. Soedin.* 434 (1975); *CA* **84**, 40726 (1976).
103. J. K. Bhatnagar and R. K. Puri, *Lloydia* **37**, 318 (1974).
104. R. K. Puri and J. K. Bhatnagar, *Phytochemistry* **14**, 2096 (1975).
105. L. A. Mitscher, J. V. Juvarkar, and J. L. Beal, *Experientia* **32**, 415 (1976).
106. M. Saleh and S. S. Ahmed, *CA* **79**, 39992 (1973).
107. S. M. Aslanov, *Khim. Prir. Soedin.* 264 (1975); *CA* **83**, 128652 (1975).
108. A. Usubillaga, G. de Castellano, J. Hidalgo, C. Guevara, P. Martinod, and A. Paredes, *Phytochemistry* **16**, 1861 (1977).
109. E. Ali, A. K. Chakravarty, T. K. Dhar, and S. C. Pakrashi, *Tetrahedron Lett.* 3871 (1978).
110. E. Ali, A. K. Chakravarty, S. C. Pakrashi, K. Biemann, and C. E. Hignite, *Tetrahedron* **33**, 1371 (1977).
111. D. K. Seth and R. Chatterjee, *J. Inst. Chem., Calcutta* **41**, 194 (1969); *CA* **72**, 79405 (1970).
112. K. Schreiber and H. Ripperger, *Kulturpflanze* **15**, 199 (1967); *CA* **69**, 27721 (1968).
113. S. M. Aslanov, *CA* **77**, 72556 (1972).
114. K. K. Purushothaman, S. Saradambal, and V. Narayanaswami, *CA* **71**, 73988 (1969).
115. M.-J. Shih and J. Kuć, *Phytochemistry* **13**, 997 (1974).
116. S. I. Balbaa, G. H. Mahran, and A. Y. Zaky, *CA* **73**, 59239 (1970).
117. W. Döpke, I. L. Mola, and U. Hess, *Pharmazie* **31**, 656 (1976).
118. G. Adam, H. T. Huong, and N. H. Khoi, *Phytochemistry* **19**, 1002 (1980).

119. V. R. Dnyansagar and A. R. Pingle, *Planta Med.* **31**, 21 (1977).
120. G. Kusano, J. Beisler, and Y. Sato, *Phytochemistry* **12**, 397 (1973).
121. J. F. Verbist and R. Mounet, *CA* **83**, 75340 (1975).
122. J. F. Verbist and R. Mounet, *CA* **83**, 75341 (1975).
123. M. M. Shabana, T. S. M. A. El-Alfy, and G. H. Mahran, *CA* **85**, 37127 (1976).
124. M. S. Karawya, A. M. Rizk, F. M. Hammouda, A. M. Diab, and Z. F. Ahmed, *Acta Chim. Acad. Sci. Hung.* **72**, 317 (1972); *CA* **77**, 58861 (1972).
125. M. S. Karawya, A. M. Rizk, F. M. Hammouda, A. M. Diab, and Z. F. Ahmed, *Planta Med.* **20**, 363 (1971).
126. S. Singh, N. M. Khanna, and M. M. Dhar, *Phytochemistry* **13**, 2020 (1974).
127. H. Mitsuhashi, U. Nagai, and T. Endo, *CA* **72**, 47321 (1970).
128. K. Kaneko, U. Nakaoka, M. W. Tanaka, N. Yoshida, and H. Mitsuhashi, *Tetrahedron Lett.* 2099 (1978).
129. R. N. Nuriddinov and S. Yu. Yunusov, *Khim. Prir. Soedin.* 600 (1969); *CA* **73**, 25742 (1970).
130. R. N. Nuriddinov and S. Yu. Yunusov, *Khim. Prir. Soedin.* 601 (1969); *CA* **73**, 15098 (1970).
131. R. Shakirov, R. N. Nuriddinov, and S. Yu. Yunusov, *Khim. Prir. Soedin.* 605 (1969); *CA* **73**, 25813 (1970).
132. R. N. Nuriddinov, B. Babaev, and S. Yu. Yunusov, *Khim. Prir. Soedin.* 168 (1968); *CA* **69**, 87319 (1968).
133. K. Samikov, R. Shakirov, and S. Yu. Yunusov, *Khim. Prir. Soedin.* 537 (1974); *CA* **82**, 14013 (1975).
134. A. M. Khashimov, R. Shakirov, and S. Yu. Yunusov, *Khim. Prir. Soedin.* 343 (1970); *CA* **73**, 99170 (1970).
135. R. Shakirov and S. Yu. Yunusoc, *Khim. Prir. Soedin.* 265 (1975); *CA* **83**, 128653 (1975).
136. J. Zwolinski, B. Wojciechowska, and A. Sas-Nowosielska, *CA* **84**, 14684 (1976).
137. A. Vassová, Z. Votický, and J. Tomko, *Collect. Czech. Chem. Commun.* **42**, 3643 (1977).
138. J. Tomko, V. Brázdová, and Z. Votický, *Tetrahedron Lett.* 3041 (1971).
139. A. Vassová and J. Tomko, *Collect. Czech. Chem. Commun.* **40**, 695 (1975).
140. D. Grančai, V. Suchý, J. Tomko, and L. Dolejš, *Chem. Zvesti* **32**, 120 (1978); *CA* **89**, 39410 (1978).
141. A. M. Khashimov, R. Shakirov, and S. Yu. Yunusov, *Khim. Prir. Soedin.* 339 (1970); *CA* **73**, 77442 (1970).
142. A. M. Khashimov, R. Shakirov, and S. Yu. Yunusov, *Khim. Prir. Soedin.* 779 (1971); *CA* **76**, 141109 (1972).
143. R. Shakirov and S. Yu. Yunusov, *Khim. Prir. Soedin.* 501 (1973); *CA* **80**, 60078 (1974).
144. R. Shakirov, A. M. Khashimov, K. Samikov, and S. Yu. Yunusov, *Khim. Prir. Soedin.* 44 (1974); *CA* **80**, 121205 (1974).
145. R. F. Keeler, *Phytochemistry* **13**, 2336 (1974).
146. K. Kaneko, M. Watanabe, Y. Kawakoshi, and H. Mitsuhashi, *Tetrahedron Lett.* 4251 (1971).
147. K. Kaneko, H. Seto, C. Motoki, and H. Mitsuhashi, *Phytochemistry* **14**, 1295 (1975).
148. K. Kaneko, M. W. Tanaka, E. Takahashi, and H. Mitsuhashi, *Phytochemistry* **16**, 1620 (1977).
149. S. Itô, M. Miyashita, Y. Fukazawa, and A. Mori, *Tetrahedron Lett.* 2961 (1972).
150. K. Kaneko, N. Kawamura, T. Kuribayashi, M. Tanaka, and H. Mitsuhashi, *Tetrahedron Lett.* 4801 (1978).
151. O. Aurich, S. Danert, A. Romeike, H. Rönsch, K. Schreiber, and G. Sembdner, *Kulturpflanze* **15**, 205 (1967).
152. W. J. Griffin, W. R. Owen, and J. E. Parkin, *Planta Med.* **16**, 75 (1968).
153. D. Vágujfalvi, *Herba Hung.* **9**, 17 (1970); *CA* **75**, 59793 (1971).

154. T. E. H. Aplin and J. R. Cannon, *Econ. Bot.* **25**, 366 (1971); *CA* **76**, 138183 (1972).
155. E. Alemán Frías, O. Aurich, L. Ezcurra Ferrer, M. Gutiérrez Vázquez, C. Horstmann, J. López Rendueles, E. Rodríguez Graquitena, E. Roquel Casabella, and K. Schreiber, *Kulturpflanze* **19**, 359 (1972).
156. H. H. S. Fong, M. Trojánkova, J. Trojánek, and N. R. Farnsworth, *Lloydia* **35**, 117 (1972).
157. T. G. Hartley, E. A. Dunstone, J. S. Fitzgerald, S. R. Johns, and J. A. Lamberton, *Lloydia* **36**, 217 (1973).
158. S. J. Smolenski, H. Silinis, and N. R. Farnsworth, *Lloydia* **37**, 30 (1974).
159. A. Urzúa and B. K. Cassels, *Phytochemistry* **11**, 3548 (1972).
160. P. Bite and T. Rettegi, *CA* **67**, 54397 (1967).
161. L. H. Briggs, R. C. Cambie, and D. M. Hyslop, *J. Chem. Soc., Perkin Trans. I* 2455 (1975).
162. D. M. van Niekerk and B. H. Koeppen, *Experientia* **28**, 123 (1972).
163. R. Kuhn, I. Löw, and H. Trischmann, *Chem. Ber.* **88**, 1492 (1955).
164. L. Ferenczy, M. M. Shabana, and P. Bite, *Acta Chim. Acad. Sci. Hung.* **65**, 101 (1970); *CA* **73**, 127930 (1970).
165. A. P. Swain, T. J. Fitzpatrick, E. A. Talley, S. F. Herb, and S. F. Osman, *Phytochemistry* **17**, 800 (1978).
166. A. Zitnak, *Proc. Can. Soc. Hortic. Sci.* **3**, 81 (1964).
167. A. Zitnak, *Proc. Can. Soc. Hortic. Sci.* **7**, 75 (1968).
168. H. Ripperger, H. Budzikiewicz, and K. Schreiber, *Chem. Ber.* **100**, 1725 (1967).
169. K. Schreiber and H. Ripperger, *Tetrahedron Lett.* 5997 (1966).
170. S. F. Herb, T. J. Fitzpatrick, and S. F. Osman, *J. Agric. Food Chem.* **23**, 520 (1975); *CA* **83**, 4106 (1975).
171. R. N. Nuriddinov, B. Babaev, and S. Yu. Yunusov, *Khim. Prir. Soedin.* 604 (1969); *CA* **73**, 15049 (1970).
172. G. Kusano, T. Takemoto, Y. Sato, and D. F. Johnson, *Chem. Pharm. Bull.* **24**, 661 (1976).
173. G. Adam, D. Voigt, and K. Schreiber, *Z. Chem.* **14**, 96 (1974).
174. G. A. Tolstikov, S. M. Vasilynk, V. P. Yur'ev, and M. I. Goryaev, *Dokl. Akad. Nauk SSSR* **182**, 611 (1968); *CA* **70**, 29193 (1969).
175. G. A. Tolstikov, V. P. Yur'ev, S. M. Vasilynk, G. N. Romachenko, and M. I. Goryaev, *Izv. Akad. Nauk Kaz. SSR, Ser. Khim.* **20**(2), 33 (1970); *CA* **73**, 35615 (1970).
176. R. Radeglia, G. Adam, and H. Ripperger, *Tetrahedron Lett.* 903 (1977).
177. R. J. Weston, H. E. Gottlieb, E. W. Hagaman, and E. Wenkert, *Aust. J. Chem.* **30**, 917 (1977).
178. S. Terada, K. Hayashi, and H. Mitsuhashi, *Tetrahedron Lett.* 1995 (1978).
179. J. C. Espie, H. Lemaire, and A. Rassat, *Bull. Soc. Chim. Fr.* 399 (1969).
180. H. Ripperger and K. Schreiber, *J. Prakt. Chem.* **313**, 825 (1971).
181. H. Ripperger, K. Schreiber, G. Snatzke, and K. Ponsold, *Tetrahedron* **25**, 827 (1969).
182. H. Ripperger, *Tetrahedron* **25**, 725 (1969).
183. H. Ripperger, *Z. Chem.* **17**, 177 (1977).
184. G. Adam, K. Schreiber, R. Tümmler, and K. Steinfelder, *J. Prakt. Chem.* **313**, 1051 (1971).
185. E. Höhne, I. Seidel, G. Adam, D. Voigt, and K. Schreiber, *J. Prakt. Chem.* **313**, 51 (1971).
186. E. Höhne, I. Seidel, G. Adam, D. Voigt, and K. Schreiber, *Tetrahedron* **29**, 747 (1973).
187. E. Höhne, H. Ripperger, and K. Schreiber, *Tetrahedron* **26**, 3569 (1970).
188. E. Höhne, I. Seidel, G. Reck, H. Ripperger, and K. Schreiber, *Tetrahedron* **29**, 3065 (1973).
189. E. Höhne, *J. Prakt. Chem.* **314**, 371 (1972).
190. P. M. Boll and W. von Philipsborn, *Acta Chem. Scand.* **19**, 1365 (1965).
191. H. Ripperger and K. Schreiber, unpublished observations (1979).
192. H. Ripperger and K. Schreiber, *Justus Liebigs Ann. Chem.* **723**, 159 (1969).
193. H. Budzikiewicz, *Tetrahedron* **20**, 2267 (1964).

194. H. Rönsch and K. Schreiber, *J. Chromatogr.* **30**, 149 (1967).
195. K.-E. Rozumek, *J. Chromatogr.* **40**, 97 (1969).
196. G. A. Tolstikov, V. P. Yur'ev, and M. I. Goryaev, *Khim. Prir. Soedin.* 286 (1967); *CA* **68**, 3109 (1968).
197. C. Coune and A. Denoël, *CA* **83**, 93827 (1975).
198. M. B. E. Fayez and A. A. Saleh, *Planta Med.* **15**, 430 (1967).
199. J. Tomko, G. Adam, and K. Schreiber, *J. Pharm. Sci.* **56**, 1039 (1967); *CA* **67**, 117023 (1967).
200. G. Adam, K. Schreiber, and J. Tomko, *Justus Liebigs Ann. Chem.* **707**, 203 (1967).
201. K. Schreiber and G. Adam, *Tetrahedron* **20**, 1707 (1964).
202. K. Schreiber and G. Adam, *Justus Liebigs Ann. Chem.* **666**, 176 (1963).
203. G. Adam and K. Schreiber, *Z. Chem.* **9**, 227 (1969).
204. K. Schreiber and H. Ripperger, *Justus Liebigs Ann. Chem.* **655**, 136 (1962).
205. K. Schreiber and H. Ripperger, *Justus Liebigs Ann. Chem.* **672**, 232 (1964).
206. E. Höhne, K. Schreiber, H. Ripperger, and H. H. Worch, *Tetrahedron* **22**, 673 (1966).
207. S. V. Kessar, A. L. Rampal, S. S. Gandhi, and R. K. Mahajan, *Tetrahedron* **27**, 2153 (1971).
208. S. V. Kessar, R. K. Mahajan, S. S. Gandhi, and A. L. Rampal, *Tetrahedron Lett.* 1547 (1968).
209. S. V. Kessar, Y. P. Gupta, M. Singh, and R. K. Mahajan, *Tetrahedron* **27**, 2869 (1971).
210. S. V. Kessar, P. J. S. Bhatti, and R. K. Mahajan, *Tetrahedron Lett.* 603 (1969).
211. G. Kusano, N. Aimi, and Y. Sato, *J. Org. Chem.* **35**, 2624 (1970).
212. G. Adam, D. Voigt, and K. Schreiber, *J. Prakt. Chem.* **313**, 45 (1971).
213. R. Tschesche and M. Spindler, *Chem. Ber.* **111**, 801 (1978).
214. H. Ripperger, F.-J. Sych, and K. Schreiber, *Tetrahedron* **28**, 1619 (1972).
215. S. V. Kessar, A. Sharma, M. Singh, and R. K. Mahajan, *Indian J. Chem.* **12**, 1245 (1974); *CA* **82**, 156576 (1975).
216. H. Ripperger and K. Schreiber, *Chem. Ber.* **102**, 4080 (1969).
217. H. Ripperger, F.-J. Sych, and K. Schreiber, *Tetrahedron* **28**, 1629 (1972).
218. H. Ripperger, F.-J. Sych, and K. Schreiber, *Tetrahedron Lett.* 5251 (1970).
219. M. Nagai and Y. Sato, *Tetrahedron Lett.* 2911 (1970).
220. F.-J. Sych, H. Ripperger, and K. Schreiber, *Tetrahedron* **28**, 1645 (1972).
221. G. A. Tolstikov, V. P. Yur'ev, V. M. Potapov, and M. I. Goryaev, *Biol. Akt. Soedin.* 100 (1968); *CA* **71**, 124776 (1969).
222. A. Poláková and K. Syhora, *Collect. Czech. Chem. Commun.* **34**, 3118 (1969).
223. A. Poláková and K. Syhora, Czech. Patent 126138; *CA* **70**, 4428 (1969).
224. A. Poláková and K. Syhora, Czech. Patent 126137; *CA* **70**, 4427 (1969).
225. F. C. Uhle, *J. Org. Chem.* **32**, 792 (1967).
226. G. Kusano, T. Takemoto, N. Aimi, H. J. C. Yeh, and D. F. Johnson, *Heterocycles* **3**, 697 (1975); *CA* **84**, 59856 (1976).
227. G. A. Tolstikov, V. P. Yur'ev, G. N. Romachenko, and M. I. Goryaev, *Izv. Akad. Nauk Kaz. SSR, Ser. Khim.* **21**(5), 42 (1971); *CA* **76**, 59857 (1972).
228. G. N. Romachenko, M. I. Goryaev, and M. P. Irismetov, *Izv. Akad. Nauk Kaz. SSR, Ser. Khim.* **22**(3), 67 (1972); *CA* **77**, 102015 (1972).
229. V. V. Kuril'skaya, M. P. Irismetov, M. I. Goryaev, V. S. Bazalitskaya, and L. G. Mikhaleva, *Izv. Akad. Nauk Kaz. SSR, Ser. Khim.* **27**(3), 46 (1977); *CA* **88**, 51090 (1978).
230. G. N. Romachenko, M. P. Irismetov, and M. I. Goryaev, *Izv. Akad. Nauk Kaz. SSR, Ser. Khim.* **23**(4), 76 (1973); *CA* **80**, 3708 (1974).
231. G. N. Romachenko, M. I. Goryaev, and M. P. Irismetov, *Izv. Akad. Nauk Kaz. SSR, Ser. Khim.* **24**(5), 42 (1974); *CA* **82**, 73306 (1975).
232. M. P. Irismetov, M. I. Goryaev, and V. V. Kuril'skaya, *Izv. Akad. Nauk Kaz. SSR, Ser. Khim.* **28**(2), 58 (1978).

233. M. I. Goryaev, M. P. Irismetov, and G. N. Romachenko, *Izv. Akad. Nauk Kaz. SSR, Ser. Khim.* **23**(1), 70 (1973); *CA* **78**, 136522 (1973).
234. G. S. Khatamkulova, M. I. Goryaev, and M. P. Irismetov, *Izv. Akad. Nauk Kaz. SSR, Ser. Khim.* **24**(4), 71 (1974); *CA* **81**, 152509 (1974).
235. M. P. Irismetov, V. V. Kuril'skaya, and M. I. Goryaev, *Izv. Akad. Nauk Kaz. SSR, Ser. Khim.* **26**(4), 41 (1976); *CA* **86**, 16840 (1977).
236. G. Piancatelli and A. Scettri, *Gazz. Chim. Ital.* **106**, 167 (1976).
237. V. P. Yur'ev, G. A. Tolstikov, and M. I. Goryaev, *Dokl. Akad. Nauk SSSR* **176**, 122 (1967); *CA* **68**, 22138 (1968).
238. G. Adam, D. Voigt, and K. Schreiber, *J. Prakt. Chem.* **315**, 739 (1973).
239. G. Adam, D. Voigt, and K. Schreiber, *Tetrahedron* **27**, 2181 (1971).
240. I. Belič and H. Sočič, *Experientia* **27**, 626 (1971).
241. I. Belič and H. Sočič, *J. Steroid Biochem.* **3**, 843 (1972); *CA* **78**, 26208 (1973).
242. I. Belič, V. Gaberc-Porkekar, and H. Sočič, *Vestn. Slov. Kem. Drus.* **22**, 49 (1975); *CA* **86**, 68109 (1977).
243. I. Belič, M. Mervič, T. Kastelic-Suhadolc, and V. Kramer, *J. Steroid Biochem.* **8**, 311 (1977); *CA* **87**, 35636 (1977).
244. I. Belič, V. Kramer, and H. Sočič, *J. Steroid Biochem.* **4**, 363 (1973); *CA* **79**, 144832 (1973).
245. I. Belič, V. Hiršl-Pintarič, H. Sočič, and B. Vranjek, *J. Steroid Biochem.* **6**, 1211 (1975); *CA* **84**, 3278 (1976).
246. I. Belič, R. Komel, and H. Sočič, *Steroids* **29**, 271 (1977).
247. Y. Sato, Y. Sato, and K. Tanabe, *Steroids* **9**, 553 (1967).
248. G. G. Malanina, L. I. Klimova, L. M. Morozovskaya, O. S. Anisimova, L. M. Alekseeva, and N. N. Suvorov, *Khim.-Farm. Zh.* **8**(5), 18 (1974); *CA* **81**, 63874 (1974).
249. L. M. Morozovskaya, L. I. Klimova, G. G. Malanina, N. K. Genkina, L. N. Finyakin, and N. N. Suvorov, *Zh. Org. Khim.* **10**, 2125 (1974); *CA* **82**, 86496 (1975).
250. L. M. Morozovskaya, E. S. Belen'kaya, L. I. Klimova, and G. S. Grinenko, *Khim.-Farm. Zh.* **10**(11), 64 (1976); *CA* **87**, 53483 (1977).
251. E. S. Belen'kaya, L. M. Morozovskaya, L. I. Klimova, and G. S. Grinenko, USSR Patent 514848; *CA* **85**, 78271 (1976).
252. Y. Sato and M. Nagai, *J. Org. Chem.* **37**, 2629 (1972).
253. C. G. Bakker and P. Vrijhof, *Tetrahedron Lett.* 4699 (1978).
254. G. Adam and K. Schreiber, *Chem. Ind.* (*London*) 989 (1965).
255. G. Adam and K. Schreiber, *Tetrahedron* **22**, 3581 (1966).
256. G. Adam, D. Voigt, K. Schreiber, M. von Ardenne, R. Tümmler, and K. Steinfelder, *J. Prakt. Chem.* **315,** 125 (1973).
257. G. Adam and K. Schreiber, *Justus Liebigs Ann. Chem.* **709**, 191 (1967).
258. H.-H. Worch, E. Höhne, G. Adam, and K. Schreiber, *J. Prakt. Chem.* **312**, 1043 (1970).
259. G. Adam, *Chem. Ber.* **101**, 1 (1968).
260. H. Ripperger and K. Schreiber, *Tetrahedron* **23**, 1841 (1967).
261. G. Adam and K. Schreiber, *Justus Liebigs Ann. Chem.* 2048 (1973).
262. G. Adam and K. Schreiber, *Tetrahedron Lett.* 923 (1965).
263. G. Adam and K. Schreiber, *Chem. Ber.* **102**, 878 (1969).
264. R. T. Li and Y. Sato, *Steroids* **13**, 451 (1969).
265. R. T. Li and Y. Sato, *J. Org. Chem.* **33**, 3635 (1968).
266. J. A. Beisler and Y. Sato, *J. Chem. Soc. C* 149 (1971).
267. R. Franzmair, *Monatsh. Chem.* **107**, 501 (1976).
268. Österr. Stickstoffwerke AG, Fr. Patent 2,065,099; *CA* **77**, 48719 (1972).
269. K. Schreiber and C. Horstmann, *Chem. Ber.* **99**, 3183 (1966).
270. G. Schramm, Austrian Patent 280, 494; *CA* **73**, 35635 (1970).
271. G. Schramm, Ger. Patent 1,900,060; *CA* **73**, 77488 (1970).

272. G. Schramm, Austrian Patent 280,497; *CA* **73**, 66825 (1970).
273. C. Schlögl, G. Schramm, and W. Obendorf, Ger. Patent 1,908,518; *CA* **73**, 110007 (1970).
274. Österr. Stickstoffwerke AG, Br. Patent 1311307; *CA* **79**, 18954 (1973).
275. K. Schreiber and C. Horstmann, unpublished data. (1966).
276. E. Heftmann, *Lloydia* **30**, 209 (1967); **31**, 293 (1968).
277. E. Heftmann, "Steroid Biochemistry." Academic Press, New York, 1970.
278. E. Heftmann, *Lipids* **6**, 128 (1971); *CA* **74**, 121291 (1971); *Lipids* **9**, 626 (1974); *CA* **82**, 40660 (1975).
279. K. Schreiber, *Abh. Dtsch. Akad. Wiss. Berlin.* 69 (1969).
280. H. R. Schütte, *in* "Biosynthese der Alkaloide" (K. Mothes and H. R. Schütte, eds.), p. 616. VEB Dtsch. Verlag Wiss., Berlin, 1969.
281. H. R. Schütte, *Prog. Bot.* **37**, 133 (1975).
282. L. J. Goad, *in* "Terpenoids in Plants" (I. B. Pridham, ed.), p. 159. Academic Press, New York, 1967.
283. T. W. Goodwin, ed., *Biochem. Soc. Symp.* **29** (1970).
284. T. W. Goodwin, *Biochem. J.* **123**, 293 (1971).
285. A. K. Chakravarty, T. K. Dhar, and S. C. Pakrashi, *Phytochemistry* **19**, 1249 (1980).
286. A. K. Chakravarty, C. R. Saha, and S. C. Pakrashi, *Phytochemistry* **18**, 902 (1979).
287. A. K. Chakravarty, T. K. Dhar, and S. C. Pakrashi, *Tetrahedron Lett.* 3875 (1978).
288. W. Döpke, E. Sewerin, U. Hess, and C. Nogueiras, *Z. Chem.* **16**, 104 (1976).
289. A. G. González, C. G. Francisco, R. Freire, R. Hernández, J. A. Salazar, E. Suárez, A. Morales, and A. Usubillaga, *Phytochemistry* **14**, 2483 (1975).
290. K. Kaneko, S. Terada, N. Yoshida, and H. Mitsuhashi, *Phytochemistry* **16**, 791 (1977).
291. K. Kaneko, K. Hirayama, and H. Mitsuhashi, *Pap. Symp. Nat. Prod., 14th*, 358 (1970).
292. K. Kaneko, M. W. Tanaka, and H. Mitsuhashi, *Phytochemistry* **16**, 1247 (1977).
293. K. Kaneko, M. W. Tanaka, U. Kawakoshi, Y. Suzuki, and H. Mitsuhashi, *Pap. Symp. Nat. Prod., 15th*, 177 (1971).
294. H. Ripperger, M. Moritz, and K. Schreiber, *Phytochemistry* **10**, 2699 (1971).
295. E. Heftmann and M. L. Weaver, *Phytochemistry* **13**, 1801 (1974).
296. R. Tschesche, B. Goossens, and A. Töpfer, *Phytochemistry* **15**, 1387 (1976).
297. F. Ronchetti, G. Russo, G. Ferrara, and G. Vecchio, *Phytochemistry* **14**, 2423 (1975).
298. F. Ronchetti and G. Russo, *J. Chem. Soc., Chem. Commun.* 785 (1974).
299. F. Ronchetti and G. Russo, *Tetrahedron Lett.* 85 (1975).
300. R. Tschesche and G. Piestert, *Phytochemistry* **14**, 435 (1975).
301. R. Tschesche and J. Leinert, *Phytochemistry* **12**, 1619 (1973).
302. R. Tschesche and R. Fritz, *Z. Naturforsch., Teil B* **25**, 590 (1970).
303. A. Töpfer, Ph.D. Thesis, Univ. Bonn, 1975.
304. R. Tschesche, H. Hulpke, and R. Fritz, *Phytochemistry* **7**, 2021 (1968).
305. R. Tschesche and M. Spindler, *Phytochemistry* **17**, 251 (1978).
306. R. Tschesche and H. R. Brennecke, *Chem. Ber.* **112**, 2680 (1979).
307. R. Tschesche and H. R. Brennecke, *Phytochemistry* **19**, 1449 (1980).
308. K. Kaneko, M. W. Tanaka, and H. Mitsuhashi, *Phytochemistry* **15**, 1391 (1976).
309. R. Tschesche and M. Spindler, *Phytochemistry* **17**, 251 (1978).
310. K. Kaneko, H. Seto, C. Motoki, and H. Mitsuhashi, *Phytochemistry* **14**, 1295 (1975).
311. L. Canonica, F. Ronchetti, and G. Russo, *J. Chem. Soc., Chem. Commun.* 286 (1977).
312. J. G. Roddick, *Phytochemistry* **16**, 805 (1977).
313. J. G. Roddick, *Phytochemistry* **15**, 475 (1976).
314. J. G. Roddick, *in* "The Biology and Taxonomy of the Solanaceae" (J. G. Hawkes, R. N. Lester, and A. D. Skelding, eds.), Linnean Society Symposium Series, No. 7, p. 223. Academic Press, New York, 1979.

315. J. G. Roddick, *Phytochemistry* **13**, 1459 (1974).
316. J. G. Roddick and D. N. Butcher, *Phytochemistry* **11**, 2019 (1972).
317. M. R. Heble, S. Narayanaswami, and M. S. Chadha, *Naturwissenschaften* **55**, 350 (1968).
318. M. R. Heble, S. Narayanaswami, and M. S. Chadha, *Phytochemistry* **10**, 2393 (1971).
319. P. Khanna, A. Uddin, G. L. Sharma, S. K. Manot, and A. K. Rathore, *Indian J. Exp. Biol.* **14**, 694 (1976); *CA* **86**, 40211 (1977).
320. P. Khanna, G. L. Sharma, A. K. Rathore, and S. K. Manot, *Indian J. Exp. Biol.* **15**, 1025 (1977).
321. M. R. Heble, S. Narayanaswami, and M. S. Chadha, *Science* **161**, 1145 (1968).
322. M. R. Heble, S. Narayanaswami, and M. S. Chadha, *Phytochemistry* **10**, 910 (1971).
323. J. H. Supniewska and B. Dohnal, *Diss. Pharm. Pharmacol.* **24**, 187 (1972).
324. J. H. Supniewska and B. Dohnal, *Diss. Pharm. Pharmacol.* **24**, 193 (1972); *CA* **77**, 58820 (1972).
325. S. J. Stohs and H. Rosenberg, *Lloydia* **38**, 181 (1975).
326. D. Vágujfalvi, M. Maróti, and P. Tétényi, *Phytochemistry* **10**, 1389 (1971).
327. E. Heftmann and S. Schwimmer, *Phytochemistry* **11**, 2783 (1972).
328. E. Heftmann and S. Schwimmer, *Phytochemistry* **12**, 2661 (1973).
329. K. Miyahava, Y. Ida, and T. Kawasaki, *Chem. Pharm. Bull.* **20**, 2506 (1972).
330. A. G. González, C. G. Francisco, R. F. Barreira, and E. S. Lopez, *An. Quim.* **67**, 433 (1971); *CA* **75**, 115896 (1971).
331. K. Kaneko, M. Watanabe, and H. Mitsuhashi, *Phytochemistry* **12**, 1509 (1973).
332. G. Adam, H. T. Huong, and N. H. Khoi, *Phytochemistry* **17**, 1802 (1978).
333. T. Nohara, H. Yabuta, M. Suenobu, R. Hida, K. Miyahara, and T. Kawasaki, *Chem. Pharm. Bull.* **21**, 1240 (1973).
334. S. Kiyosawa and T. Kawasaki, *Chem. Pharm. Bull.* **25**, 163 (1977).
335. S. Kiyosawa, K. Goto, R. Owashi, and T. Kawasaki, *Tetrahedron Lett.* 4599 (1977).
336. A. G. González, R. Freire, C. G. Francisco, J. A. Salazar, and E. Suárez, *Tetrahedron* **29**, 1731 (1973).
337. D. R. Liljegren, *Phytochemistry* **10**, 3061 (1971).
338. N. Lavintman, J. Tandecarz, and C. E. Cardini, *Plant Sci. Lett.* **8**, 65 (1977); *CA* **86**, 67260 (1977).
339. R. D. Durbin and T. F. Uchytil, *Biochim. Biophys. Acta* **191**, 176 (1969).
340. K.-H. Meyer, S. Schütz, K. Stoepel, and H.-G. Kroneberg, U.S. Patent 3,328,387; *CA* **68**, 105467 (1968).
341. M. I. Goryaev, G. N. Romachenko, and M. P. Irismetov, *Izv. Akad. Nauk Kaz. SSR, Ser. Khim.* **22**(6), 83 (1972); *CA* **78**, 84645 (1973).
342. M. P. Irismetov, M. I. Goryaev, V. V. Kuril'skaya, and G. D. Tsepochkin, *Vestn. Akad. Nauk Kaz. SSR* No. 8, p. 64 (1978); *CA* **90**, 39119 (1979).
343. G. A. Tolstikov, V. P. Yur'ev, and M. I. Goryaev, *Tr. Inst. Khim. Nauk, Akad. Nauk Kaz. SSR* **19**, 64 (1967); *CA* **68**, 114823 (1968).
344. E. Bianchi, C. Djerassi, H. Budzikiewicz, and Y. Sato, *J. Org. Chem.* **30**, 754 (1965).
345. M. Havel and V. Černý, *Collect. Czech. Chem. Commun.* **40**, 3199 (1975).
346. M. Havel and V. Černý, *Collect. Czech. Chem. Commun.* **40**, 1579 (1975).
347. Österr. Stickstoffwerke AG; Fr. Patent 2,003,261; *CA* **72**, 111727 (1970).
348. R. Shakirov, R. N. Nuriddinov, and S. Yu. Yunusov, *Khim. Prir. Soedin.* 384 (1965); *CA* **64**, 12746 (1966).
349. G. Adam, H. T. Huong, and N. H. Khoi, unpublished data (1980).
350. K. Samikov, R. Shakirov, and S. Yu. Yunusov, *Khim. Prir. Soedin.* 815 (1978).
351. D. Voigt, K. Schreiber, and G. Adam, *Adv. Mass Spectrom. Biochem. Med.* **2**, 183 (1976); *CA* **86**, 135798 (1977).

352. D. Voigt, G. Adam, and K. Schreiber, *Pharmazie* **30**, 214 (1975).
353. G. P. Moiseeva, R. Shakirov, M. R. Yagubaev, and S. Yu. Yunusov, *Khim. Prir. Soedin.* 623 (1976); *CA* **87**, 136144 (1977).
354. I. Mathe and I. Mathe, *Acta Pharm. Hung.* **44**, Suppl., 19 (1974); *CA* **82**, 13993 (1975).
355. G. J. Bird, D. J. Collins, F. W. Eastwood, and J. M. Swan, *Aust. J. Chem.* **32**, 597 (1979).
356. G. J. Bird, D. J. Collins, F. W. Eastwood, and J. M. Swan, *Aust. J. Chem.* **32**, 611 (1979).
357. G. J. Bird, D. J. Collins, F. W. Eastwood, R. H. Exner, M. L. Romanelli, and D. D. Small, *Aust. J. Chem.* **32**, 783 (1979).
358. G. J. Bird, D. J. Collins, F. W. Eastwood, and R. H. Exner, *Aust. J. Chem.* **32**, 797 (1979).
359. B. M. Gatehouse and A. J. Jozsa, *Acta Crystallogr., Sect. B* **33**, 3782 (1977); *CA* **88**, 144570 (1978).
360. G. Adam and H. T. Huong, *Tetrahedron Lett.* 1931 (1980).
361. A. K. Rathore, K. P. Sharma, and G. L. Sharma, *CA* **92**, 55119 (1980).
362. K. Samikov, R. Shakirov, and S. Yu. Yunusov, *Khim. Prir. Soedin.* 350 (1979); *CA* **91**, 207406 (1979).
363. S. C. Pakrashi, A. K. Chakravarty, E. Ali, T. K. Dhar, and S. Dan, *CA* **92**, 6806 (1980).
364. M. P. Irismetov, M. I. Goryaev, V. V. Kuril'skaya, *CA* **91**, 211671 (1979).
365. K. Samikov, R. Shakirov, K. A. Ubaidullaev, and S. Yu. Yunusov, *Khim. Prir. Soedin.* 183 (1975); *CA* **83**, 79473 (1975).
366. K. Kaneko, N. Kawamura, H. Mitsuhashi, and K. Ohsaki, *Chem. Pharm. Bull.* **27**, 2534 (1979).
367. K. Kaneko, U. Nakaoka, M. Tanaka, N. Yoshida, and H. Mitsuhashi, *Phytochemistry* **20**, 157 (1981).
368. K. Kaneko, M. Tanaka, U. Nakaoka, Y. Tanaka, N. Yoshida, and H. Mitsuhashi, *Phytochemistry* **20**, 327 (1981).
369. J. D. Mann, *Adv. Agron.* **30**, 307 (1978).
370. J. D. Phillipson, S. R. Hemingway, and C. E. Ridsdale, *Lloydia* **41**, 503 (1978).
371. E. W. Weiler, H. Krüger, and M. H. Zenk, *Planta Med.* **39**, 112 (1980).
372. V. Ahmad, S. F. Ali, and R. Ahmad, *Planta Med.* **39**, 186 (1980).
373. I. R. Hunter, M. K. Walden, J. R. Wagner, and E. Heftmann, *J. Chromatogr.* **119**, 223 (1976).
374. T. W. Goodwin, *Annu. Rev. Plant Physiol.* **30**, 369 (1979).
375. N. Hosoda and M. Yatazawa, *Agric. Biol. Chem.* **43**, 821 (1979).
376. C. K. Kokate and S. S. Radwan, *Z. Naturforsch., Teil C* **34**, 634 (1979).
377. A. Uddin and H. C. Chaturvedi, *Planta Med.* **37**, 90 (1979).
378. J. G. Roddick, *Phytochemistry* **18**, 1467 (1979).
379. J. G. Roddick, *Phytochemistry* **19**, 2455 (1980).
380. F. Coll, unpublished (1981).
381. Y. M. El Kheir and M. H. Salih, *Fitoterapia* **50**, 255 (1979.
382. G. Indrayanto and E. Sundrawati, *CA* **92**, 160519 (1980).
383. S. B. Mahato, N. P. Sahu, A. N. Ganguly, R. Kasai, and O. Tanaka, *Phytochemistry* **19**, 2017 (1980).
384. H. T. Huong, Ph.D. Thesis, Academy of Sciences of the GDR, Halle, 1980.
385. J. Garnero and D. Joulain, *8th Int. Congr. of Essential Oils, Cannes*, 1980.
386. S. F. Osman, R. M. Zacharius, and D. Naglak, *Phytochemistry* **19**, 2599 (1980).
387. K. Tori, S. Seo, Y. Terui, J. Nishikawa, and F. Yasuda, *Tetrahedron Lett.* **22**, 2405 (1981).

—CHAPTER 3—

PHENANTHROINDOLIZIDINE AND PHENANTHROQUINOLIZIDINE ALKALOIDS

I. RALPH C. BICK AND WANNEE SINCHAI

Chemistry Department, University of Tasmania, Hobart, Australia

I. Introduction

The closely related phenanthroindolizidine and phenanthroquinolizidine groups of alkaloids have been the subject of detailed studies in recent years, particularly by Indian chemists. Various aspects of their chemistry and pharmacology have been reviewed by Govindachari (*1*, *56*, *80*) and by Wiegrebe (*2*). Short reports of progress in the chemistry (*3*) and biosynthesis (*4*) of the phenanthroindolizidine alkaloids, and on the synthesis (*5*) of the phenanthroquinolizidines, have also been published.

II. Structures and Configurations

In this section structures are shown for the 16 phenanthroindolizidine and two phenanthroquinolizidine alkaloids that have so far been described, together with the stereochemistry where known. References to their isolation

THE ALKALOIDS, VOL. XIX

ISBN 0-12-469519-1

and structural and stereochemical determination are provided; the literature concerned with other aspects of their chemistry and their pharmacology is cited in subsequent sections.

A. Phenanthroindolizidine Alkaloids

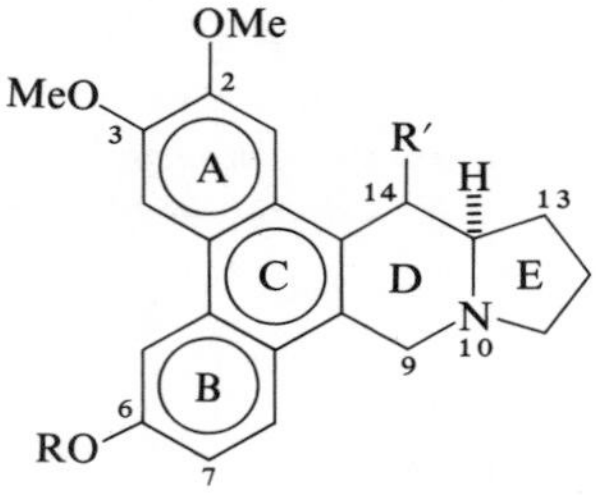

1 R = R′ = H; 7-demethoxydemethyltylophorine (*7*) (C-13a: *R*)
2 R = Me, R′ = H; antofine (7-demethoxytylophorine) (*6–11*, *41*) (C-13a: *R*)
3 R = Me, R′ = OH; 14-hydroxy-2,3,6-trimethoxyphenanthroindolizidine (*12*) (C-13a: *R*)

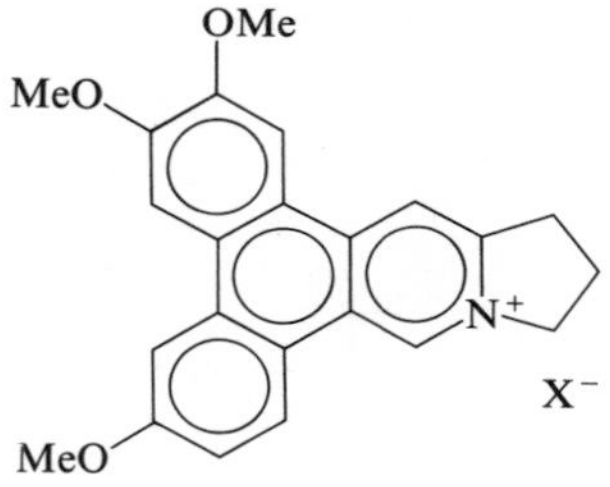

4 Dehydroantofine* (*6*)

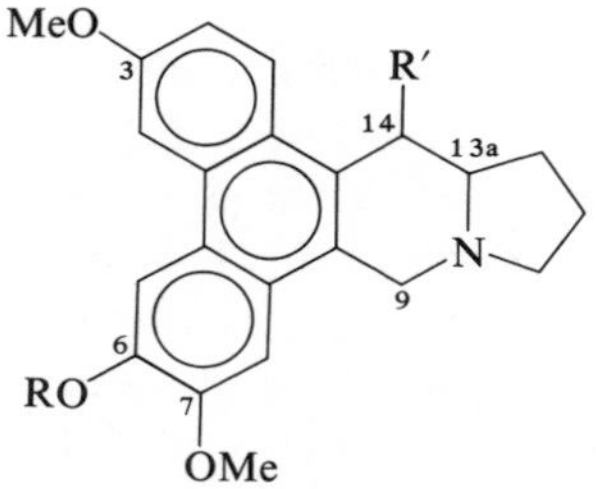

5 R = Me, R′ = H; deoxypergularinine (*14*, *15*) (C-13a: *S*)
6 R = Me, R′ = OH; pergularinine (*14*, *15*) (C-13a: *S*, C-14: *R*). Tylophorinine (*16–18*, *32*, *44*) is the corresponding racemic base.
7 R = H, R′ = OH; tylophorinidine (*14*, *15*, *19*, *20*, *23*, *36*, *44*) (C-13a: *S*, C-14: *S*)

* This base may be an artifact produced by atmospheric oxidation of antofine (**2**).

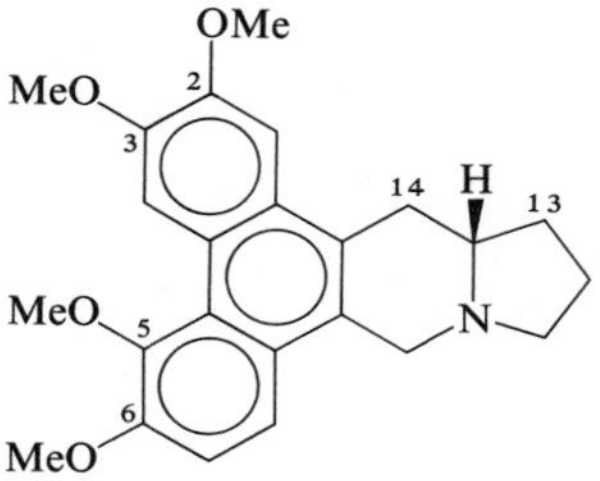

OMe, MeO, MeO, MeO, 2, 3, 5, 6, 14, H, 13, N

8 Tylocrebrine (*21, 22, 37, 43*) (C-13a: *S*)

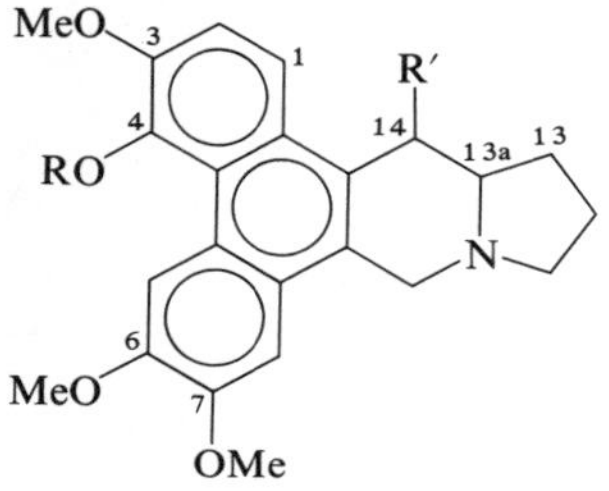

MeO, RO, MeO, OMe, R′, 1, 3, 4, 6, 7, 14, 13a, 13, N

9 R = Me, R′ = H (C-13a: *R*); (+)-isotylocrebrine (*23, 37, 42*)
10 R = Me, R′ = OH; Rao's alkaloid A (*22, 33*) from *Tylophora crebriflora*
11 R = H, R′ = H; Rao's alkaloid B (*22, 33*) from *T. crebriflora*
12 R = H, R′ = OH; Rao's alkaloid C (*22, 33*) from *T. crebriflora*

OMe, MeO, MeO, OMe, 2, 3, 6, 7, 9, 14, H, 13, N

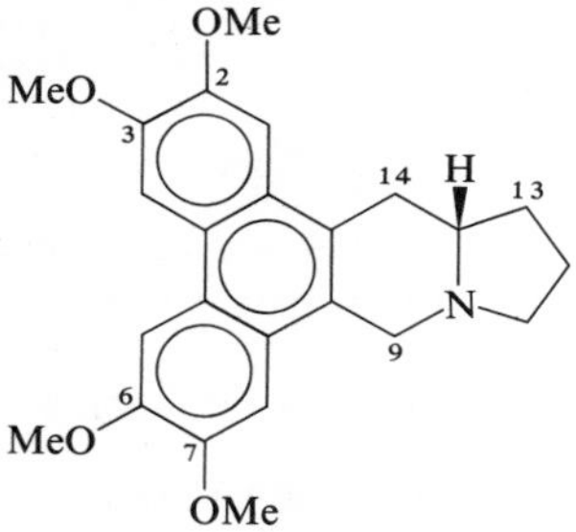

13 Tylophorine (*7, 10, 14–18, 21, 22, 24, 30, 31, 42*) (C-13a: *S*)

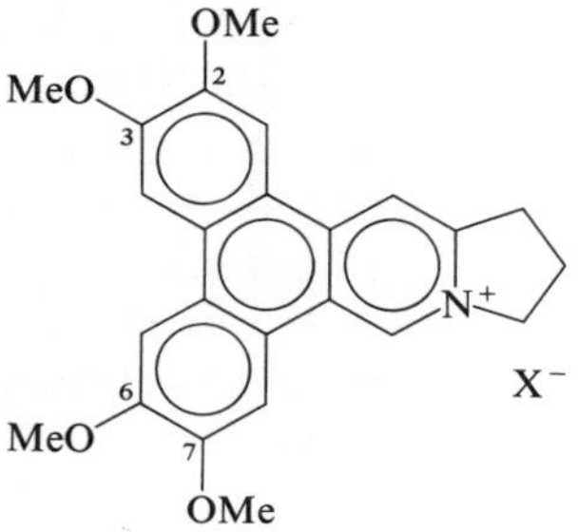

OMe, MeO, MeO, OMe, 2, 3, 6, 7, N^+, X^-

14 Dehydrotylophorine* (*13*)

* This base may be an artifact produced by atmospheric oxidation of tylophorine (**13**).

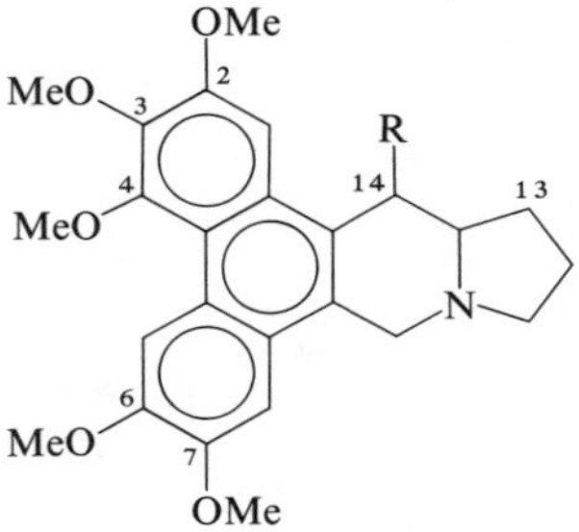

OMe, MeO, MeO, MeO, OMe, R, N, 2, 3, 4, 6, 7, 13, 14

15 R = H; Rao's alkaloid E (*22*, *33*) from *Tylophora crebriflora*
16 R = OH; Rao's alkaloid D (*22*, *33*) from *T. crebriflora*

B. Phenanthroquinolizidine Alkaloids

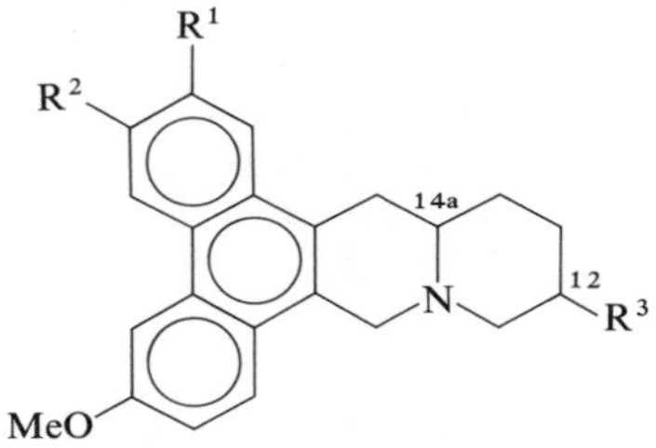

R^1, R^2, R^3, MeO, N, 14a, 12

17 $R^1 = R^2 = OCH_2O$, $R^3 = OH$; cryptopleuridine (*71*)
18 $R^1 = R^2 = OMe$, $R^3 = H$ (C-14a:*R*); cryptopleurine (*25–27*, *34*, *43*)

C. Minor Alkaloids of Uncertain Structure

Rao *et al.* (*18*) have isolated three minor alkaloids from *Tylophora asthmatica*, one of which, C, may be identical (*56*) with tylophorinidine (**7**), the difference in melting points being due to solvation. This base also occurs in *T. dalzellii* (*18*). Another *T. asthmatica* alkaloid, B, has a structure similar to tylophorine (**13**), with one methoxyl replaced by a hydroxyl group (*18*). The third minor alkaloid, A, may be a bisnortylophorine (*18*) with two methoxyls replaced by phenolic groups. A minor alkaloid of uncertain structure isolated from *Pergularia pallida* may be 14-hydroxytylophorine (*15*, *80*). Evidence for the presence of dehydrodeoxytylophorinine and dehydrodeoxytylophorinidine in extracts of *T. asthmatica* has been found, but as in the case of **4** and **14**, these bases may be artifacts (*13*).

III. Occurrence

The name *Tylophora* alkaloids is often applied to the phenanthroindolizidines from the frequency with which they occur in this genus. They are also found in other genera of the Asclepidaceae, as well as in the Moraceae and Urticaceae, which are phylogenetically remote from the latter family but related to one another; the Lauraceae, which produce both phenanthroindolizidines and phenanthroquinolizidines, are likewise considered much less advanced than the Asclepidaceae. The occurrence of known alkaloids is shown in the following table.

Family	Genus and species	Alkaloids	Reference
Asclepiadaceae	*Antitoxicum funebre* Boiss. and Kotschy	**2**	(*8*)
	Cynanchum vincetoxicum (L.) Pers. (syn. *Vincetoxicum officinale* Monch)	**1–3, 13**	(*7, 9, 10, 12*)
	Pergularia pallida	**5–7, 13**	(*14, 15, 20*)
	Tylophora asthmatica Wight et Arn. (syn. *T. indica* (Burm) Merrill.)	**6–9, 13, 14**	(*13, 16, 17–19, 23, 24, 28*)
	T. cordifolia	**6**	(*28*)
	T. crebriflora S. T. Blake	**8, 10–13, 15, 16**	(*22*)
	T. dalzellii Hook. f.	**7**?	(*18*)
	T. flava	**6, 13**	(*28*)
Moraceae	*Ficus septica*	**2, 8, 13**	(*11, 21*)
Lauraceae	*Cryptocarya phyllostemon* Kostermans	**2, 4**	(*6*)
	C. pleurosperma White and Francis	**17, 18**	(*29, 71*)
Urticaceae	*Boehmeria cylindrica* (L.) Sw.	**18**	(*27*)
	B. platyphylla Don	**18**	(*26*)

IV. Degradative Experiments and Structural Determination

Tylophorine (**13**), the most common of these alkaloids and the first phenanthroindolizidine base to be isolated (*16*), was studied by Govindachari and collaborators (*30, 31*), who found that after two Hofmann degradations the product still contained nitrogen: evidently, the latter is common to two ring systems. One of these was identified as a five-membered ring as a result of an Emde degradation on tylophorine methochloride, followed by dehydrogenation. The nonbasic nitrogen-containing product gave a positive Ehrlich test, indicating the presence of a pyrrole nucleus. Vigorous oxidation of tylophorine methiodide produced metahemipic acid (**19**) as the only identifiable product; oxidation of the isomethohydroxide gave an imide, **20**, and the corresponding dicarboxylic acid, whose structures were established by synthesis. Structure **13**, put forward on the basis of these and other data for tylophorine, has been confirmed by several syntheses (see Section IX).

In the case of tylophorinine (**6**), the presence of the 14-hydroxyl group prevented any useful results from being obtained by Emde or Hofmann

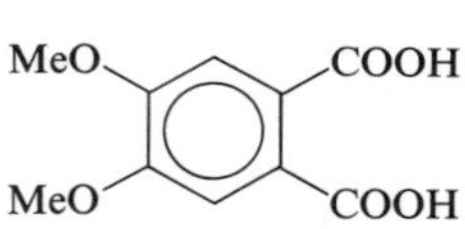

19

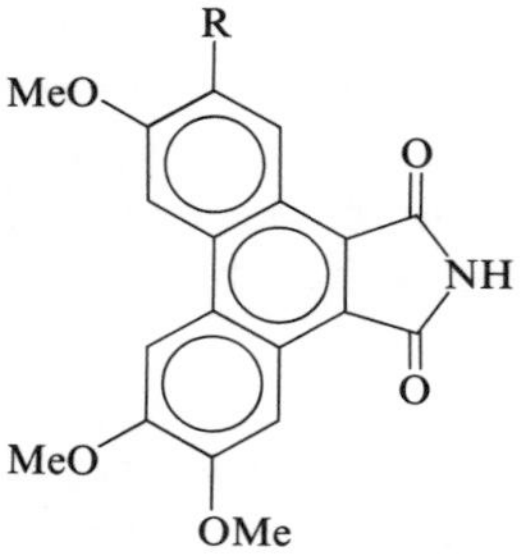

20 R = OMe
21 R = H

degradations (*32*); however, direct oxidation gave **19** and a phenanthrene imide, **21**, similar to that obtained from tylophorine but with one less methoxyl group. The hydroxyl group in tylophorinine can be easily removed by hydrogenolysis, indicating its benzylic nature; on the other hand, tylophorinine does not behave as a carbinolamine, so that the hydroxyl group must be located at C-14 rather than C-9. This leaves two possible structures for the alkaloid, either **6** or the isomer with a methoxyl at C-2 instead of C-7; a decision between these alternatives was made by synthesis.

These methods, or extensions of them, have enabled the structures of other phenanthroindolizidine alkaloids to be established. In some cases it has been possible to relate the structure of a newly discovered base to a known structure by simple transformations; thus, hydrogenolysis of Rao's alkaloid C (**12**) gives his alkaloid B (**11**), which can be methylated with diazomethane to the same isomer of tylocrebrine (**8**) as is formed on hydrogenolysis of Rao's alkaloid A (**10**) (*33*). As another example, the quaternary base dehydrotylophorine (**14**) can be converted into tylophorine (**13**) by catalytic hydrogenation (*13*). These correlations with known structural types have been greatly assisted by spectroscopic methods (See Sections IV–VII.)

The structure of cryptopleurine as **18** was established by X-ray crystallographic analysis of its racemic methiodide (*34*, *35*). The structure of tylophorinidine as **7** was confirmed in a similar manner, and its absolute configuration was also determined in this case (*36*).

V. Electronic Spectra

Phenanthroindolizidine and phenanthroquinolizidine alkaloids may be recognized by their UV spectra (*70*), which correspond to a substituted phenanthrene chromophore. There is a distinct increase in intensity of the absorption maximum around 260 nm as the number of oxy substituents increases from three (e.g., **5**, **18**) to four (e.g., **10**, **11**, **13**) and five (e.g., **15**, **16**); this increase has been used diagnostically (*23*, *32*, *37*). A careful comparison of the spectra of newly discovered alkaloids with those from bases of known structure, or from model compounds, has helped to limit the number of structural possibilities, and in some instances has allowed a tentative assignment of the location of oxy functions to be made (e.g., tylocrebrine, **8**) (*37*).

The quaternary alkaloid dehydrotylophorine (**14**), with an extra aromatic ring, has a characteristic spectrum with intense absorptions (*13*).

UV Spectroscopy has also been applied to distinguish alcoholic and phenolic groups by observing the spectra before and after addition of alkali (e.g., tylophorinine, **6**) (*32*). Examples of phenanthroindolizidine alkaloids with an aliphatic hydroxyl group which have so far been encountered have proved to be benzylic alcohols from the ready removal of the —OH group by hydrogenolysis. The two possible locations for the hydroxyl resulting from this hydrogenolysis—at C-14 or C-9—can be distinguished by a comparison of the UV spectra in neutral and acid media: If a carbinolamine group is present, a pseudo base with a 3,4-dihydroisoquinolinium structure is formed in acid solution, which can be recognized by the pronounced bathochromic shift as compared to the spectrum in neutral solution (*2*).

VI. Infrared Spectra

IR Spectroscopy has been useful in detecting and distinguishing phenolic and alcoholic groups by observation of the carbonyl absorption bands after acetylation; e.g., tylophorinidine (**7**) gives a diacetate with absorption bands at 1760 (phenolic acetate) and 1730 cm^{-1} (alcoholic acetate), respectively (*23*).

Phenanthroquinolizidine alkaloids would normally be expected to exhibit Bohlmann bands (*65*, *66*) in their IR spectra, and a published spectrum of cryptopleurine, **18**, (*67*) in fact shows absorptions around 2800 cm^{-1}. These bands may prove useful in helping to recognize other alkaloids of this group. Phenanthroindolizidine alkaloids should also show Bohlmann bands, because a trans-fused ring system is more stable than the corresponding cis conformation for both quinolizidines and indolizidines (*68*), and should result in the lone-pair electrons on the nitrogen having at least two trans-oriented hydrogens on α-carbons. The spectrum of antofine (**2**) shows distinct bands in the 2800 cm^{-1} region (*6*), as does a published spectrum (*46*) of tylophorinine, **6**. On the other hand, Bohlmann bands appear to be absent from the spectrum recorded for tylocrebrine (**8**) (*37*).

VII. Mass Spectra

The phenanthroindolizidines give easily recognizable mass spectra owing to the intense ion at $M - 69$, usually the base peak, formed by loss of ring E as a pyrroline unit by a retro-Diels–Alder fission (*cf.* Scheme 1, for tylophorine, **13**) (*10*).

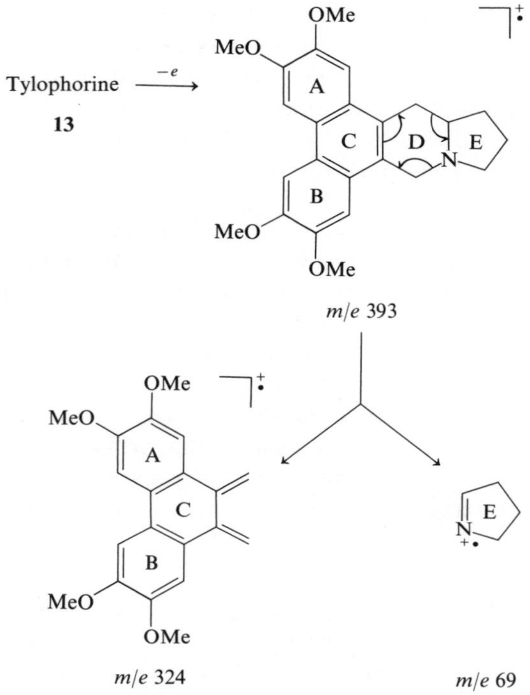

Tylophorine $\xrightarrow{-e}$

13

m/e 393

m/e 324

m/e 69

SCHEME 1

In the case of a phenanthroquinolizidine alkaloid, the corresponding tetrahydropyridine unit would be lost on electron impact, but with cryptopleuridine (**17**) there is a loss of 99 mass units from the molecular ion (*71*), which indicates that a hydroxyl group is located in ring E. Mass spectrometry can also be useful in locating a hydroxyl group in ring D. Tylophorinidine (**7**) exhibits a base peak at *m/e* 296 formed as in Scheme 1, which then readily eliminates carbon monoxide to give a fragment ion at *m/e* 268 (*38*, *56*):

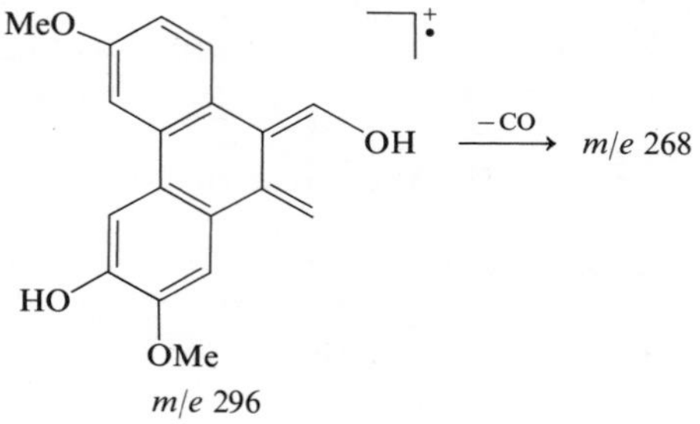

$\xrightarrow{-CO}$ *m/e* 268

m/e 296

This behavior is consistent with the presence of a hydroxyl group at position 14 or 9. However, the latter alternative can be eliminated on other grounds: in particular, if the hydroxyl group is located at C-14, the *O*-acetyl derivative can undergo a McLafferty rearrangement on electron impact with loss of acetic acid, as shown in Scheme 2 in the case of 14-hydroxy-2,3,6-trimethoxy-phenanthroindolizidine (**3**) (*12*). The product forms an intense ion that can lose hydrogen to form the base peak at *m/e* 360; this fragment can further lose HCN to give another strong peak at *m/e* 333. In the presence of a 14-*O*-acetate grouping, this sequence takes precedence over the retro-Diels–Alder fragmentation; the product from the latter forms a weaker ion which, however, can lose ketene to give a strong ion at *m/e* 310 (*12*).

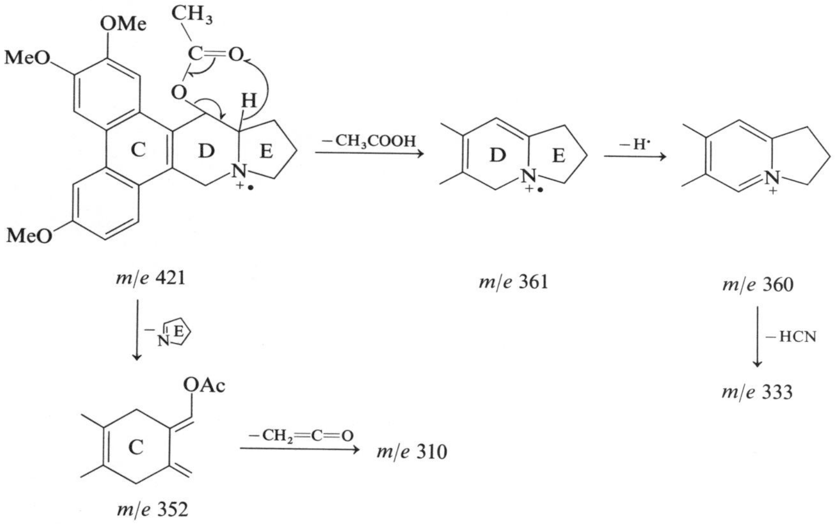

OMe CH_3
MeO C=O
O H
C D E $\xrightarrow{-CH_3COOH}$ D E $\xrightarrow{-H^{\cdot}}$
N+• N+• N+
MeO

m/e 421 *m/e* 361 *m/e* 360

↓ −(E) N ↓ −HCN

OAc *m/e* 333

C $\xrightarrow{-CH_2=C=O}$ *m/e* 310

m/e 352

SCHEME 2

VIII. Nuclear Magnetic Resonance Spectra

NMR Spectroscopy has been valuable in locating the oxy functions in phenanthroindolizidine alkaloids. In the case of antofine (**2**), analysis of the spectroscopic data coupled with degradative experiments did not allow a distinction to be drawn between structure **2** and the 3,6,7-trimethoxy analog **5** (*10*), so the alkaloid was oxidized with mercuric acetate to an immonium salt unsaturated between positions 9 and 10 (Scheme 3) (*7*). Nucleophilic substitution of a nitrile group at C-9 produced a cyanoantofine whose NMR spectrum showed the proton at C-8, which could be clearly distinguished

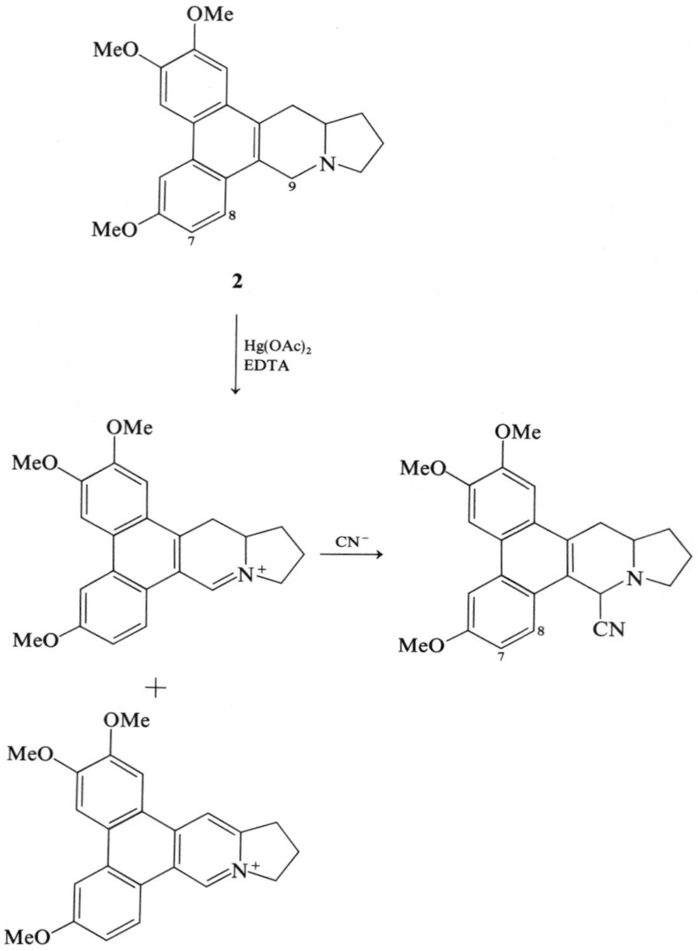
OMe
MeO
N
9
8
7
2
Hg(OAc)2
EDTA
CN⁻
N+
CN

SCHEME 3

as a doublet, to be shifted downfield (*7*). The other aromatic protons were unaffected except for a slight shift of the C-7 proton, which appeared as a double doublet with a meta and an ortho coupling. These data established structure **2** as the correct alternative for antofine (*7*).

Structures such as **2** and **5** can also be distinguished by a nuclear Overhauser experiment (*6*). In this type of alkaloid, the protons attached to C-9 form a pair of doublets, one of which is easily distinguished because it resonates around δ 4.7, being deshielded by the aromatic rings B and C.

When this proton is irradiated, a 25% increase in the intensity of one aromatic signal is observed, but the rest of the spectrum is practically unaffected. This signal must come from the proton attached to C-8; it shows an ortho coupling to the adjacent proton on C-7 in the case of **2** (*6*), but would appear as a singlet for a base such as **5**.

In Rao's alkaloid A from *Tylophora crebriflora* (**10**), there is an alcoholic hydroxyl at C-14 which deshields the proton at C-1 and shifts its signal to lower fields. This signal appears as a doublet, and shows vicinal coupling to the proton at C-2; therefore, these two positions are unsubstituted, and the two methoxyls in ring A must be located at positions 3 and 4. Because the remaining aromatic protons give singlets, the other two methoxyls must be at C-6 and C-7 (*33*). Rao's alkaloid D (**16**), from the same plant, has a similar structure except for an extra methoxyl group, which must be located at position 2 as the proton at C-1 produces a singlet in this case (*33*). In these alkaloids, the singlet at lowest field can be correlated with the proton at C-5 by analogy with values recorded by other workers (*39*, *40*) for "bay" protons (i.e., at C-4 or C-5) in various substituted phenanthrenes.

In Rao's alkaloids B (**11**) and C (**12**), the C-5 protons resonate at slightly higher field than in Rao's alkaloids A (**10**), D (**16**), and E (**15**), as a result of the influence of the phenolic hydroxyl (*33*). This effect is consistent with the presence of a 4-hydroxyl group in **11** and **12**; an ortho-substituted hydroxyl (at C-6) would make little difference in chemical shift to the C-5 proton as compared to a methoxyl group. The location of the phenolic hydroxyls in Rao's alkaloids B and C is confirmed by a positive test with Gibb's reagent, because position 4 is the only one with a free para position (*33*).

In contrast to the above-mentioned effect of a C-14 hydroxyl in shifting the C-1 proton signal to lower field, the effect after acetylation of the hydroxyl is to cause an upfield shift, which has been made use of in a structural study of 14-hydroxy-2,3,6-trimethoxyphenanthroindolizidine (**3**) (*12*). In this case, the shift affects not only the C-1 proton, but the protons of the methoxy group attached to C-2.

O-Acetylation can also be used to assist in locating the position of a phenolic group. One of the "bay" protons of tylophorinidine (**7**) is deshielded in the spectrum of the *O*-acetyl derivative as compared to that of the *O*-methylated tylophorinidine, but the remaining aromatic protons are practically unaffected (*23*). Because *O*-methyltylophorinidine is known to have the same substitution pattern as tylophorinine (**6**), it is clear that the deshielded "bay" proton in *O*-acetyltylophorinidine is attached to C-5, and that the deshielding is due to an ortho acetyl group; the hydroxyl group must, therefore, be located at C-6 (*23*).

Base-catalyzed deuterium exchange of ortho protons can also be used to assist in locating phenolic groups (*2*). The deuteration should affect para protons as well in cases such as **11** and **12**, where the para positions are unsubstituted.

NMR Spectroscopy has also been used to study the stereochemistry of tylophorinine (**6**) and the *O*-methyl derivative of tylophorinidine (**7**). These two bases are diastereomeric, and the NMR spectrum of the former indicates that it forms dimers in various solvents by hydrogen bonding between the hydroxyl group of one molecule and the nitrogen of another (*23*). This is shown by an unusually low-field signal for the proton at C-1 (δ 8.38) and an unusually high-field resonance for that at C-8 (δ 6.20) (*28*). In order to form such a dimer, it is argued that tylophorinine (**6**) must assume a flattened boat conformation for the six-membered ring of the indolizidine system, with the C-14 hydroxyl group pseudoequatorial and the C-13a hydrogen cis-oriented (*28*). The latter orientation is in accord with the very low coupling constant between the C-13a and C-14 hydrogens (*23*), and also with other evidence discussed below (Section VIII) concerning the configurations of these diastereomeric bases, which indicates a trans-diaxial arrangement of the C-14 hydroxyl and the C-13a hydrogen in the case of the *O*-methyl derivative of tylophorinidine (**7**), and a cis disposition for tylophorinine (**6**).

The location of the hydroxyl group in cryptopleuridine (**17**) was determined largely by a careful comparison of the NMR spectra of the base and its *O*-acetyl derivative with that of cryptopleurine (**18**) (*71*).

IX. Stereochemistry

It was shown by Wiegrebe *et al.* (*41*) that antofine (**2**) on vigorous ozonolysis gives D-proline (**22**), which was identified by degrading it with D-amino acid

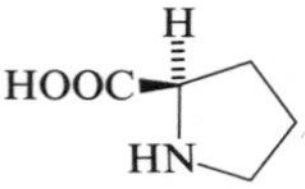

H HOOC HN **22** H HOOC—CH_2 HN **23**

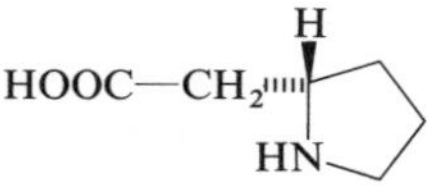

oxidase. Thus antofine (**2**) must have the *R*-configuration at C-13a, as must 14-hydroxy-2,3,6-trimethoxyphenanthroindolizidine (**3**) and 7-demethoxy-demethyltylophorine (**1**), which can be correlated with it. On the other hand, exhaustive ozonolysis of tylophorine (**13**) yielded *S*-pyrrolidine-2-acetic acid (**23**) (*42*); thus, this base must belong to the enantiomeric series to antofine (**2**). The ORD curve of tylophorine (**13**) shows a negative Cotton effect in the region 200–280 nm, whereas that of antofine has a positive one (*42*). Observations of this kind have allowed absolute configurations to be assigned to various alkaloids provided they have no other chiral center. Tylocrebrine

(**8**) (*43*) shows a negative Cotton effect, as did tylophorine, whereas (+)-isotylocrebrine (**9**) (*42*) and cryptopleurine (**18**) (*43*) show a positive effect and must belong to the *R*-series.

Tylophorinidine (**7**) has an extra chiral center, but hydrogenolysis gives the 14-deoxy derivative with a negative Cotton effect (*44*); therefore, tylophorinidine belongs to the *S*-series as far as the chiral center at C-13a is concerned. The facile loss of the benzylic hydroxyl group from *O*-methyltylophorinidine by hydrogenolysis under acidic conditions to give a racemic product may be interpreted as a dehydration followed by reduction; this in turn suggests a trans diaxial arrangement of the C-14 hydroxyl and the C-13a hydrogen (*23*), and indicates the absolute stereochemistry shown in **7** for tylophorinidine. As mentioned above, this configuration was confirmed by X-ray crystallographic analysis of its diacetate methiodide (*36*).

Tylophorinine (**6**) is a diastereomer of *O*-methyltylophorinidine (*23*), and a careful ORD study of the former compound and its derivatives has revealed that it is racemic (*44*); pergularinine is the levorotary form. Evidence from hydrogenolysis experiments, in which an optically pure, levorotary deoxy product was obtained (*15*, *23*), and from an analysis of the NMR data (*23*, *28*), accords with the assignment to tylophorinine (**6**) of a flattened boat conformation for ring D, with a pseudoequatorial hydroxyl group at C-14, and a cis-oriented proton at C-13a.

The configurations deduced by physical and degradative methods have been confirmed in certain cases by synthesis (see Section X).

For various purposes, such as the identification of a racemic product of synthesis with an optically active naturally occurring alkaloid, it is advantageous to be able to racemize the latter before the comparison is made. This can be done in the case of phenanthroindolizidines by dehydrogenation of ring D to an isoquinolinium derivative, as shown in Scheme 3 in the case of antofine (**2**) with mercuric acetate (*2*, *7*) or NBS (*72*), followed by borohydride reduction. For the reverse process of resolution, (+)-camphorsulfonic acid has proven effective for (+)-tylophorine, **13** (*61*).

X. Synthesis

Phenanthroindolizidine and phenanthroquinolizidine alkaloids have attracted considerable attention as far as synthesis is concerned, largely owing to their interesting pharmacological properties. A summary of the main synthetic methods follows.

Method 1. The first reported synthesis in the field gave racemic cryptopleurine (*45*).

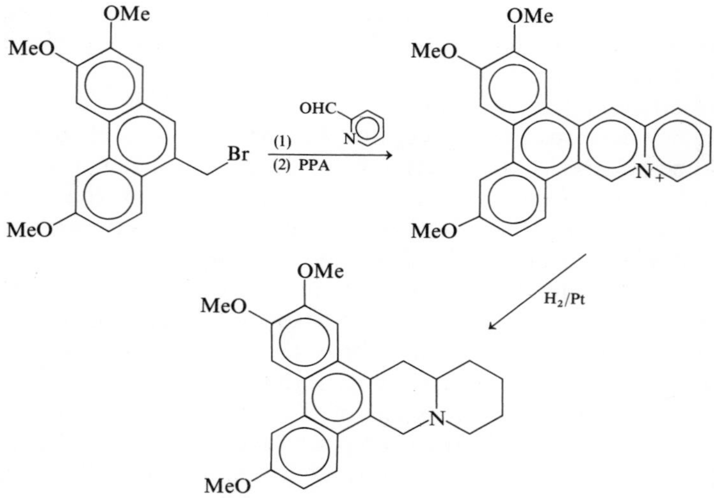

OMe MeO MeO Br (1) OHC N (2) PPA OMe MeO MeO N+ H_2/Pt OMe MeO MeO N

18 (±)-Cryptopleurine

Method 2. This method has been used widely in various modifications since its early introduction for the synthesis of cryptopleurine (**18**) (*47*). It is related to method 1, but instead of picolinic aldehyde, it employs an ester of

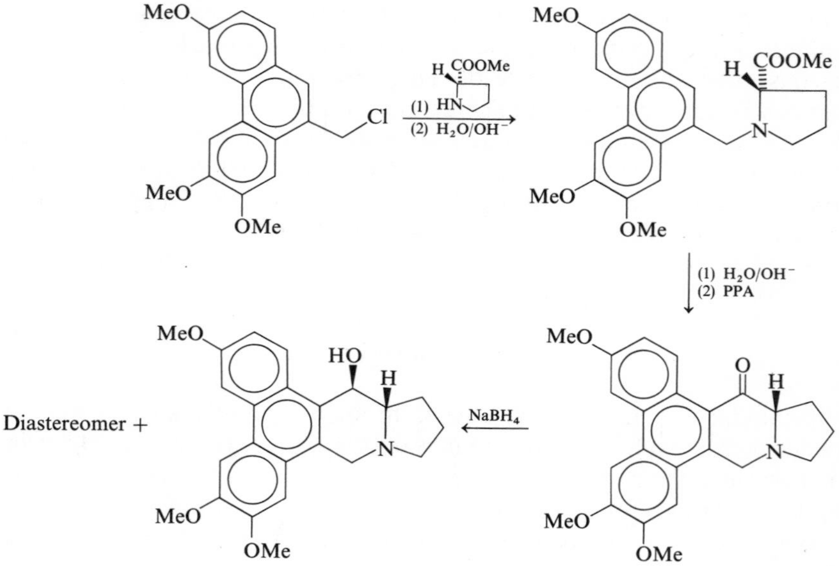

MeO MeO OMe Cl H COOMe (1) HN (2) H_2O/OH^- MeO MeO OMe H COOMe N (1) H_2O/OH^- (2) PPA MeO MeO OMe O H N NaBH$_4$ Diastereomer + MeO MeO OMe HO H N

(−)-Tylophorinine
(Pergularinine, **6**)

pipecolic acid (*47*) or proline (*46*, *48*), depending on whether a quinolizidine or an indolizidine nucleus is required. The modification shown here helped to confirm the *S*-configuration at C-13a of (−)-tylophorinine (*46*). If the C-14 hydroxyl is not required, the borohydride reduction may be carried out on the tosylhydrazone formed from the penultimate stage (*48*), or the oxygen function may be removed in other ways (*38*, *47*).

Method 3. This method was used for preparing the first synthetic phenanthroindolizidine alkaloids, tylophorine and tylocrebrine (*61*).

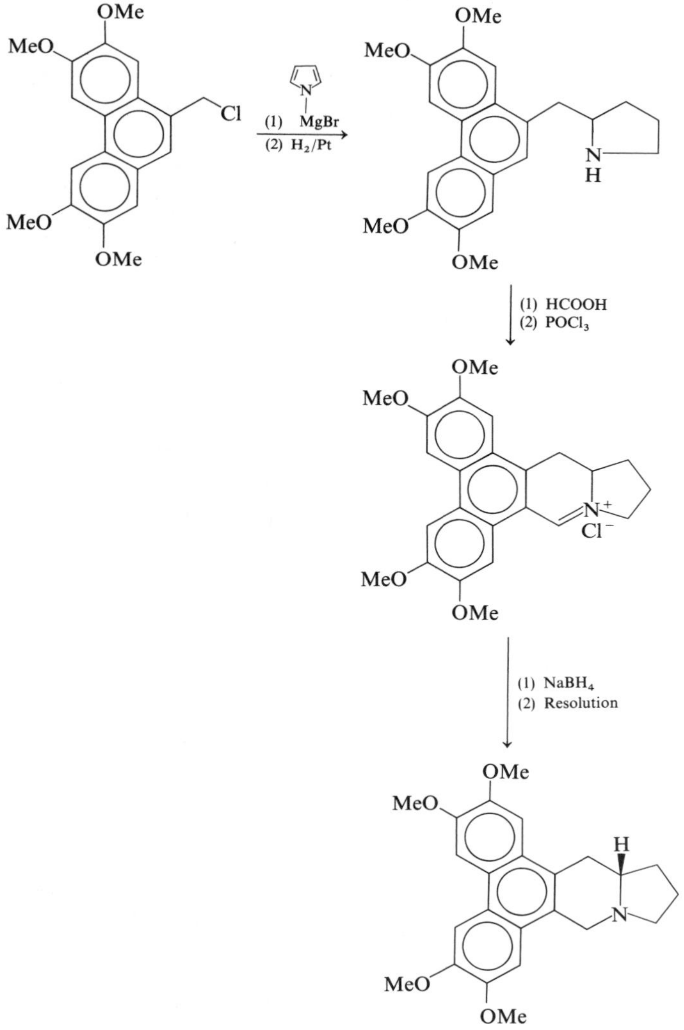

OMe MeO MeO OMe Cl

(1) MgBr (2) H_2/Pt

OMe MeO MeO OMe N H

(1) HCOOH (2) $POCl_3$

OMe MeO MeO OMe N^+ Cl^-

(1) $NaBH_4$ (2) Resolution

OMe MeO MeO OMe H N

13 (−)-Tylophorine

Method 4. A photochemical cyclization is employed to produce an amidophenanthrene derivative, which is subsequently further cyclized. In a variation of this method, the amidophenanthrene is formed from the corresponding acid and methyl prolinate by dehydration with DCC (*38*).

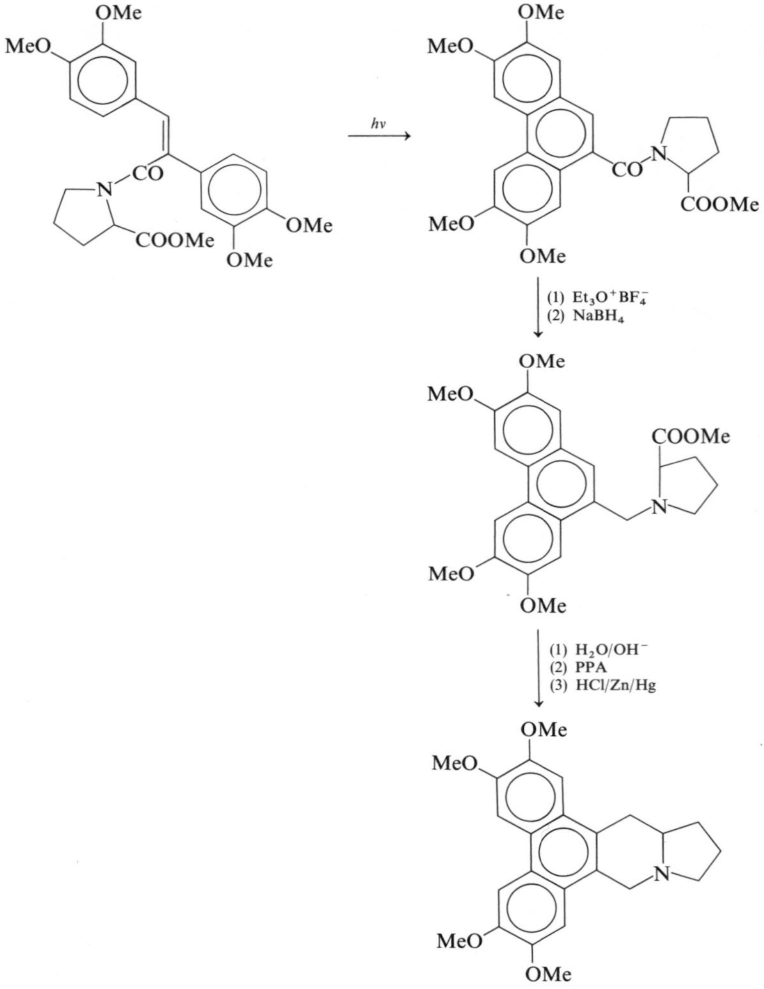

$h\nu$

(1) $Et_3O^+BF_4^-$
(2) $NaBH_4$

(1) H_2O/OH^-
(2) PPA
(3) HCl/Zn/Hg

13 (±)-Tylophorine

Method 5. This stereoselective synthesis resulted in the preparation of the enantiomer of naturally occurring antofine, and helped to confirm its absolute stereochemistry (*49*, *60*).

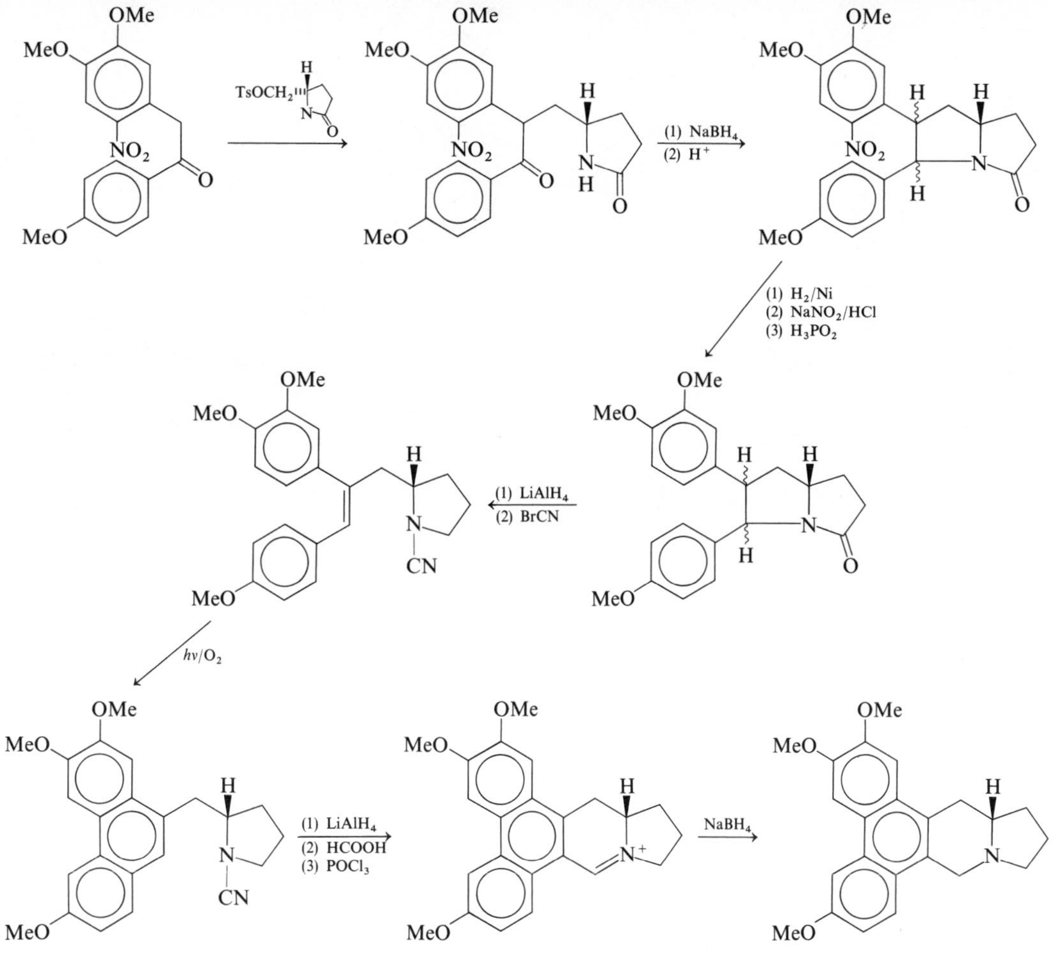

$TsOCH_2$-(pyrrolidinone) →

(1) $NaBH_4$
(2) H^+

(1) H_2/Ni
(2) $NaNO_2/HCl$
(3) H_3PO_2

(1) $LiAlH_4$
(2) BrCN

$h\nu/O_2$

(1) $LiAlH_4$
(2) HCOOH
(3) $POCl_3$

$NaBH_4$

(+)-Antofine (enantiomer of **2**)

Method 6. This synthesis is biomimetic (*52*), and the product of the first reaction, **33**, occurs naturally along with cryptopleurine (**18**) in *Boehmeria platyphylla* and *B. cylindrica* (*26, 27*). The penultimate stage is the alkaloid julandine (**24**), which also occurs in these two plants (*26, 27*). The synthesis can be modified (*52*) to afford tylophorine (**13**) through the secophenanthroindolizidine alkaloid septicine (**30**) (*21, 33, 56*).

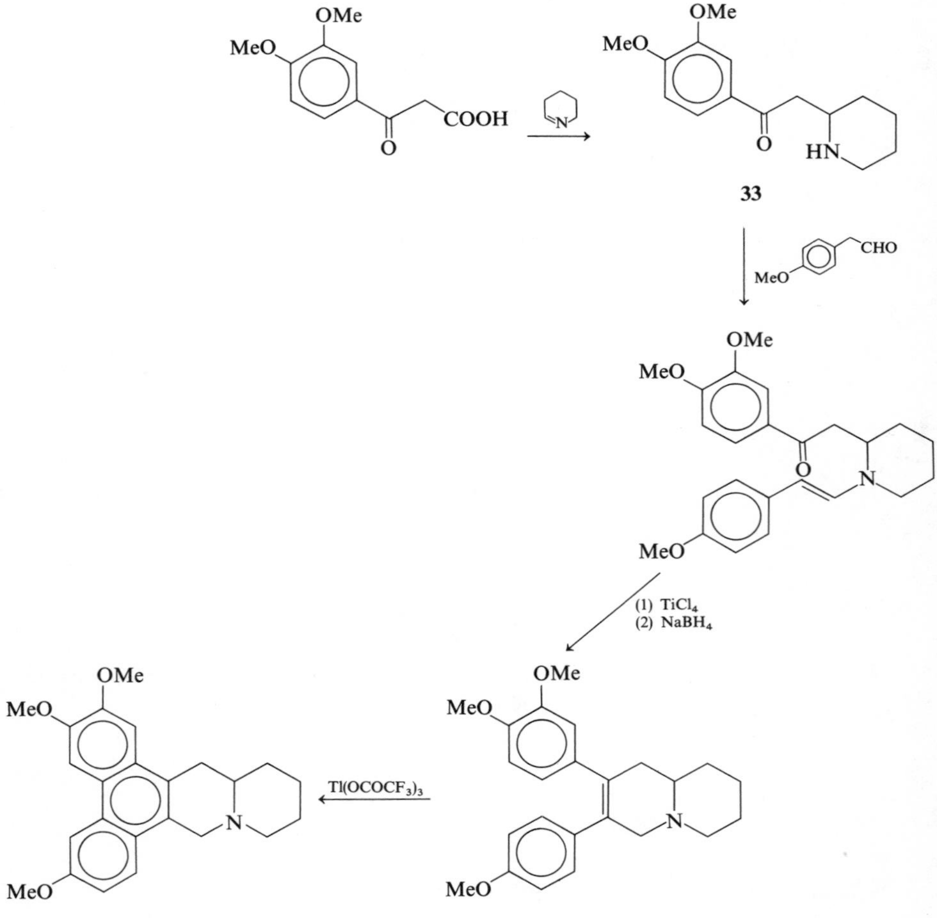

33

(1) $TiCl_4$
(2) $NaBH_4$

$Tl(OCOCF_3)_3$

18 (±)-Cryptopleurine

24 Julandine

Method 7. This method follows similar lines to the probable biosynthetic pathway to cryptopleurine (*50*).

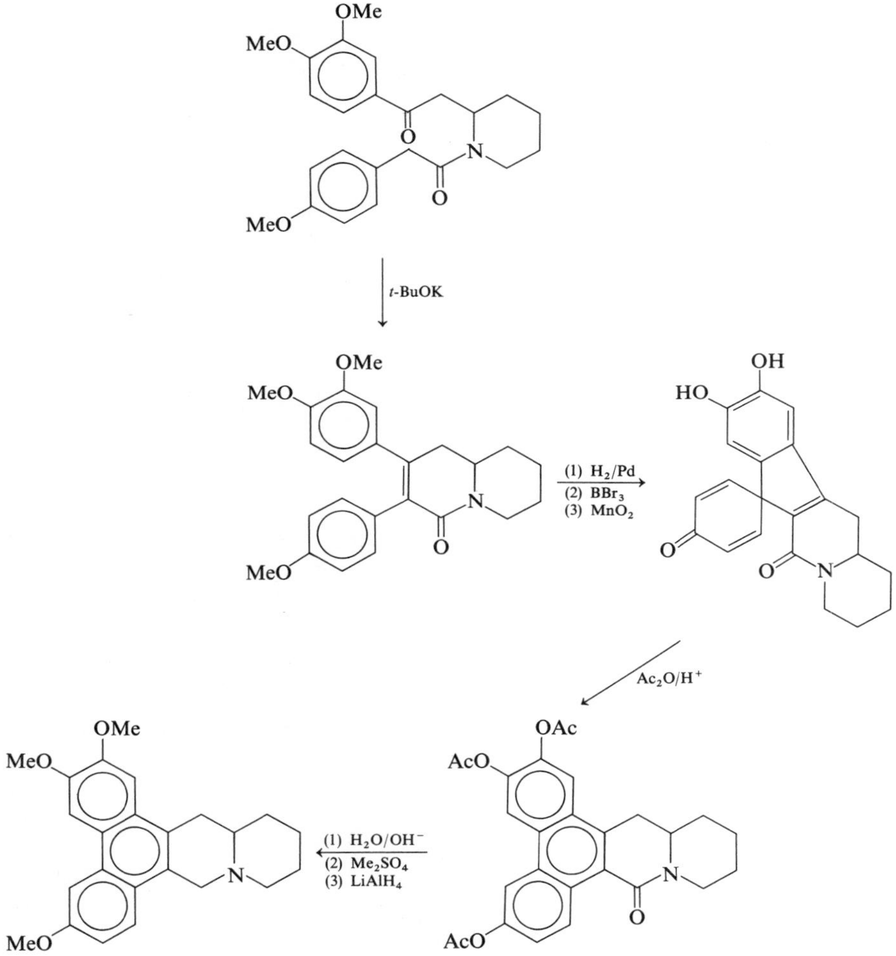

OMe
MeO
O
N
O
MeO
t-BuOK
OMe
MeO
N
O
MeO
(1) H_2/Pd
(2) BBr_3
(3) MnO_2
OH
HO
O
O
N
Ac_2O/H^+
OMe
MeO
N
MeO
(1) H_2O/OH^-
(2) Me_2SO_4
(3) $LiAlH_4$
OAc
AcO
N
O
AcO

18 (±)-Cryptopleurine

An improved synthesis employs anodic oxidation of the diaryl quinolizidone derivative to give the corresponding *O,O*-dimethylspirodienone, which is converted to cryptopleurine by essentially the same sequence of reactions as that shown above (*51*).

Method 8. The previous syntheses start with a ready-made heterocyclic nucleus for ring E; by contrast, in this recent method (*69*), ring E is assembled during the synthesis. As in methods 4–7, ring C is formed by an oxidative coupling between rings A and B, in this case with vanadium trifluoride oxide.

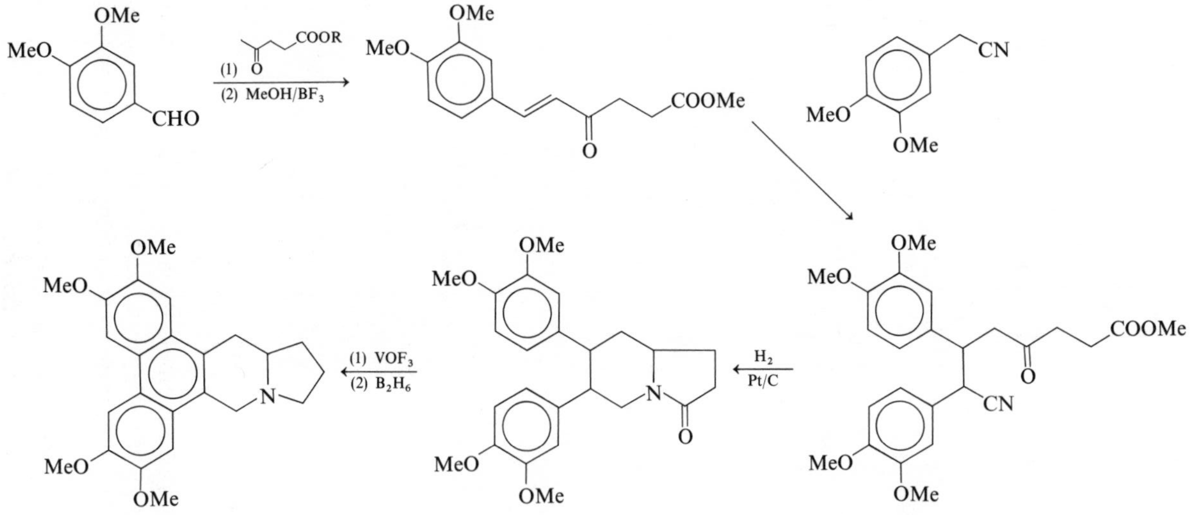

(1) COOR, O; (2) $MeOH/BF_3$

H_2, Pt/C

(1) VOF_3; (2) B_2H_6

13 (±)-Tylophorine

Method 9. This method gives the phenanthroindolizidine skeleton, and may be used for synthesizing tylophorine (**13**) and other alkaloids with suitable modifications (*53*).

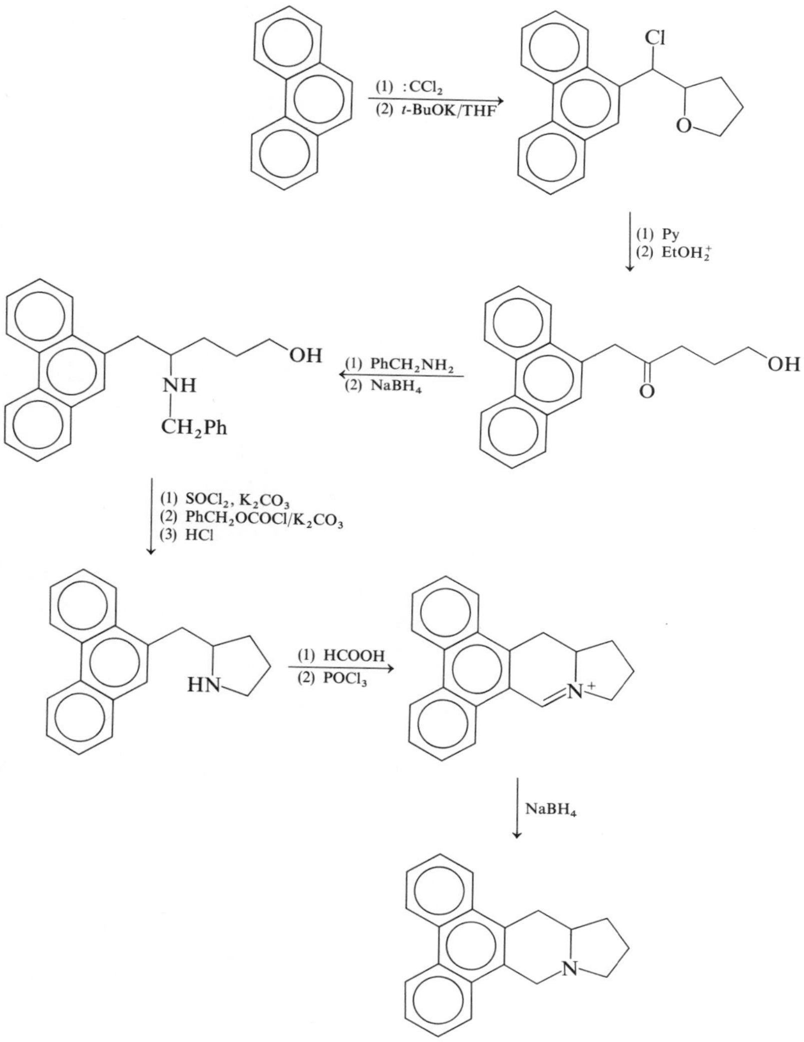

(1) :CCl_2
(2) *t*-BuOK/THF

Cl

O

(1) Py
(2) $EtOH_2^+$

OH

O

(1) $PhCH_2NH_2$
(2) $NaBH_4$

OH

NH

CH_2Ph

(1) $SOCl_2$, K_2CO_3
(2) $PhCH_2OCOCl/K_2CO_3$
(3) HCl

HN

(1) HCOOH
(2) $POCl_3$

N^+

$NaBH_4$

N

Method 10. An ingenious recent synthesis employs an imino Diels–Alder reaction to form rings D and E of tylophorine simultaneously (*82*).

OMe
MeO
CHO
MeO
OMe

$\xrightarrow{LiCH{=}CH_2}$

OMe
MeO
OH
MeO
OMe

$\xrightarrow[EtCOOH]{CH_3C(OEt)_3}$

OMe
MeO
O
COEt
MeO
OMe

(1) Me_2AlNH_2
(2) CH_2O/OH^-
(3) Ac_2O/Py

OMe
MeO
O
C
$NHCH_2OAc$
MeO
OMe

$\xleftarrow[\Delta]{-HOAc}$

OMe
MeO
N
O
MeO
OMe

OMe
MeO
N
O
MeO
OMe

$\xrightarrow{LiAlH_4}$

OMe
MeO
N
MeO
OMe

13 (±)-Tylophorine

XI. Biosynthesis

As a result of feeding experiments with [2-^{14}C]phenylalanine (**25**) (*55*) and [2-^{14}C]tyrosine (**26**) (*54*) on *Tylophora aromatica* plants, it has been shown that radioactive tylophorine (**13**) is formed; the labeling pattern, deduced by degradation, is consistent with Scheme 4 (*54, 55*). [2-^{14}C]Cinnamic acid was incorporated efficiently in place of phenylalanine (*62*); presumably, the cinnamic acid is first transformed into benzoylacetic acid and its *p*-hydroxy derivative, both of which are also incorporated (*73*). In a subsequent stage, the keto acids are converted to phenacyl pyrrolidines, and feeding experiments with doubly labeled substances of the latter type have shown that not only the unsubstituted compound and its *p*-hydroxy derivative, but the corresponding 4-hydroxy-3-methoxy derivative, are incorporated intact. These substances are thus key intermediates, and their role is underlined by the recent isolation of phyllostone (**31**) (*6*) from *Cryptocarya phyllostemon*, a plant which also produces antofine (**2**) (*6*). Three alkaloids with structures closely analogous to phyllostone have also been isolated from the acanthaceous plant *Ruspolia hypercraterformis* (*81*).

The pyrrolidine ring in these compounds evidently originates in ornithine (**27**), as suggested by labeling experiments (*55*), although the details of the incorporation are not yet clear.

In a further stage of the proposed scheme, a phenylacylpyrrolidine condenses with an arylpyruvic acid derived from tyrosine to give an amino acid analogous to those formed during isoquinoline biosynthesis (*74*). The diarylindolizidine **28** postulated as the decarboxylation product of the amino acid is a close analog of septicine (**30**), an alkaloid whose occurrence along with tylophorine (**13**) and tylocrebrine (**8**) in *Ficus septica* (*21*) as well as in *T. crebriflora* (*33*) and *T. asthmatica* (*56*) lends support to the proposed scheme. Furthermore, septicine is transformed into a mixture of these two bases on UV irradiation, with retention of the configuration at C-13a (*56*).

The final steps shown in Scheme 4 include oxidative phenol coupling (*63, 64, 75*) and other reactions analogous to those which occur during aporphine biosynthesis (*74, 76*). In addition to the final products (**13** and **6**) indicated in Scheme 4, any of the other phenanthroindolizidine alkaloids so far isolated could be produced by simple modifications of the scheme similar to those that take place in the biogenesis of aporphines. It is interesting that in spite of the apparent symmetry in substitution pattern of rings A and B of tylophorine (**13**), they are formed by separate pathways from phenylalanine and tyrosine, respectively.

No biosynthetic work has yet been carried out on the phenanthroquinolizidine alkaloids, but it has been suggested (*54, 77*) that they are formed in a corresponding way to the phenanthroindolizidines with the substitution of a

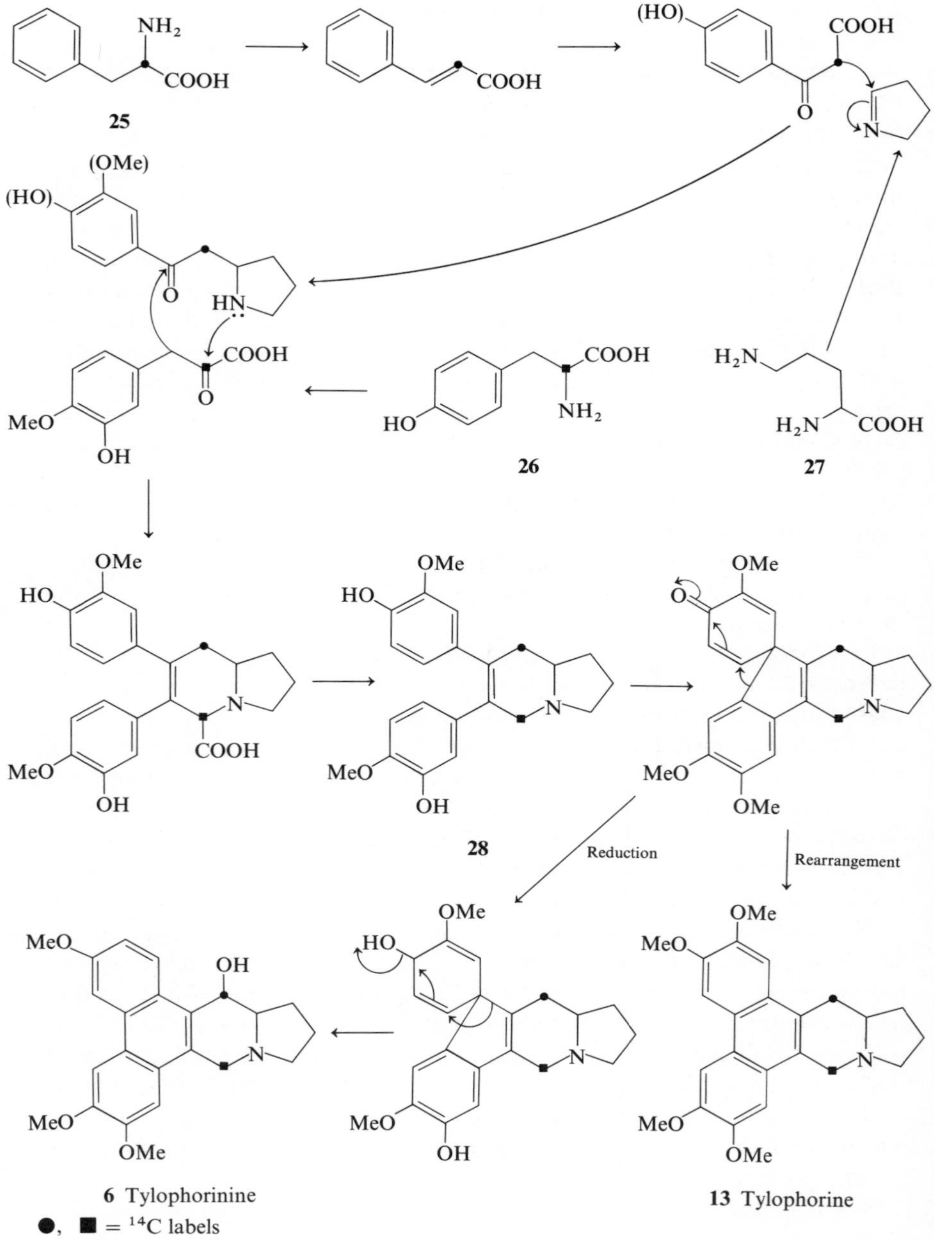

25

26

27

28

Reduction

Rearrangement

6 Tylophorinine

13 Tylophorine

●, ■ = ^{14}C labels

SCHEME 4

unit of lysine for ornithine. The occurrence of pleurospermine (**32**), an isomer of phyllostone (**31**) in *C. pleurosperma* (*57*), and of its *O*-methyl derivative (**33**) together with julandine (**24**) in *Boehmeria cylindrica* and *B. platyphylla* (*26*, *27*), accompanied in each case by cryptopleurine (**18**), lends some support to this proposal.

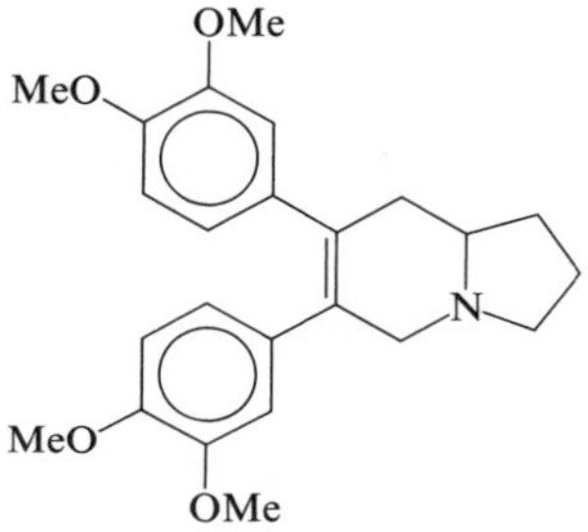

30 Septicine

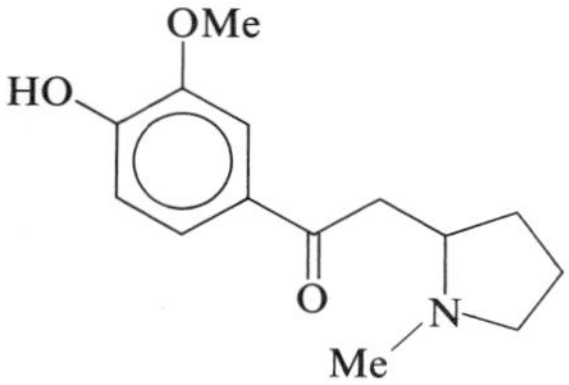

31 Phyllostone

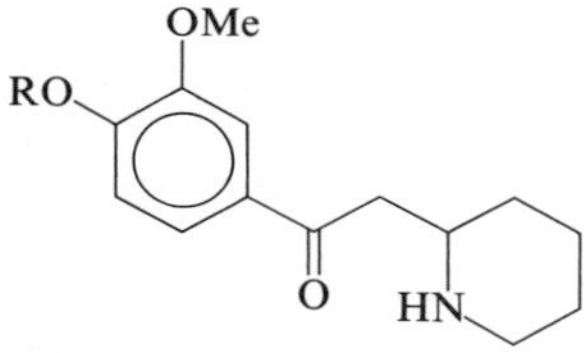

32 Pleurospermine R = H
33 3,4-Dimethoxy-ω-(2′-piperidyl)acetophenone R = Me

XII. Biological Activity

Interest in this group of alkaloids has centered in particular around reports of their antitumor activity (*27*, *58*, *59*, *79*). At the same time, their extremely toxic and vesicant nature has been noted by several investigators (*16*, *25*, *32*, *37*); however, the toxicity appears to be variable. Tylophorine (**13**) is toxic to *Paramecium caudatum* in a dilution of 1:50,000 (*24*), and is lethal to frogs at a dose of 0.4 mg/kg; on the other hand, toxicity to mice and guinea pigs is very small (*24*). The administration of tylophorine produces an initial drop in blood pressure, which subsequently rises to a level above normal; this is caused by the paralyzing action of the alkaloid on the heart muscles, accompanied by a stimulating effect on muscles of the blood vessels (*24*). Tylocrebine (**8**) has high activity against lymphoid leukemia L1210 in mice (*58*), but showed irreversible central nervous system toxicity in clinical trials; it inhibits protein biosynthesis irreversibly in HeLa cells (*83*).

Cryoptopleurine (**18**) is also a highly active inhibitor of protein synthesis (*59*). It shows highly specific and potent cytotoxic activity against a particular carcinoma of the nasopharynx in cell culture (*27*). On the other hand, it has been shown to stimulate the growth of nerve tissue (*78*).

The inhibiting effect of these alkaloids on protein biosynthesis seems to be associated with the translocution phase (*84*) and chain elongation (*83*).

An alkaloid, possibility identical with tylophorinidine, **7**, which was isolated by Rao from *Tylophora asthmatica* (*18*) and from *T. dalzellii* (*18*), showed significant activity in murine leukemia (L-1210 system).

References

1. T. R. Govindachari, *in* "The Alkaloids" (R. H. F. Manske, ed.), Vol IX, p. 517. Academic Press, 1967.
2. W. Wiegrebe, *Pharm. Z.* **117**, 1509 (1972).
3. J. E. Saxton, *Alkaloids* (*London*) **1**, 81 (1971); **2**, 74 (1972); **3**, 94 (1973); **4**, 100 (1974); **5**, 89 (1975); J. A. Lamberton, *Alkaloids* (*London*) **6**, 88 (1976); **7**, 66 (1977); **8**, 62 (1978).
4. R. P. Herbert, *Alkaloids* (*London*) **1**, 15 (1971); J. Staunton, **2**, 27 (1972); R. P. Herbert, **8**, 6 (1978); E. Leete, *Biosynthesis* **5**, 148 (1977).
5. J. E. Saxton, *Alkaloids* (*London*) **1**, 93 (1971); M. F. Grundon, *ibid.* **6**, 98 (1976).
6. I. R. C. Bick, W. Sinchai, T. Sévenet, A. Ranaivo, M. Nieto, and A. Cavé, *Planta Med.* **39**, 205 (1980); also unpublished data.
7. W. Wiegrebe, L. Faber, H. Brockmann, Jr., H. Budzikiewicz, and U. Krüger, *Justus Liebigs Ann. Chem.* **721**, 154 (1969).
8. T. F. Platonova, A. D. Kuzovkov, and P. S. Massagetow, *Zh. Obshch. Khim.* **28**, 3131 (1958); *CA* **53**, 7506d (1959).
9. A. Haznagy, L. Tóth, and K. Szendrei, *Pharmazie* **20**, 649 (1965).
10. M. Pailer and W. Streicher, *Monatsh. Chem.* **96**, 1094 (1965).
11. R. B. Herbert and C. J. Moody, *Phytochemistry* **11**, 1184 (1972).
12. W. Wiegrebe, H. Budzikiewicz, and L. Faber, *Arch. Pharm. Ber. Dtsch. Pharm. Ges.* **303**, 1009 (1970).
13. T. R. Govindachari, N. Viswanathan, J. Radhakrishnan, R. Charubala, N. Nityanandra Rao, and B. R. Pai, *Indian J. Chem.* **11**, 1215 (1973).
14. N. B. Mulchandani and S. R. Venkatachalam, *At. Energy Comm., India, Bhabha At. Res. Cent., Rep.* **BARC-764**, 8 (1974); see also *Alkaloids* (*London*) **6**, 89 (1976).
15. N. B. Mulchandani and S. R. Venkatachalam, *Phytochemistry* **15**, 1561 (1976).
16. A. N. Ratnagiriswaran and K. Venkatachalam, *Indian J. Med. Res.* **22**, 433 (1935); *CA* **29**, 8229 (1935).
17. T. R. Govindachari, B. R. Pai, and K. Nagarajan, *J. Chem. Soc.* 2801 (1954).
18. K. V. Rao, R. A. Wilson, and B. Cummings, *J. Pharm. Sci.* **60**, 1725 (1971).
19. N. B. Mulchandani, S. S. Iyer, and L. P. Badheka, *Chem. Ind.* (*London*) 505 (1971).
20. N. B. Mulchandani and S. R. Venkatachalam, *Int. Symp. Chem. Nat. Prod., 8th, New Delhi*, Abstr., **A-10**, 13 (1972); see also *Alkaloids* (*London*) **4**, 102 (1974).
21. J. H. Russel, *Naturwissenschaften* **50**, 443 (1963).
22. K. V. Rao, R. Wilson, and B. Cummings, *J. Pharm. Sci.* **59**, 1501 (1970).
23. T. R. Govindachari, N. Viswanathan, J. Radhakrishnan, B. R. Pai, S. Natarajan, and P. S. Subramaniam, *Tetrahedron* **29**, 891 (1973).
24. R. N. Chopra, N. N. Ghosh, J. B. Bose, and S. Ghosh, *Arch. Pharm. Ber. Dtsch. Pharm. Ges.* **275**, 236 (1937).

25. I. S. de la Lande, *Aust. J. Exp. Biol. Med. Sci.* **26**, 181 (1948).
26. N. K. Hart, S. R. Johns, and J. A. Lamberton, *Aust. J. Chem.* **21**, 1397, 2579 (1968).
27. N. R. Farnsworth, N. K. Hart, S. R. Johns, J. A. Lamberton, and W. Messmer, *Aust. J. Chem.* **22**, 1805 (1969).
28. J. D. Phillipson, I. Tezcan, and P. J. Hylands, *Planta Med.* **25**, 301 (1974).
29. E. Gellert, *Aust. J. Chem.* **9**, 489 (1956).
30. T. R. Govindachari, M. V. Lakshmikantham, K. Nagarajan, and B. R. Pai, *Tetrahedron* **4**, 311 (1958).
31. T. R. Govindachari, M. V. Lakshmikantham, B. R. Pai, and S. Rajappa, *Tetrahedron* **9**, 53 (1960).
32. T. R. Govindachari, B. R. Pai, I. S. Ragade, S. Rajappa, and N. Viswanathan, *Tetrahedron* **14**, 288 (1961).
33. K. V. Rao, *J. Pharm. Sci.* **59**, 1608 (1970).
34. J. Fridrichsons and A. McL. Mathieson, *Nature* (*London*) **173**, 132 (1954).
35. J. Fridrichsons and A. McL. Mathieson, *Acta Crystallogr.* **8**, 761 (1955).
36. V. K. Wadhawan, S. K. Sikka, and N. B. Mulchandani, *Tetrahedron Lett.* 5091 (1973).
37. E. Gellert, T. R. Govindachari, M. V. Lakshmikantham, I. S. Ragade, R. Rudzats, and N. Viswanathan, *J. Chem. Soc.* 1008 (1962).
38. R. B. Herbert and C. J. Moody, *Chem. Commun.* 121 (1970).
39. J. Reisch, M. Băthory, K. Szendrei, E. Minker, and I. Novăk, *Tetrahedron Lett.* 67 (1969).
40. K. D. Bartle and J. A. S. Smith, *Spectrochim. Acta, Part A* **23**, 1689, (1967).
41. W. Wiegrebe, L. Faber, and T. Breyhan, *Arch. Pharm. Ber. Dtsch. Pharm. Ges.* **304**, 188 (1971).
42. T. R. Govindachari, T. G. Rajagopalan, and N. Viswanathan, *J.C.S. Perkin I* 1161 (1974).
43. E. Gellert, R. Rudzats, J. C. Craig, S. K. Roy, and R. Woodard, *Int. Symp. Chem. Nat. Prod., 11th, Golden Sands, Bulgaria*, Abstr., **2**, 1 (1978).
44. T. R. Govindachari, N. Viswanathan, and B. R. Pai, *Indian J. Chem.* **12**, 886 (1974).
45. C. K. Bradsher and H. Berger, *J. Am. Chem. Soc.* **80**, 930 (1958).
46. T. R. Govindachari, P. R. Pai, S. Prabhakar, and T. S. Savitri, *Tetrahedron* **21**, 2573 (1965).
47. P. Marchini and B. Belleau, *Can. J. Chem.* **36**, 581 (1958).
48. B. Chauncy and E. Gellert, *Aust. J. Chem.* **23**, 2503 (1970).
49. L. Faber and W. Wiegrebe, *Helv. Chim. Acta* **56**, 2882 (1973).
50. J. M. Paton, P. L. Pauson, and T. S. Stevens, *J. Chem. Soc. C* 1309 (1969).
51. E. Kotani, M. Kitazawa, and S. Tobinaga, *Tetrahedron* **30**, 3207 (1974).
52. R. B. Herbert, *Chem. Commun.* 794 (1978).
53. S. Takano, K. Yuta, and K. Ogaswara, *Heterocycles* **4**, 947 (1976).
54. N. B. Mulchandani, S. S. Iyer, and L. P. Badheka, *Phytochemistry* **8**, 1931 (1969).
55. N. B. Mulchandani, S. S. Iyer, and L. P. Badheka, *Phytochemistry* **10**, 1047 (1971).
56. T. R. Govindachari, *J. Indian Chem. Soc.* **50**, 1 (1973).
57. E. Gellert, *Aust. J. Chem.* **12**, 90 (1959).
58. E. Gellert and R. Rudzats, *J. Med. Chem.* **7**, 361 (1964).
59. G. R. Donaldson, M. R. Atkinson, and A. W. Murray, *Biochem. Biophys. Res. Commun.* **31**, 104 (1968).
60. L. Faber and W. Wiegrebe, *Helv. Chim. Acta* **59**, 2201 (1976).
61. T. R. Govindachari, M. V. Lakshmikantham, and S. Rajadurai, *Tetrahedron* **14**, 284 (1961).
62. N. B. Mulchandani, S. S. Iyer, and L. P. Badheka, *Phytochemistry* **15**, 1697 (1976).
63. D. H. R. Barton, *Proc. Chem. Soc., London* 293 (1963).
64. D. H. R. Barton and T. Cohen, *Festschr. Prof. Dr. Arthur Stoll Siebzigsten Geburtstag* 117 (1957).
65. F. Bohlmann, *Angew. Chem.* **69**, 641 (1957).

66. F. Bohlmann, D. Schumann, and H. Schulz, *Tetrahedron Lett.* 173 (1965).
67. E. Gellert and N. V. Riggs, *Aust. J. Chem.* **7**, 113 (1954).
68. T. A. Crabb, R. F. Newton, and D. Jackson, *Chem. Rev.* **71**, 109 (1971).
69. A. J. Liepa and R. E. Summons, *Chem. Commun.* 826 (1977).
70. A. W. Sangster and K. L. Stuart, *Chem. Rev.* **65**, 69 (1965).
71. S. R. Johns, J. A. Lamberton, A. A. Sioumis, and R. I. Willing, *Aust. J. Chem.* **23**, 353 (1970).
72. K. V. Rao and L. S. Kapicak, *J. Heterocycl. Chem.* **13**, 1073 (1976).
73. R. B. Herbert, F. B. Jackson, and I. T. Nicholson, *Chem. Commun.* 865 (1976).
74. J. Staunton, *Alkaloids (London)* **2**, 10 (1972).
75. A. R. Battersby, *in* "Oxidative Coupling of Phenols" (W. I. Taylor and A. R. Battersby, eds.), p. 119. Arnold, London, 1967.
76. R. B. Herbert, *Alkaloids (London)* **1**, 19 (1971).
77. I. R. C. Bick and W. Sinchai, *Heterocycles* **9**, 903 (1978).
78. H. Hofmann, *Aust. J. Exp. Biol. Med. Sci.* **30**, 541 (1952).
79. J. Hartwell and B. J. Abbott, *Adv. Pharmacol. Chemother.* **7**, 117 (1969).
80. T. R. Govindachari and N. Viswanathan, *Heterocycles* **11**, 587 (1978).
81. F. Roessler, D. Ganzinger, S. Johne, E. Schöpp, and M. Hesse, *Helv. Chim. Acta* **61**, 1200 (1978).
82. S. M. Weinreb, N. A. Khatri, and J. Shringarpure, *J. Am. Chem. Soc.* **101**, 5073 (1979).
83. M.-T. Huang and A. P. Grollman, *Mol. Pharmacol.* **8**, 538 (1972).
84. P. Grant, L. Sanchez, and A. Jimenez, *J. Bacteriol.* **120**, 1308 (1974).

INDEX

E

F

G

H

I

J

K

L

M

N

O

P

R